T0331944

Peirce on Inference

Peirce on Inference

*Validity, Strength, and
the Community of Inquirers*

RICHARD KENNETH ATKINS

OXFORD
UNIVERSITY PRESS

OXFORD
UNIVERSITY PRESS

Oxford University Press is a department of the University of Oxford. It furthers
the University's objective of excellence in research, scholarship, and education
by publishing worldwide. Oxford is a registered trade mark of Oxford University
Press in the UK and certain other countries.

Published in the United States of America by Oxford University Press
198 Madison Avenue, New York, NY 10016, United States of America.

Library of Congress Control Number: 2023909277

ISBN 978–0–19–768906–6

DOI: 10.1093/oso/9780197689066.001.0001

Printed by Integrated Books International, United States of America

For Dedo Petar and Baba Cvetanka Kiprovski
With gratitude for your love and support

За Дедо Петар и Баба Цветанка Кипровски
Со благодарност за Вашата љубов и поддршка

Contents

Acknowledgments

This book is dedicated to my parents-in-law, Petar and Cvetanka Kiprovski, who live with my wife and me. Their love and support over the past fifteen years have been vital to our family, to the lives of our children, and to my ability to maintain an active research agenda. Completing this book, and in fact all of my books, would not have been possible without them. Of course, they would not be in my life were it not for my wife, Božanka. Her patience, love, and understanding are beyond measure.

I have been making steady progress on this book for the past six years. Along the way, many of my students at Boston College have heard me lecture on several subjects covered in this book. I am thankful for their good humor and excellent questions as I muddled through the issues. I made much headway during my sabbatical over the Fall semester of 2020. I am deeply appreciative for the support I have received at Boston College from the administration, staff, and faculty. In particular, Junhee Han has been an excellent and diligent research assistant.

Two anonymous reviewers provided detailed comments on an earlier version of this book and on the final typescript. One of those reviewers, who is no longer anonymous, is Jeff Kasser. He rescued me from numerous errors, corrected several infelicities of expression, and encouraged me to improve the typescript in many ways. I am deeply in his debt. The other reviewer, who remains anonymous, counseled me to reconsider the structure and focus of the book. The result is a text that is much clearer and more focused. This book would be much worse without the insight of these two reviewers. Any errors, of which surely many remain, are mine alone.

It has been a pleasure to work again with Lucy Randall at Oxford University Press. Her support of the project has been instrumental to seeing it through to publication, and I am immensely grateful for it. Thanks are also due to Brent Matheny for shepherding the book through production.

Introduction

This book is an essay on Peirce's theory of inference. Peirce's writings on inference are voluminous, and they span from the start of his philosophical
career to its end, over fifty years. One way in which to examine Peirce's theory
of inference is to detail the chronological development of Peirce's ideas from
start to finish. No doubt such an enterprise would have its merits. Yet it would
also take many volumes to complete, involve much repetition, and require
glossing over confusions in Peirce's earlier writings only to clear them up in
later chapters. This book pursues an alternative strategy. Although I shall
draw on the full scope of Peirce's writings from start to finish, my aim shall be
to construct a coherent account of inference faithful to Peirce's ideas overall.
I shall endeavor to put Peirce's work on inference into its best light by fitting
together core insights found throughout diverse writings into a cogent and
coherent theory of inference. That is, my strategy is to present a charitable
retrospective perspective on Peirce's theory of inference.

That such is my strategy does not imply I shall be insensitive to key
developments in Peirce's theory of inference. Much to the contrary, I shall
argue that a cogent and coherent presentation of a Peircean theory of inference requires appreciating why Peirce changes his mind with respect to various topics. For instance, Peirce defines logical validity in different ways in his
earlier writings and in his later writings. Understanding why the shift in his
conception of validity occurs requires being sensitive to the shortcomings of
his earlier theory of inference. Similarly, Peirce's classification of inferences
changes from his earlier writings to his later, as well as his terminology.
Tracking those changes and why Peirce makes them is an aid to developing a
charitable retrospective perspective on Peirce's theory of inference.

Of course, it would not be very interesting to develop such a Peircean perspective on inference unless there are some valuable lessons to draw from it.
Peirce's theory of inference sits at the intersection of many different areas of
philosophy, including philosophy of logic, epistemology (especially formal
epistemology), and philosophy of science. To give just a taste of the sorts
of insights that can be mined from Peirce's lifelong reflections, allow me to

Peirce on Inference. Richard Kenneth Atkins, Oxford University Press. © Oxford University Press 2023.
DOI: 10.1093/oso/9780197689066.003.0001

make a few brief comments on each of these areas. With respect to philosophy of logic, Peirce's theory of inference explains why logicians can and should take an interest in how people reason without acquiescing in psychologism, the thesis that the laws of logic are somehow grounded in how we in fact think. Moreover, Peirce offers a distinctive theory of validity according to which inferences are valid just in case the procedure involved in the inference has some truth-producing virtue and the inference is in fact made in accordance with that procedure. For deduction, that truth-producing virtue is truth-preservation, and so Peirce's theory of deductive inference is broadly aligned with contemporary accounts of validity. Yet Peirce, influenced by the medieval philosophers, holds that inductive and abductive inferences can be valid, too. In the case of inductive inference, its truth-producing virtue is being truth-approximating in the long run. In the case of abductive inference, its truth-producing virtue is being truth-conducive, provided there is any truth to be ascertained. Accordingly, Peirce's theory of inference promises us a unified theory of logical validity on which both inductions and abductions can be valid.

As to epistemological insights, Peirce develops and defends a distinctive reply to the Humean problem of induction. Peirce denies that acceptance of the principle that nature is uniform is warranted. Against those who argue it is a regulative assumption of inquiry, Peirce contends that an appeal to the principle of uniformity as a regulative assumption only warrants its use in cases when we must assume it in order to win the truth. Yet induction is used more widely than in such cases. He also denies that the principle of uniformity can itself be rationally justified, even if circularly. Instead, Peirce argues that induction makes no use of such a principle. Induction does not turn on the assumption that nature is uniform. Moreover, if induction did turn on the assumption that nature is uniform, so much the worse for induction. Nature is not uniform, Peirce argues. Even if nature were uniform, we could not know it. And even if we could know that nature is uniform, the principle of uniformity would be worthless. Instead, Peirce contends that induction uses more specific guiding principles, such as that what is true of one sample of copper with respect to its magnetic properties is true of any other. These specific guiding principles can, in turn, be confirmed or refuted in the ordinary process of inquiry.

Turning to philosophy of science, Peirce is sometimes presented as an arch-frequentist staunchly opposed to various forms of subjective Bayesianism. This is a mistake. Peirce is quite explicit that the procedure of assigning priors and updating our beliefs "is supported, under abstract conditions, by the doctrine of chances, and that there are cases in which it is useful" (HP 2:711, 1901). What Peirce denies is that the conditions under which such a procedure is useful are always satisfied. Moreover, he argues that the procedure as such is consistent with the method of tenacity. If we keep all counterevidence at bay, we will never have to change our beliefs. Peirce does not reject subjective Bayesianism. He argues that, unless the procedure of assigning priors and updating our beliefs is supplemented with sound practices of inquiry, it is of little value. Further, he holds that there are certain domains of inquiry in which the procedure is useless, such as in drawing inferences on the basis of testimonia in ancient history. Peirce does not reject subjective Bayesianism but identifies its limits. As such, his account offers a rapprochement between frequentist-type and belief-type accounts of inductive inference.

The preceding comments are just a taste of some of the key insights we can glean from Peirce's writings. In addition to these, Peirce has perceptive things to tell us about matters bearing on Goodman's new riddle of induction, the problem of underdetermination, axiomatic approaches to deductive inference, why we should trust our instinct to guess at the answers to questions, and the like. Anyone interested in such questions will be well-rewarded by grappling with Peirce's lifelong investigations into these topics, even if one ultimately disagrees with him.

Most scholars, if they know much of anything about Peirce, know that he distinguishes among deduction, induction, and abduction as genera of inference. From there, matters become murkier. Accordingly, some distinctions and terminological ground clearing are in order from the start. The first distinction to be drawn is between deduction, induction, and abduction as genera of inference and deduction, induction, and abduction as stages of inquiry. These are not the same. Some stages of inquiry will draw on more than one species of inference within that genus. For example, at an inductive stage of inquiry, we may make both (to use Peirce's mature terminology) qualitative and quantitative inductions. Suppose I wish to know whether a standard cuboidal die is fair. I decide to toss it one hundred times, recording whether it turns up odd or even more frequently. I find that forty-eight of those times, the die turned up even. I infer the die is fair. The statistic ascertained by rolling the die is a quantitative induction. The inference from *if the die is fair, then it will show an even number about half of the rolls of one hundred rolls* and *the die showed an even number about half of the rolls* to *the die is fair* is a qualitative induction. Both inferences, however, are used at the inductive stage of inquiry. In addition to the inductive stage of inquiry, I also deduced that *if the die is fair, it will show an even number on about half of one hundred rolls of the die*. And if we imagine I was losing at a dice game when I came to suspect some die was not fair, then at the abductive stage I came to suspect that the die was not fair (rather than the gods were thwarting my chances, say) and opted to put that hypothesis to the test.

A whole cluster of confusions concerns Peirce's theory of abduction. One confusion is that many scholars have thought Peircean abduction is the same as inference to the best explanation. Depending on what we mean by 'inference to the best explanation,' this is wrong. Abduction invents, selects, and entertains hypotheses (see HP 2:895, 1901). It does not eliminate, corroborate, or confirm them. Nonetheless, abductive inferences can be (but are not always!) (a) inferences to the best explanation given the evidence we have or (b) inferences to the best explanation to pursue. A second confusion concerning abduction is that many scholars have thought what Peirce calls hypothesis in his early work and what he calls abduction in his late work are the same. This is a mistake. What Peirce calls hypothesis in his early work is what he later calls qualitative induction. There is no equivalent of abduction in Peirce's earlier works. In those earlier works, Peirce thinks that new ideas are introduced by observation, not by a genus of inference. Of course, all of these assertions need defense, which will be forthcoming in this book. The first confusion just mentioned is cleared up in Chapter 7. The second confusion is cleared up in Chapter 2.

The mention of hypothesis brings us to another terminological confusion. In his early works, Peirce identifies two species of induction, hypothesis and induction. The terminology is infelicitous. As Murray Murphey remarks, "[i]t is something of a historical tragedy that Peirce should have used the terms 'hypothesis' and 'induction' as names for his two forms of ampliative inference, for his choice of terminology has helped to obscure his meaning" (1961, 113). Peirce uses 'induction' in a narrow sense in his earlier works. It is specifically statistical inference. Hypothesis, in contrast, is inference from a conditional and its consequent to its antecedent. It is akin to (but not identical to) what today we call the hypothetico-deductive method or induction by confirmation. Having deduced a prediction from a hypothesis, we test whether the prediction is borne out. Our contemporary terminology, however, is also unhappy. If the 'hypothetico-' is supposed to be the proposal of a hypothesis, then that is properly an abduction and not an induction. And deduction is a different genus of inference as well. Reference to the 'hypothetico-deductive method' confuses induction as a method of inquiry involving abductive and deductive stages with hypothetic induction as a species of inference. Furthermore, hypothetic inductions do not always confirm the hypothesis; sometimes, the hypothesis is eliminated or merely corroborated. Consequently, calling such an inference an induction by confirmation is misleading. In his earlier works, Peirce sometimes calls hypothesis hypothetic inference, but this terminology risks confusing hypothetic inference with abductive inference in his later works since both concern hypotheses. I have chosen instead to term 'hypothesis' hypothetic induction and 'induction' statistical induction.

In his later works, Peirce identifies three species of induction rather than just two. These three species of induction are crude induction, qualitative induction, and quantitative induction. Quantitative induction is clearly the heir of statistical induction. Whereas in his early work sampling is Peirce's model for quantitative induction, his later work expands quantitative induction to include all sorts of measurement. Qualitative induction is the heir of hypothetic induction, though that is only apparent when we get clear on what hypothetic induction is. Crude induction, in contrast, is inference from the mass of past experience involving no established proof of some claim to the denial of that claim. For instance, it is inference from the fact that *it has never been established someone has quantum tunneled through a wall* to *no one has ever quantum tunneled through a wall*. Throughout this book, I shall use 'hypothetic induction' and 'qualitative induction' interchangeably. Also, I shall use 'statistical induction' and 'quantitative induction' interchangeably. The difference is that when I am discussing Peirce's earlier works, I shall use the language which reflects his earlier terminology, viz. hypothetic induction and statistical induction. When I am discussing Peirce's mature views, I shall use his mature terminology. And at other times, I will use whichever terminology seems most apt in the context.

In addition to all the distinctions just made, I have already noted that Peirce uses 'validity' in a way broader than we use it today. For Peirce, inductive and abductive inferences can be valid. Moreover, inductive and abductive inferences can be strong or weak. Peirce holds that deduction alone does not admit of a distinction between validity and strength (see EP 2:232, 1903). Accordingly, while Peirce's terminology does track the manner in which we ordinarily evaluate inductive inferences by strength, his conception of validity does not.

That these distinctions are called for and well-motivated by Peirce's writings can only be established in the course of a fuller examination of this theory of inference. Without further ado, I turn to that now. The book roughly divides into two parts. Chapters 1 through 3 set the stage by explaining Peirce's account of the nature of inference in general, his classification of inferences, and his theory of logical validity. In the first chapter, I unpack Peirce's conceptions of inference and of argument. In Chapter 2, I examine how Peirce's classification of inferences changes from his works in the 1860s to his works in the 1900s. Chapter 3 explains Peirce's general theory of logical validity. Chapters 4 through 7 take a more detailed look at each of the genera of inference. For each genus of inference, I (a) examine its species, (b) explain why each species has the truth-producing virtue claimed for it and so is valid, (c) discuss how such inferences can be made strong, and (d) consider objections to the validity of such inferences and Peirce's replies to those objections. The nature of deduction is discussed in Chapter 4, along with objections to its validity and replies. Chapters 5 and 6 explore Peirce's theory of induction. Chapter 5 is an examination of Peirce's account of the species of induction, why they are valid, and how they can be made strong. Chapter 6 considers objections and replies to the validity of induction. Finally, Chapter 7 subjects Peirce's theory of abduction to scrutiny. I examine what abduction is, why it is valid, how abductions can be made strong, and objections and replies to the validity of abduction.

1

Inference and Argument

A farmer, standing in her field, looks up at the clouds. She sees they are dark
and heavy. She looks at her barometer. She concludes it is likely to rain and
proceeds to prepare for the storm. A doctor, reviewing a patient's medical
reports and seeing he has high cholesterol, prescribes a statin. A grocer
receives an order of avocados. She opens a few boxes, sees that all but a couple
of the avocados inside look ripe, and accepts the order.

In each of these cases, a person makes an inference based on observations.
The premises of that inference may be more or less explicit to the person who
makes it. Nonetheless, if pressed, the person could likely make the premises
of that inference explicit. The farmer might say she reasoned like so:

(1) The clouds are dark and heavy.
(2) The atmospheric pressure has dropped.
(3) Therefore, it will likely rain.

The doctor might say she reasoned like so:

(1) This patient has high cholesterol, which if left untreated will negatively impact his health.
(2) There are no contraindications to this patient being prescribed statins
to manage his cholesterol.
(3) Therefore, this patient ought to be prescribed statins.

And the grocer may say she reasoned like so:

(1) Most of the avocados which I have seen in this order are ripe and
salable.
(2) Therefore, most of the avocados in this order are ripe and salable.

According to Peirce, the logician's task is to subject such inferring to logical scrutiny. As such, logic is a science both of thought and of the contents

Peirce on Inference. Richard Kenneth Atkins, Oxford University Press. © Oxford University Press 2023.
DOI: 10.1093/oso/9780197689066.003.0002

of thought. Logic is a science of thought insofar as it studies the process of inferring. Logic is a science of the contents of thought insofar as it studies the arguments made in the process of inferring. Peirce distinguishes between self-controlled inference (reasoning proper) and inference that is not self-controlled. Sometimes, he restricts inference to self-controlled reasoning alone, as when he writes, "one opinion does produce another opinion. The former is a *premise*, the latter a *conclusion*. The process, provided it is deliberate, is called *inference*" (LoF 1:350, 1898). But usually, he treats self-controlled inferencing as reasoning proper and admits some inferencing is not self-controlled. In 1911, he distinguishes between reasoning as mental action (which he regards as synonymous with inference) and reasoning as a mental occupation in which "one casts about for arguments, considers them, and draws a conclusion from them" (R 852:2, 1911). These comments suggest that inference is a process of thought in which one draws a conclusion from other claims or commitments, whereas an argument is what is thought, the premises and conclusion of the inference. As I shall explain, the difference between inference and argument is that the former is a category of mental action, whereas the latter is a category of semiotic relations.

Peirce states that he uses the word 'logic' in two distinct senses, broadly as "the science of the necessary laws of thought" and narrowly as the "necessary conditions of the attainment of truth" (CP 1.444, 1896). In other passages, Peirce claims that logic consists of three subdisciplines:

(I) One of which studies "those properties of beliefs which belong to them as beliefs, irrespective of their stability";

(II) The second of which "consider[s] to what conditions an assertion must conform in order that it may correspond to the 'reality,' that is, in order that the belief it expresses may be stable. This is what is more particularly understood by the word *logic*. It must consider, first, *necessary*, and second, *probable* reasoning";

(III) And the third of which is the "study of those general conditions under which a problem presents itself for solution and those under which one question leads on to another" (CP 3.430, 1896).

This threefold division of logic corresponds with the study of those things on which we base our inferences, our inferences themselves, and finally the process of inquiry as a whole. Plainly, inquiry requires investigators to make inferences. Those investigators have various commitments or beliefs.

Their inferences on the basis of those beliefs may be better or worse. And the procedures the investigators follow may or may not have truth-producing virtues, such as being truth-preserving, approximating the truth in the long run, or conducing toward the truth in an efficient way.

Peirce's claim that logic is concerned with belief and with inference as a process of thought, however, seems to run afoul of his stated aim to develop an "Unpsychological View of Logic" (W 1:305, 1865). Peirce criticizes logicians (he specifically targets Edmund Husserl) who "after underscored protestations that their discourse shall be of logic exclusively and not by any means of psychology . . . forthwith become intent upon those elements of the process of thinking which seem to be special to a mind like that of the human race, *as we find it*" (CP 4.7, 1906). But people who live in glass houses shouldn't throw stones. Peirce often reflects on the ways in which people in fact think, so much so that scholars have wondered whether Peirce, too, is guilty of some form of psychologism, the thesis that the laws of logic are somehow grounded in or derived from psychological laws about how the mind works.[1]

I shall argue in this chapter that Peirce's curious intermingling of psychological and logical claims is not a threat to his development of an unpsychological view of logic. To be certain, Peirce regards the study of how people in fact reason to be a source of valuable insights into the nature and structure of inferences and arguments. Nonetheless, he denies that logic is merely concerned with the inferences that we in fact make; rather, his broader target is the sorts of inferences that could be made and how far they are good. Logic is concerned with both the possible forms of inference and the possible forms of argument.

Inference as Mental Action

Peirce's aim is to develop an unpsychological view of logic. But his classic essay "The Fixation of Belief" appears to confound this very task when he depends so clearly on psychological claims. Various methods of fixing belief, he tells us, fail because the "social impulse" and "a wider sort of social feeling" are against them (ILS 59, 62, 1877). He characterizes belief and doubt in terms of feelings, motivations, and dispositions to action. And he claims that the settlement of opinion—not the truth!—is "the sole object of inquiry" (ILS 58, 1877). Is not Peirce's practice of investigating how people think so as to glean logical insights inconsistent with his stated aim to develop an unpsychological view of logic? This apparent tension in Peirce's stances can be resolved if we draw a distinction between two different conceptions of logic and the role that empirical or experimental inquiries can play in identifying principles we do and ought to use in reasoning.

In order to better understand Peirce's commitments, we need to get clearly before us his account of inference. Inference, on Peirce's view, is a psychological process. Ordinarily, when we make inferences, we have some set of beliefs or commitments (the premises),[2] and from those we draw a conclusion. On Peirce's view, making an inference involves three core components.

The Three Core Components of Inference

The first component of an inference is the premises. These premises may be explicit to the person making the inference or implicit to her as she makes the inference. For example, suppose a person sets a copper disk spinning and discovers it stops spinning when placed between the poles of a magnet. She infers this will happen to all copper disks. What is explicit to her may only be the commitment that *this copper disk stopped spinning when placed between the poles of a magnet* together with the conclusion that *all copper disks will stop spinning when placed between the poles of a magnet*. Such an inference, however, presumably involves the implicit commitment that *what is true of one sample of copper with respect to its magnetic properties is true of all samples of copper*. This, too, may be regarded as a premise of her inference.

The second component of an inference is a habit, method, or way of proceeding which the person making the inference follows. For instance, in making the inference from *If it is raining, then the ground is wet* and *it is raining* to *the ground is wet*, a person employs a habit represented by the argument form modus ponens. If a person infers from a sample of beans that *50% of the beans in the bag are white*, she makes an inference. Her way of proceeding is to draw a sample. That sample might be drawn in different ways. She may pick through the beans to get her sample (in which case it will clearly be a bad induction). Or she may mix the beans thoroughly and take several large samples from different places in the bag (in which case her induction will be strong). The manner of proceeding is the habit or procedure. The habit or procedure may be only dimly apprehended by the person making the inference.

These habits or procedures are at least implicitly supposed by the person making the inference to have some truth-producing virtue. Though the settlement of opinion is the sole object of inquiry, opinion tends to be most settled when the opinion is true. Consequently, in primarily aiming at the settlement of opinion, inquiry also derivatively aims at the truth. Peirce identifies three truth-producing virtues of procedures: truth-preservation; truth-approximation over the long run; and truth-conduciveness, provided there is any truth to be ascertained. The supposition that any given procedure has these virtues may be false: The procedure may or may not in fact have the truth-producing virtue the person making the inference supposes it has. For example, a person may reason from *If it is raining, then the ground is wet* and *the ground is wet* to *it is raining*. If she supposes this inference to be a truth-preserving inference, then plainly she errs. The inference is a classic case of affirming the consequent. She has made an invalid deductive inference.

The procedure followed can be described. The description of the procedure can be treated as an implicit premise of the inference. We may call the descriptions of these procedures guiding principles of the inference. In the example given earlier about copper disks, *what is true of one sample of copper with respect to its magnetic properties is true of all samples of copper* is a guiding principle of the inference. It presupposes the procedure of treating one sample of copper as having the same properties as any given sample of copper. This may be clearer once we recognize that guiding principles can have both formal and material components. The formal component of the guiding principle is that *if S is a representative sample of population P, then the properties of S are the properties of the individuals comprising P*. The material principle is that *this sample of copper is a representative sample of all coppers*.

The third component of an inference is the conclusion. Conclusions are, of course, propositions. However, they are accepted under different modalities. In the example from the rain case given earlier where the argument is valid, the person concludes *it is the case that the ground is wet*. In contrast, in the case of sampling beans, the conclusion might be accepted under diverse modalities. For example, if the sample is small, the person may merely conclude *so far as I have ascertained, 50% of the beans in the bag are white*. But if she has taken several large samples from a well-mixed bag of beans, she may conclude *I am confident approximately 50% of the beans in the bag are white*.

When we take all of these components and put them together, we can state the total effect of such a process of inferring in the form of a conditional. That conditional will have (a) as its antecedent the premises and a description of the procedure used in drawing the conclusion and (b) as its consequent the conclusion of the inference under the modality in which it has been accepted. Suppose a person intends to make a deductively valid inference about the ground being wet. The effect of the entire processes might be:

> If (1) it is raining, then the ground is wet and (2) it is raining and (3) modus ponens is truth-preserving, then (4) it is the case that the ground is wet.

But in the case where the inference is intended to be deductive but is invalid, the effect of the entire process is:

> If (1) it is raining, then the ground is wet and (2) the ground is wet and (3) affirming the consequent is truth-preserving, then (4) it is the case that it is raining.

Valid deductive inferences are truth-preserving. Necessarily, if the premises are true, then the conclusion is true. But it is worth noting that 'necessarily' modifies the entire conditional and not the consequent. It would be a mistake to make the following inference:

> If (1) it is raining, then the ground is wet and (2) it is raining and (3) modus ponens is truth-preserving, then (4) necessarily, it is the case that the ground is wet.

In cases in which the inference is inductive, the effect of the entire process might be:

> If (1) half the beans in my samples are white and (2) I have taken several large samples from a thoroughly mixed bag of thousands of beans and counted them carefully and accurately, then (3) I am confident that approximately half of the beans in the bag are white.

But one might also make this inductive inference, which is also acceptable albeit weak:

> If (1) half the beans in my sample are white and (2) I have a small, randomly selected sample of ten beans from a bag of thousands of beans, carefully and

accurately counted, then (3) so far as I have ascertained, half of the beans in the bag are white.

Whereas this inductive inference is not acceptable:

If (1) half the beans in my sample are white and (2) I have a small sample of ten beans from a bag of thousands of beans, carefully and accurately counted, then (3) I am confident that approximately half of the beans in the bag are white.

This last inference is not acceptable because the sample is contemptibly small and may not have been randomly selected. To accept confidently the conclusion as being approximately true is a mistake.

Peirce's Four Conceptions of the Leading Principle

In Peirce's mature work, he calls the entire conditionals just stated the leading principle of the inference. Yet Peirce mentions leading principles throughout his early works as well. It is possible to discern four distinct conceptions of the "leading principle" in Peirce's writings.

First, in his earliest work, a leading principle is a conditional. These conditionals have one of two forms. The first form, for deductive reasoning, is *necessarily, if P, then C*, where P is the conjunction of the premises and C is the conclusion. The second form, for inductive reasoning, is *probably, C on condition of A*, where A is the conjunction of antecedent circumstances on which some event C is conditional. I call these sorts of leading principles modalized conditionals. They will not play much of a role in this chapter, but they will be relevant in the next chapter.

Second, in the 1870s, Peirce begins regarding implicit commitments used in inferences—such as *what is true of the magnetic properties of one sample of copper is true of the magnetic properties of all samples of copper*—to be leading principles. He calls these guiding principles in "The Fixation of Belief," but scholars have taken 'guiding principle' to be synonymous with 'leading principle.'

Third, in the 1880s, Peirce distinguishes between formal and material components of these guiding principles. As noted earlier, the formal component of the guiding principle about copper is that *if S is a representative sample of population P, then the properties of S are the properties of the individuals comprising P.* The material principle is that *this sample of copper is a representative sample of all coppers.* He calls these logical and factual leading principles, respectively. I shall call them logical and factual *guiding* principles, to distinguish them from the first and fourth conceptions of a leading principle.

Fourth, in the 1900s, Peirce seems to return to a position akin to his earlier conception of a leading principle. Instead of regarding leading principles as being of two sorts of modalized conditionals, however, he thinks of them as conditionals such that (a) the antecedent (i) states the explicit and implicit premises and (ii) describes the procedure used and its putative truth-producing virtue, whereas (b) the consequent of the conditional states the conclusion as accepted under some modality, such as *being the case* or *being so far as has been ascertained.*

At various points throughout this book, I shall have need to recur to each of these different conceptions of a leading principle. I shall call the first sort of leading principle a modalized conditional. I shall call the second and third sorts of leading principles guiding principles, sometimes referring specifically to the logical or formal component of the guiding principle and sometimes to the factual or material component. I shall call the fourth conception a leading principle proper.

Peirce is never as clear about these distinctions as one would hope him to be, no doubt because they are not clear to him. Indeed, his entry for 'leading principle' in *Baldwin's Dictionary* reveals that Peirce has not cleared up the doctrine for himself. In the next section, I shall take a closer look at his account of leading principles in *Baldwin's Dictionary* with the aim of dispelling these confusions attendant on Peirce's doctrine of leading principles.

Clearing Up Peirce's Doctrine of Leading Principles

It is very nearly a platitude that every inference can be stated as a conditional where the premises are the antecedent of the conditional and the conclusion is the consequent. This is the leading principle proper. Conditionals are suited to represent such processes of inference because they are asymmetric and they are true if and only if, provided the premises are true, the conclusion is true as well. With respect to asymmetry, whereas the premises have as a consequence the conclusion, the conclusion may not have as a consequence the premises. With respect to truth, material conditionals are false just in case the antecedent is true and the consequent is false.[3]

One question that arises when we consider the role of the leading principle proper is whether it is also a premise of the inference. If it is, then an account of leading principles proper is threatened by an infinite regress. The threat of such a regress is familiar from Lewis Carroll's "What the Tortoise Told Achilles." If we posit the leading principle proper as itself a premise, a sort of super-premise needed to connect the premises and the conclusion, then we run the risk of an infinite regress. We will have to postulate an additional super-super-premise to connect the first super-premise, the premises, and the conclusion, and so on ad infinitum.

Peirce's response to the problem of an infinite regress is to argue that once we analyze the inference as far as possible such that the same basic formal, logical principle repeats (say, of the valid argument form modus ponens), then analysis comes to an end.[4] As already indicated, such a principle is the habit or procedure used in the inference. It is in fact the logical guiding principle. We are already starting to glimpse a confusion in Peirce's doctrine of the leading principle. On the one hand, there are conditionals which are supposed to represent the total effect, the total process, of making the inference. On the other hand, there are the logical or formal principles that "lead" in the process of making an inference, such as modus ponens. Is the leading principle the total conditional, that is, the leading principle proper? Or is the leading principle the description of the procedure used in making the inference, that is, the logical guiding principle?

This confusion is evident in Peirce's entry for 'leading principle' in *Baldwin's Dictionary*. On the one hand, Peirce claims:

> [i]t is of the essence of reasoning that the reasoner should proceed, and should be conscious of proceeding, according to a general habit. . . . The effect of this habit or method could be stated in a proposition of which the antecedent should describe all possible premises upon which it could operate, while the consequent should describe how the conclusion to which it would lead would be determinately related to those premises. Such a proposition is called the 'leading principle' of the reasoning. (BD 2:1, 1901)

This passage suggests that Peirce has in mind the leading principle proper.

On the one hand, Peirce claims that:

(a) two persons may infer the same conclusion from the same premises using "different, or even conflicting, leading principles";

(b) that the person who makes the inference need not have a "distinct apprehension of the leading principle of the habit" which she follows;

(c) that a person may think she makes an inference in conformity with some leading principle when in fact she makes it in conformity with a different leading principle; and

(d) that "[f]rom the effective leading principle, together with the premises, the propriety of accepting the conclusion in such sense as it is accepted [i.e., as true, as approximating the truth in the long run; or as conducive to the ascertainment of the truth] follows necessarily in every case" (BD 2:1, 1901).

In addition, he proceeds to schematically present an argument wherein a leading principle consists of two components L and L'. He then imagines that some of the components of the leading principle (L or L') are themselves premises:

Suppose that the leading principle involves two propositions, L and L', and suppose that there are three premises, P, P', P''; and let C signify the acceptance of the conclusion, as it is accepted, either as true, or as a legitimate approximation to the truth, or as an assumption conducive to the ascertainment of the truth. Then, from the five premises L, L', P, P', P'', the inference to C would be necessary; but it would not be so from L, L', P', P'' alone, for, if it were, P would not really act as a premise at all. From P' and P'' as the sole premises, C would follow, if the leading principle consisted of L, L', and P. Or from the four premises L', P, P', P'', the same conclusion would follow if L alone were the leading principle. What, then, could be the leading principle of the inference of C from all five propositions L, L', P, P', P'', taken as premises? It would be something already implied in those premises; and it might be almost any general proposition so implied. Leading principles are, therefore, of two classes; and any leading principle whose truth is implied in the premises of every inference which it governs is called a 'logical' (or, less appropriately, a *formal*) leading principle; while a leading principle whose truth is not implied in the premises is called a 'factual' (or *material*) leading principle. (BD 2:1, 1901)

But this suggests that a leading principle may be either an explicit premise (as when one is cognizant of using the argument form of modus ponens) or an implicit premise. It also risks conflating the procedure with a premise. The act of sampling and the procedure of sampling are not themselves premises, even though the procedure can be described and that description can be included among the premises.

How are we to square these claims with Peirce's (apparently) straightforward account of a leading principle as a conditional with the conjunction of the premises as the antecedent and the conclusion as the consequent? I have already indicated that, in my judgment, Peirce's doctrine of the leading principle confuses different ways in which a person may be "led" in making an inference. In what follows, I shall make seven clarifications with respect to Peirce's doctrine of the leading principle.

Clarification One: First, even though every inference may be stated as a conditional, insofar as principles such as the argument form modus ponens are guiding principles of an inference, these principles are not the same as that conditional.[5] I have already been employing some terminology Peirce uses in "The Fixation of Belief" to help mark the distinction. In that essay, Peirce refers to guiding principles rather than leading principles. If we think of leading principles proper as the conditionals which state the effect of the entire process, then we may think of guiding principles as commitments which function somewhat like implicit premises of inferences or as descriptions of procedures used in making inferences.[6]

Peirce's example of an inference involving a guiding principle illustrates this point. He remarks that a person may observe a rotating disk of copper quickly come to rest when placed between the poles of a magnet. The person concludes that "this will happen with every disk of copper" (ILS 51, 1877). What is the leading principle of this inference? If leading principles are the entire effect of the process, then the leading principle is along the lines of *if this disk of copper quickly comes to rest when placed between the poles of a magnet, then this will happen with every disk of copper.* The guiding principle, however, is "what is true of one piece of copper is true of another" (ILS 51, 1877). This guiding principle should be added to the antecedent of the leading principle just stated.

This example shows how it is possible for (a) two people, from the same premise, to draw the same conclusion, even though they use different leading principles, for these leading principles are really guiding principles.

Guiding principles, such as the one about copper, can be false, and they can be more or less specific. Peirce remarks, "[s]uch a guiding principle with regard to copper would be much safer than with regard to many other substances—brass, for example" (ILS 51, 1877). Brass is a nonmagnetic alloy of copper. A disk of brass will not stop spinning when placed between the two poles of a magnet. Hence, the guiding principle that *what is true of any disk with copper in it is true of any other disk with copper in it* is false (for which reason Peirce claims it is not safe). Brass disks are disks with copper in them but not disks of copper. A copper disk will stop spinning if placed between the poles of a magnet, but a brass disk will not. The guiding principle about disks with copper in them is also more general than the principle concerning all copper disks, since the former but not the latter applies to copper and any of its alloys. The guiding principle may be even more general yet, such as *what is true of any metal disk is true of any other metal disk.*

Nonetheless, the inference from *this disk of copper quickly comes to rest when placed between the poles of a magnet* to *this will happen with every disk of copper* may be made in accordance with any of the three guiding principles just mentioned. The guiding principles, however, are different. In fact, they may even conflict. Suppose that one employs the guiding principle that *copper is the only metal such that the magnetic properties of any of its samples are the magnetic properties of any other.* This guiding principle (false though it is) would certainly license the inference from observation of one spinning copper disk coming to rest when placed between two magnets to any other copper disk doing the same. And yet that guiding principle is plainly inconsistent with the more general guiding principle that *what is true of any metal disk is true of any other metal disk,* which would license the same inference (again, false though the principle is).

The preceding explains how (a) two people may draw the same conclusion from the same premises though they use different "leading" (i.e., guiding) principles. It also helps to explain how it is that (b) a person may not have a distinct apprehension of the leading principle she uses when drawing the inference. These leading principles are in fact guiding principles. Did the person make the inference based on a guiding principle about metals in general, about metallic elements (as opposed to alloys), or about the specific metal in question (copper rather than mercury, say)? A person may make an inference without distinctly formulating the guiding principle she is using.

Consider another argument:

(1) All humans are mortal.
(2) All humans are rational animals.
(3) Therefore, all rational animals are mortal.

What is the guiding principle of this inference? If a person has studied syllogistic logic, she may initially think that the guiding principle she used in making the inference concerns a relation of inclusion: *human* is included in the scope of *mortal things* as well as in the scope of *rational animals*. So *rational animals* is included in the scope of *mortal things*. But such an inference is invalid. She may then wonder whether she is really susceptible to making such fallacious inferences. Upon further reflection, she may realize that she in fact takes *rational animal* to define *human*, and so the inference was validly made by substituting the definition of human in (2) for the subject term of (1). Only by more carefully probing her process of inference (i.e., by subjecting it to logical scrutiny) does she come to a more distinct apprehension of the guiding principle she in fact used.

These considerations show not only how (b) a person may make an inference without a distinct apprehension of the guiding principle she used but also how (c) a person could think she made an inference in conformity with some single guiding principle when in fact she made it in accordance with a different guiding principle. Peirce notes that one may even draw an illogical consequence when one "is not prepared to insist upon the leading principle as a premise" (NEM 3:756, 1893). Taking a leading principle in this context to be a guiding principle, on reflection one may realize that she perhaps should not endorse the guiding principle employed in her inferring.

Clarification Two: The second clarification is that guiding principles may be analyzable into different propositions. Peirce states that a guiding principle may be analyzed into different propositions L and L', and some of the propositions will be formal or logical and others will be material or factual. Consider again the guiding principle that *what is true of one piece of copper is true of another*. As already noted, one way of regarding this guiding principle is as constituted of two separate propositions, one formal and one material. The logical guiding principle is that *if S is a representative sample of population P, then the properties of S are the properties of the individuals comprising P*. The factual guiding principle is that *this sample of copper is a representative sample of all copper*.[7]

Contrast this with the more general guiding principle that *what is true of any metal disk is true of any other metal disk*. Analyzed in the same way, it would consist of the logical guiding principle that *if S is a representative sample of population P, then the properties of S are the properties of the individuals comprising P* together with the factual guiding principle that *this sample of copper is a representative sample of all metals*. Notice that the logical guiding principle is the same in both cases. However, the factual guiding principle is true in the first case but false in the second.

Furthermore, note that the logical guiding principle suggests a procedure, a way of proceeding in one's inquiry. In this case, the procedure is taking some sample one has as representative of the individuals comprising an entire class of the things, such that one can infer from the properties of the sample one has to the properties of the individuals comprising the entire class of things. As we shall see, the putative truth-producing virtues of the procedure determine the validity of the inference.

Clarification Three: The third point in need of clarification is that a proposition which is an implicit guiding principle in one inference may be an explicit premise in another inference. To continue with the copper example, a person may make the following inference:

(1) This sample of copper is representative of all copper.
(2) This copper disk stops spinning when placed between the poles of a magnet.
(3) Therefore, all copper disks will stop spinning when placed between the poles of a magnet.

Notice that the inference just given is different from the following:

(1) This copper disk stops spinning when placed between the poles of a magnet.
(2) Therefore, all copper disks will stop spinning when placed between the poles of a magnet.

In the former inference, the factual guiding principle is a premise of the inference, whereas in the latter the factual guiding principle is not a premise.

What, then, is the difference between a premise and a guiding principle? One might suspect that guiding principles are merely suppressed premises, so that Peirce's doctrine of the guiding principle amounts to little more than the observation that many times our inferences are enthymematic. But Peirce demurs. He has doubts about "the doctrine of a suppressed premise" (W 2:24n1, 1867). Either the suppressed premise is "conveyed" or not. If it is conveyed, then the premise is not suppressed. If it is not conveyed, then it is not a premise at all because it serves no function in the inference.

Rather, Peirce claims that one who makes an inference may only indistinctly apprehend the guiding principle of the inference. When a guiding principle (or its constituents) is clearly recognized, it "amounts to another premise" (W 2:25n1, 1867). But if a guiding principle or its constituents is only obscurely recognized, then while the person may become aware of the principle "by an act of attention," when the inference was made the person making the inference "did not *actually* understand this premise to be contained in it" (W 2:25n1, 1867).

Guiding principles or their constituent propositions may be more or less clearly recognized by the person making the inference when the actual inference is made. If the guiding principle or one of its constituents is clearly recognized by the person making the inference, then it is not only a guiding principle but also a premise of the inference. If the guiding principle is not clearly recognized by the person making the inference, then it is not a premise of the inference, even though it functions like a premise. The guiding principle functions like a premise because to draw the conclusion from the premises that were distinctly apprehended we must presume the person who made the inference to be committed to it, that is, that the information in that guiding principle was somehow conveyed.

The preceding suggests that when we subject an inference to logical scrutiny, we will endeavor to make all of the premises and guiding principles used in the inference explicit. When we do so, we will have stated the complete argument of the inference. Doing so will give us the leading principle proper, that is, the effect of the entire process. In fact, and as already suggested, it is even possible that the inference when actually made is complete. In such a case, the person who drew the inference has distinctly apprehended the guiding principle of the inference while making the inference. When this happens, Peirce asks, what shall we say is the leading principle? The question belies his confusion between leading principles proper and guiding principles. Peirce tells us that "[i]t would be something already implied in those premises; and it might be almost any general proposition so implied" (BD 2:1, 1901). That is, the leading principle is a proposition akin to: *If these premises are true, then the conclusion is true as well.* This is the leading principle proper, abstractly stated. The guiding principle, though, consists either of implicit premises or of descriptions of habits or procedures used to make the inference. These implicit premises and descriptions can be included as premises, included in the antecedent of the leading principle proper. I shall return to this point later.

I want to stress that guiding principles can function as premises. Note, however, that a habit is a procedure, a way of preceding, in making an inference. It is not properly a premise because it is not a proposition. Nonetheless, the habit or procedure can be described and treated as though it were a premise in the inference. Peirce writes, "[t]he particular habit of mind which governs this or that inference may be formulated in a proposition" (ILS 51, 1877). Such a description is a description of what Peirce calls the reasoner's logica utens, or logic in use: "Every reasoner, therefore, since he approves certain habits, and consequently methods, of reasoning, accepts a logical doctrine, called his *logica utens*" (CP 2.773, 1902). The person making the inference may have but a dim apprehension of what procedure she is following or what it involves. She may know no more of it than does a billiard player the laws of mechanics. In such cases, the logica utens is "like the analytical mechanics resident in the billiard player's nerves" (RLT 109, 1898). It is the reasoner's way of proceeding, even if that way of proceeding has not been established as a consequence of critical reflection on how to reason well. To be sure, a person's logica utens may be improved by a logica docens, by learned doctrines of good reasoning, but the logica utens need not be.[8]

Importantly, procedures are not propositions and so cannot be premises, even if they can be described and those descriptions can be treated as premises. Peirce writes:

> It is of the essence of reasoning that the reasoner should proceed, and should be conscious of proceeding, according to a general habit, or method, which he holds would either (according to the kind of reasoning) always lead to the truth, provided the premises were true; or, consistently adhered to, would eventually approximate indefinitely to the truth; or would be generally conducive to the ascertainment of truth, supposing there be any ascertainable truth. (BD 2:1, 1901)

Habits are ways of going about reaching conclusions from premises; they are methods or procedures. They are not premises, even though they can be described. Their descriptions can be treated as if they are premises of the inference.

Clarification Four: Peirce also tells us that the person who makes an inference accepts the conclusion "either as true, or as a legitimate approximation to the truth, or as an assumption conducive to the ascertainment of the truth" (BD 2:1, 1901). Peirce holds that we accept conclusions under certain modalities because the procedure is supposed to have "one kind of virtue or another in producing truth" (CP 2.780, 1901). These truth-producing virtues are of three kinds: truth-preservation; truth-approximation in the long run; and truth-conduciveness provided there is any truth to be ascertained. (Henceforth for readability, I shall drop the additional clauses "in the long run" and "provided there is any truth to be ascertained," but the reader should always regard them as implied.) Every time we make an inference, we do so in accordance with some procedure which we take to be either truth-preserving, truth-approximating, or truth-conducing. As I shall explain in more detail in the next chapter, these procedures and their truth-producing virtues are determinative of the species and genus of the inference, respectively.

The key point for the present is that the procedure is not a proposition and, a fortiori, not a premise of the inference. Nonetheless, the procedure can be described and the truth-producing virtue of the procedure is suggested in various modifiers, such as "probably," "necessarily," "so far as I have ascertained," and so on. Here, a distinction is needed between the truth-producing virtue the procedure in fact has (if it has any) and the virtue we take the procedure to have. A good deductive inference will be truth-preserving. However, we may wrongly take some procedure to be truth-preserving when it is not in fact truth-preserving. Similarly, a good inductive inference will be truth-approximating. However, we may wrongly take some procedure to be truth-approximating when it is not in fact truth-approximating.

Clarification Five: Peirce claims, "[f]rom the effective leading principle, together with the premises, the propriety of accepting the conclusion in such sense as it is accepted follows necessarily in every case" (BD 2:1, 1901). This statement needs to be unpacked. The fifth point of clarification concerns Peirce's statement that the conclusion follows in every case. 'Follows' here is not merely a temporal following but a logical following: "In inference, one belief not only follows *after* another, but follows *from* it" (SWS 21, 1895). To illustrate the difference, Peirce writes, "[i]f noticing my ink is bluish, I cast my eye out of the window and my mind being awakened to color remark particularly a poppy, that is no inference" (SWS 21, 1895). The difference between merely following-after and following-from is that in the latter but not the former there is both "some habit of association" and a "carrying forward of the *asserting* element of judgment" (SWS 22, 1895) from the premises to the conclusion.

A habit of association is not merely suggestion. The noting of bluish ink may suggest the color of the Niagara, but I am not disposed to think of the Niagara whenever I see bluish ink. In contrast, when I see an argument in syllogistic form, I am disposed to drop the middle term: "The operation of acquiring associations . . . is a sort of generalization by which, for example, the middle term is dropped" (RLT 191–192, 1898).[9] Moreover, that the conclusion follows from the premises also implies that commitments not pertinent to the conclusion are not to be regarded as premises of the inference. The inference does not contain any premises extraneous to the conclusion drawn. As Peirce states, to be a premise, a claim must "act as a premise" (BD 2:1, 1901).

Clarification Six: When Peirce claims that the propriety of accepting the conclusion in the modality at which we do "follows necessarily in every case," he has a distinct conception of in what this sort of necessity consists. There are varieties of necessity, including deontic, physical, and logical necessity. When considering inferences, we need to distinguish carefully between what Peirce calls the informationally necessary, on the one hand, and the logically necessary, on the other, in the contemporary sense of logical necessity.

It is easiest to start with logical necessity, since this will be most familiar to contemporary philosophers. Logical necessity pertains to logical structure and to the a priori. When we are concerned with the logically possible and necessary, we are to consider "the state of information in which we are supposed to know *nothing*, except the meanings of words, and their consequences, and in which we are supposed to know *everything*" (SWS 28, 1895). A proposition is logically possible just in case "a person who knows no facts though perfectly *au fait* at reasoning and well-acquainted with the words involved, is unable to pronounce untrue" (SWS 28, 1895). That is, granted the truth of the propositions and the meanings of the words, the conclusion is not inconsistent with the premises. In contrast, a proposition will be logically necessary just in case a person who knows no facts though perfectly *au fait* at reasoning and well-acquainted with the words involved, is unable to deny it without inconsistency.

In contrast, informational necessity concerns what we know, which may be merely logically possible but may also include knowledge of other matters. Information is "a state of knowledge, which may range from total ignorance of everything except the meanings of words up to omniscience" (SWS 27, 1895). The informationally necessary is that which is "perfectly known to be true" whereas the informationally possible is that which is "not perfectly known not to be true" (SWS 27, 1895). Plainly, these are suitable to be represented by our ordinary epistemic modal operator K and $\neg K \neg$ (or, to follow Hintikka, P). Such perfect knowledge may be the "actual information" we have "*for the present*" or it may be "some hypothetical state of knowledge" as when we imagine ourselves to know all the laws of nature (SWS 27, 1895). Accordingly, it may consist not only in what we know but in what we take ourselves to know. Furthermore, such perfect knowledge may be relativized to domains, so that metaphysical possibilities are "states of things which the most perfect . . . metaphysician does not *quâ* . . . metaphysician know not to be true" (SWS 27, 1895).[10]

With respect to logical necessity, we imagine a person to have no information except what is stated in the premises and the meanings of words (together with, presumably, the axioms of the relevant logical calculi). This is appropriate to evaluating deductive inferences, when we will logically scrutinize the propriety of accepting the conclusion given no more information than what is contained in the premises and the meanings of the words. However, we cannot evaluate inductive inferences in the same way, for the obvious reason that inductive inferences are not deductively valid, that is, truth-preserving.

Instead, when we inquire into the propriety of accepting an inductive conclusion, we will consider the procedure that the person has followed together with the premises and the conclusion. In these cases, we may ask whether given the procedure and the premises, the person making the inference is right to draw the conclusion she has. If drawn in accordance with a procedure that in fact has the truth-producing virtue it is supposed to have, such a conclusion will be an informationally necessary consequence of the procedure and premises. The conclusion will not, however, be a logically necessary consequence of them. Similar considerations apply, mutatis mutandis, to abductive inference, or inferences that follow truth-conducing procedures.

Accordingly, when Peirce claims that the conclusion is a necessary consequence of the premises, he does not mean that the conclusion is entailed by the premises in the sense of deductive entailment. He means that the premises supply sufficient information for the person making the inference to (i) know the conclusion is true, (ii) accept the conclusion has been drawn according to a procedure that approximates the truth if pursued indefinitely, or (iii) suspect that some conclusion is a good explanation given the evidence we have or that pursuit of the conclusion will be truth-conducive. With respect to (ii), notice that the person making the inference may not know the proposition stated as the conclusion. Induction thus differs from deduction in that when one knows the premises of a valid deduction one also knows the conclusion,[11] but when one knows the premises of a valid induction one does not thereby know the proposition stated in the conclusion. This is not surprising, since the sample may be small and inference on a small sample may not justify belief in the conclusion. Also, if we accept that the truth of the proposition is a necessary condition on knowledge, we may not know the conclusion of even a strong induction. Suppose a person takes several large samples from a well-mixed bag of beans and finds that half of the beans in the sample are white. She knows *half of the beans in the sample are white*. She infers *approximately half of the beans in the bag are white*. Note, though, that she may be wrong. By a bit of bad luck, her sample may be unrepresentative. Accordingly, if she is wrong, she does not know that approximately half of the beans in the bag are white since knowledge requires truth.

Before moving on to the seventh point of clarification, notice that on Peirce's account, the inference from *John is a bachelor* to *John is unmarried* will be logically necessary because one who knows the meaning of 'bachelor' knows that bachelors are unmarried men, regardless of whether that person knows whether John is in fact a bachelor. However, ordinarily the inference from *x is F* to *x is G* is not logically necessary, when *F* is not identical to *G*. Nonetheless, if we subject this inference to logical scrutiny and make all of the propositions comprising the guiding principle explicit, such an analysis will require us to regard both (i) *all bachelors are unmarried men* to be a factual guiding principle and (ii) the argument form of Barbara (*all M are P; all S are M; therefore, all S are P*) to be a logical guiding principle.

Clarification Seven: These considerations lead us to a seventh and final clarification. For there seems to be a tension in the preceding. On the one hand, Peirce claims that the person makes the inference in accordance with some habit or procedure. But as noted, the habit she follows may not in fact be truth-preserving, truth-approximating, or truth-conducing. She may think, for instance, that the conclusion is a necessary consequence of the premises when it is not. In that case, her inference will be fallacious. On the other hand, Peirce claims that the conclusion is an informationally necessary consequence of the premises: "From the effective leading principle, together with the premises, the propriety of accepting the conclusion in such sense as it is accepted follows necessarily in every case" (BD 2:1, 1901). (This was point (d) about leading principles, made earlier.) Notice that Peirce does not here claim the person making the inference *thinks* the propriety of accepting the conclusion at its modality does follow necessarily. Rather, the conclusion does in fact follow necessarily in every case. But in that case, it would seem to be impossible for a person to make a fallacious inference.

This aspect of Peirce's (confused) doctrine of the leading principle has its roots in his broader philosophical aims. In his early *Journal of Speculative Philosophy* series (1868–1869), Peirce aims to develop an alternative to the Cartesian project. As he states in "Some Consequences of Four Incapacities," his project is written in a "spirit of opposition to Cartesianism" (W 2:213, 1868).[12] All too briefly stated, Descartes maintains that all of our scientific knowledge is to be grounded in clear and distinct ideas, which are indubitable. According to Descartes, these clear and distinct ideas—such as that *I am a thinking thing* and that *God exists*—are known not by inference but by a nondiscursive faculty of intuition. Descartes also calls these innate ideas. As such, Descartes and his heirs[13] recognize two faculties, one inferential and the other intuitional.

Peirce, in contrast, argues that all of our knowledge is inferential and that we have no faculty of intuition. He does not deny that we have innate ideas, that is, ideas with which we are born. He maintains our idea of space, for example, is innate. Peirce contends, though, that these innate ideas are instilled in us through natural selection. They are vague and fallible; they are not indubitable. We may need to refine or even jettison some of them.

To meet the challenge of Cartesian skepticism, however, Peirce must develop an inferentialist account of knowledge that does not compromise our cognitive grip on the world. He does not need to show that we in fact currently possess knowledge of truths on which we can build the edifice of knowledge. Rather, what he needs to show is that his inferentialist account does not threaten the possibility of attaining knowledge of how things are. If inference is not valid, if it is not in some way truth-preserving, truth-approximating, or truth-conducing, then skeptical worries threaten his epistemological project. Inference needs to be such that it will eventually alight on the truth and root out error.

It is here that we glimpse the problem with Peirce's doctrine of the leading principle. On the one hand, he wants to maintain that mental action is of the nature of a valid inference. As valid inferential processes will eventually alight on the truth and root out error (precisely because they are truth-preserving, truth-approximating, or truth-conducing), it follows that mental action will eventually alight on the truth and root out error. On these grounds, Peirce can confidently proclaim, "[n]o modification of the point of view taken, no selection of other facts for study, no natural bent of mind even, can enable a man to escape the predestinate opinion. This great law is embodied in the conception of truth and reality. The opinion which is fated to be ultimately agreed to by all who investigate, is what we mean by the truth, and the object represented in this opinion is the real. That is the way I would explain reality" (ILS 98, 1878). If Peirce is correct, then a thoroughly inferentialist account of mental action does not threaten our cognitive grip on the world.

On the other hand, Peirce acknowledges that persons reason fallaciously. This fact is inconsistent with his claim that mental action is of the nature of a valid inference. And in fact, even in his early works, we find Peirce waffling between the claims that mental action is of the nature of a valid inference and that mental action can be modeled inferentially. He states, "every sort of modification of consciousness—Attention, Sensation, and Understanding—is an inference" (W 2:233, 1868), evidently committing himself to the thesis that mental action is in fact inferential. But he also states that it is "certainly very doubtful whether a conclusion—as something existing in the mind independently, like an image—suddenly displaces two premises existing in the mind in a similar way" to how the conclusion of an ordinary syllogism displaces the premises (W 2:214, 1868) and that he is endeavoring to reduce mental action to the "formula of valid reasoning" but only "as far as we can" (W 2:214, 1868).

In his later work, Peirce is much less sanguine about the claim that all mental action is of the nature of a valid inference. In 1902, he flatly denies that all mental action can be reduced to the form of a valid inference: "The real thinking-process presumably begins at the very percepts. But a percept cannot be represented in words, and consequently, the first part of the thinking [process] cannot be represented by any logical form of argument" (CP 2.27). Of course, the fact that mental action is not of the nature of a valid inference does not imply that mental action could not be of the nature of a valid inference. If we can figure out how to reason validly and conform our thinking to the nature of valid inference, then our reasoning will conform to the nature of valid inference. Peirce is not forced to abandon his earlier pragmatic characterization of truth, only to modify it. Truth is no longer the predestinate opinion; rather, "if we can find out the right method of thinking and can follow it out,—the right method of transforming signs,—then truth can be nothing more nor less than the last result to which the following out of this method would ultimately carry us" (EP 2:380, 1905). In 1911, he states, "I call[ed] 'truth' the predestinate opinion, by which I ought to have meant that which *would* ultimately prevail if investigation were carried sufficiently far in that particular direction" (EP 2:457).[14] That we will arrive at the predestinate opinion is no longer a great law but a "great hope" (ILS 98, 1894).

No doubt persons make fallacious inferences, and so it is not correct that all mental action is of the nature of a valid inference. But now suppose that we were to make an act of inference perfectly explicitly, so that it includes all of the premises and the guiding principles (both formal and material) used in the inference. Since it includes the logical guiding principle, it will also include a description of the procedure used in making the inference. Peirce holds that from the point of view of logical scrutiny, it will be informationally necessary that the premises and the leading principles imply the conclusion at the modality in which the conclusion is accepted.

Why should Peirce endorse this position? Since he abandons his earlier commitment to the thesis that all mental action is of the nature of a valid inference, it is not because all mental action is in fact a valid inference. But if in subjecting an inference to logical scrutiny we represent the inference as it has been made, and granted that such inferences may be fallacious, it follows that neither should we represent fallacious inferences in such a way that the conclusion is an informationally necessary consequence of the premises and guiding principles. But this is plainly inconsistent with his claim that "the propriety of accepting the conclusion" in its proper modality as an informationally necessary consequence of the premises follows in every case.

In order to understand Peirce's commitments here, we must once again disentangle leading principles proper from guiding principles. Consider again these two examples given earlier:

If (1) it is raining, then the ground is wet and (2) it is raining and (3) modus ponens is truth-preserving, then (4) it is the case that the ground is wet.

If (1) it is raining, then the ground is wet and (2) the ground is wet and (3) affirming the consequent is truth-preserving, then (4) it is the case that it is raining.

If we further posit (1), (2), and (3) of each of these leading principles as premises and (4) as the conclusion, we will get in the first case:

(0) If (1) it is raining, then the ground is wet and (2) it is raining and (3) modus ponens is truth-preserving, then (4) it is the case that the ground is wet.

(1) If it is raining, then the ground is wet.

(2) It is raining.

(3) Modus ponens is truth-preserving.

(4) Therefore, the ground is wet.

Doing the same for the second leading principle proper, both inferences will be perfectly valid. However, it is also plain that premise (3) of the second inference is false. Affirming the consequent is not truth-preserving. Premise (3), however, is not a leading principle proper. It is a guiding principle.

When we reconstruct the inference with the aim of subjecting it to logical scrutiny, we must be as charitable as possible. In being charitable, we will presume that the person making the inference is, in the main, a good reasoner. We will add to the list of premises guiding principles even when those guiding principles are known to be fallacious. Such charitable reconstructions of the inference may require us to include fallacious inference forms among the list of premises, as well as commitments to questionable methods and procedures for ascertaining the truth. For instance, with respect to deduction, if a person affirms the consequent, we will treat affirming the consequent as a logical guiding principle of the inference, even though it is false. Similarly with respect to induction, if a person uses an inappropriate sampling procedure, a description of that sampling procedure as truth-approximating in the long run will be treated as a logical guiding principle of the inference. The result is that we will have reconstructed the inference in such a way that the conclusion is an informationally necessary consequence of the premises and the guiding principles. The whole process, then, is valid when we consider validity relative to the leading principle proper but invalid when we consider validity relative to the guiding principle.

This way of looking at the matter is suggested in Peirce's discussion from "Some Consequences of Four Incapacities." There, he distinguishes between the evaluation of an inference's validity from the point of view of the psychologist and from the point of view of the logician. The psychologist, he claims, holds an "argument is valid only if the premises from which the mental conclusion is derived would be sufficient, if true, to justify it, either by themselves, or by the aid of other propositions which had previously been held for true" (W 2:222, 1868). It is those last clauses which are pertinent, for the psychologist will regard guiding principles of inference to be premises of the inference. Peirce states that a wrong rule of inference "is in fact taken as a premise, and therefore the false conclusion is owing merely to the falsity of a premise" (W 2:223, 1868).

As for the logician, fallacious reasoning occurs either "because their conclusions are absolutely inconsistent with their premises, or because they connect propositions by a species of illative conjunction, by which they cannot under any circumstances be validly connected" (W 2:222, 1868). That is, either the statements are inconsistent or the conclusion is not supported by the premises and the procedure used. The formal logician will evaluate the inference relative to the logical guiding principle, which describes the procedure used.

The upshot of these considerations is that we must distinguish the inference as made from the inference as subjected to logical scrutiny. The former will be invalid just in case either the procedure followed is not in fact truth-preserving, truth-approximating, or truth-conducing *or* the person has not in fact followed the procedure she professes to follow. The first is what occurs when one affirms the consequent, for example. The second case is what happens when a person makes an inference such as the following:

> If (1) it is raining, then the ground is wet and (2) it is raining and (3) modus ponens is truth-preserving, then (4) necessarily, it is the case that the ground is wet.

The person making such an inference has confused the procedure as involving a necessity of the consequence with the procedure as involving a necessity of the consequent.

In the case of the inference as subjected to logical scrutiny, though, we will endeavor to make as explicit as possible everything that must be assumed for the conclusion to be an informationally necessary consequence of the premises, including guiding principles. When we subject such inferences to logical scrutiny, Peirce maintains that we need to make the fallacy a premise of the leading principle proper. We must do so whether or not the person making the inference is cognizant of in fact being committed to such fallacious claims as are needed for the conclusion to be an informationally necessary consequence of the premises. The result is that even in the example just given, we should have to add a premise to the effect that the person thinks a necessary conditional implies the necessity of the consequent. The entire conditional would then be true.

Summary: Thus far we have seen that inferences consist of three core components: premises, a conclusion, and some procedure. We can state the entire process of inferring as a conditional with the premises and a description of the procedure as the antecedent and the conclusion as the consequent. This is the leading principle proper. The description of the procedure expresses the guiding principles used in making the inference. These guiding principles may be logical or factual guiding principles. The person making an inference supposes the procedure to have some truth-producing virtue, and either the procedure has that virtue or not. The conclusion is a proposition, and it is accepted under some modality given the truth of the premises.

With these distinctions made, one will rightly wonder whether they resolve the issue raised earlier of an infinite regress à la Lewis Carroll's "What the Tortoise Said to Achilles." As a psychological matter, there cannot be an infinite regress of premises. Humans have a finite mind, and as such our inferences are made from a finite set of premises. As already noted, Bellucci contends the correct Peircean solution to the problem is that the inference form is already implied by the premises and guiding principle. But it is not clear this solution will work. If a deductive inference follows the invalid inference form of affirming the consequent, then modus ponens is not implied by the inference. However, modus ponens must be accepted for the inference to be deductively valid: The truth of the premises will entail the truth of the conclusion. At best, we can get validity only on a charitable reconstruction of the argument that includes the rule of modus ponens.

A second route of reply, which Catherine Legg (2008) develops, is to contend the regress argument shows that there is something we cannot make explicit in our practices of inferring. This is what she calls the essential icon—the form or nature of the argument—that guides the inferential process. In my judgment, Legg hits the nail on the head. Although we can describe the procedures we have used to draw the inference and treat these descriptions as premises, they properly are not premises. It is the procedure which has (or is thought to have) truth-producing virtues. Although we can describe this procedure and although we can evaluate the procedure for whether it in fact has the truth-producing virtues claimed for it, the procedure is not properly a premise, for it is not a proposition. This, of course, is not to deny that the procedure can be described and treated as a logical guiding principle of the inference.

Peirce holds that when we subject an inference to logical scrutiny, we must endeavor to make the commitments involved in the inference as explicit as possible. The threat of an infinite regress arises only when we start positing the habits of the inference as premises of the inference. But when we subject an inference to logical scrutiny, we do not draw the inference yet again. We only describe the procedure used. The logical scrutiny of the inference is not the drawing of the inference. Consequently, the threat of an infinite regress is stopped so far as the logician's task is concerned. She is not drawing the inference; she is only remarking on the procedure and guiding principles used to draw the inference in the first place. She may then ask whether those procedures and guiding principles are any good.

Peirce's Unpsychological View of Logic

With our Peircean analysis of inference in place, we are now in a position to understand why Peirce's interest in inference as a psychological process does not threaten his aim to develop an unpsychological view of logic. Although the logician will be interested in inferences as they are made, she will subject those inferences to logical scrutiny. In so doing, she will make explicit the guiding principles and procedures or habits involved in the inference. Moreover, she will endeavor to charitably reconstruct the inference in such a way that it is valid. Finally, she will subject the charitably reconstructed inference to logical criticism.

All inquiry, whether scientific or otherwise, involves inference. Inference, in turn, employs guiding principles, including habits or procedures of inference and their presumed truth-producing virtues. As explained earlier, guiding principles may be more or less general. A guiding principle about copper is more specific than one about metals in general. They may also be true or false. When inferences are subjected to logical scrutiny, our habits or procedures will be described. Those descriptions will be included as implicit premises of the inference.

However, it does not follow that because some such guiding principle has been used that it is the only guiding principle that could have been used. This is true even of argument forms. For example, the inference form that *if X is R to Y and Y is R to Z, then X is R to Z*, where R is a relation and X, Y, and

Z are subjects, may be regarded as indifferent to whether R is transitive, as restricted to transitive relations alone, or as restricted to intransitive relations alone. Of course, only the second on the list will be acceptable, but the person who makes the inference may not realize this. Accordingly, we must distinguish guiding principles which *are* used from guiding principles which *could* be used. As Peirce remarks, two people may arrive at the same conclusion using two different guiding principles while being committed to the same premises. Failing to distinguish the guiding principles which could be used in some inference from the one that is used in an actual inference would incline us toward psychologism. Psychologism is the thesis that the laws of logic are somehow grounded in or derived from psychological laws about how the mind works. But Peirce makes no such identification of guiding principles with the mental habits we in fact have. What Peirce claims is that "[t]hat which determines *us*, from given premises, to draw one inference rather than another, is some habit of mind" (ILS 51, 1877, emphasis added). That is the inference as it is actually made. In our actual inferencing, we employ guiding principles.

Yet we can abstract away from any actual instance of inferencing, from what determines us in our inferences. If we do so, we might catalogue not only the guiding principles that are used in inferences but also any that could be used. As Peirce writes, "[a] book might be written to signalize all the most important of these guiding principles of reasoning" (ILS 51, 1877), but such a project would quickly spiral out of control. That is because "any fact may serve as a guiding principle" (ILS 52, 1877). This is evident from the distinction between logical and factual guiding principles. If a guiding principle is any commitment which must be granted to make the inference valid, this will include putative facts which we charitably assume the person making the inference to accept.

It is here that the role of logical scrutiny comes to the fore. When the logician charitably reconstructs a person's inference, she does so with the aim of understanding the premises and guiding principles from which the inference has been made. But she does not restrict her interest merely to those premises and inferences that have been made. They are but fodder for logical investigations, data on which she will draw but not to which her inquiries are restricted. Peirce's theory is unpsychological because the logician takes an interest in all the sorts of inferences that might be made and the conditions of their validity. To be sure, our analysis of inferences will pay heed to the actual processes of mental action or the actual practices of inquirers. But it is not restricted to those processes, and it is not restricted to an investigation of human reasoning in particular.

This leads us to the need to distinguish between two different ways of thinking about logic. First, logic may be formal logic, such as we commonly think of it today. It is mathematically informed and rigorous. Unquestionably, there is a need for formal logic; Peirce was one of the great founders of contemporary formal logic. And yet in a late work, he remarks, "[i]t is generally understood that I hold logical algebra to be the main part of logic. But that is quite a mistake. *I am in the world, but not of the world of formal logic*" (PSR 49, 1901). This leads us to a second conception of logic, according to which logic is broadly understood as the study of principles employed in all sorts of reasoning. This broader conception of logic includes formal logic but is not limited to it. Clearly not all the principles of reasoning are formal logical principles, as the copper and brass disk examples mentioned earlier show. There is in addition to formal logic informal logic. There is in addition to deductive logic inductive logic.

What are these other principles of reasoning, and to what extent are they good? Logic in the broad sense can be empirical and experimental, on the one hand, and yet not limited to how we in fact reason, on the other hand. It is not always evident which guiding principle is used in any given inference as it is actually made. As already noted, guiding principles can be stated in varying degrees of specificity and may be indistinctly apprehended. We may even think we have used some guiding principle when we have really used another. In order to ascertain which guiding principle has been used in an actual case of inference, we will have to answer the empirical question as to which guiding principles the person who made the inference committed herself. Only then can we evaluate the inference for whether it is a good inference or a bad one. But in order to ascertain which guiding principles could be used in making inferences, we can abstract away from actual inferences and consider which principles could be used in our inferential practices.

In his discussion of guiding principles in Section II of "The Fixation of Belief," Peirce remarks that though a catalogue of conceivable guiding principles would fill a book, some facts are "taken for granted in asking whether a certain conclusion follows from certain premises" whereas other facts are not (ILS 52, 1877). Among those facts taken for granted are that there are states of belief and doubt, that we can pass from a state of doubt to belief, that we can pass from belief to doubt, and that "this transition is subject to some rules which all minds are alike bound by" (ILS 52, 1877). Peirce aims to "deduce" from these processes rules of reasoning that we are all bound to obey if we wish our inferences to be truth-producing (whether truth-preserving, truth-approximating, or truth-conducting). Peirce recommends that we investigate the habits of thought, those species of guiding principles, that people in fact use. Among them, we will find that some have the virtue of being truth-producing, whereas others do not. Sorting through those guiding principles qua habits or procedures will give us clues as to which sorts of reasonings are good and which are not.

Peirce's recommendation may sound surprising, given how we standardly think of logic, especially so far as logic concerns deductive inference. Two points are in order. First, it is one thing to be able, with the power of retrospection, to formalize inferences rigorously and mathematically. It is quite another thing to strike off on such inquiries when the moorings of formalization and mathematics are uncertain. An excellent example is Aristotle, who clearly admits that his initial inquiries into logical principles were experimental. He remarks, "[o]n the subject of deduction, we had absolutely nothing else of an earlier date to mention, but were kept at work for a long time in experimental researches" (*Soph. Ref.*, 184b1–2). As Mark Greaves persuasively argues, Aristotle's claim to originality does not consist in no one having previously studied deduction. There had already been great advances in figure-based geometry, numerically based mathematics, and rhetoric. Rather, Aristotle's originality was in "laying out the framework of the universal art of deduction: general reasoning applicable to every study of substance" (2002, 97). Geometry and mathematics were restricted to their own particular kinds of substances (figures and numbers, respectively).

What I wish to emphasize is that Aristotle's researches were experimental. I take it his inquiries were experimental in that he (1) considered the different ways in which people such as geometers and number theorists actually reason, (2) abstracted from those considerations general principles of reasoning, (3) analyzed those principles into their constituent elements (quantities, terms, the copula, qualities), (4) tested different arrangements by seeking counterexamples to them, and (5) synthesized his results into a general framework, which we know today as syllogistic logic. Although he could take advantage of some of the work done in geometry, numerical mathematics, and rhetoric, he was generalizing from these results. He did not found his theory of the syllogism on mathematics but treated examples of mathematical reasoning as fodder for his own more general inquiries.

Second, although we do think of deductive inference in a strictly formal and mathematical way, it is important to bear in mind both (i) that there remains significant disagreement—if not choice—as to how we should understand logical consequence and how to address paradoxes such as the liar paradox and Curry's paradox and (ii) that it is much less clear we can give such a formal account of our inductive reasoning. As to (i), there are many-valued[15] and fuzzy logics, different accounts of logical consequence, and diverse ways of responding to semantic paradoxes. As to (ii), we have developed formal methods, such as the probability calculus and statistical tools, and accept diverse techniques of investigation, such as the use of random sampling. These formal methods and techniques of investigation, however, only apply generally to the diverse fields of inquiry in which statistical methods are used. Particular strategies need to be devised for more specific fields of inquiry. Studying human subjects is different from studying ice cores. Even though in both cases one will use statistical methods, an inquirer has to employ different strategies and implement different controls to ensure that the data and results hold. For any given inquiry, that these strategies and controls are truth-producing may only become evident in future inquiries. T. L. Short correctly argues that the right methods are discovered and improved upon over time. Scientists in pursuing their "vaguely defined aim . . . come, by degrees, to an ever more exact conception of that aim and, correspondingly, to a better idea of what their methods should be" (1998, 296).

As I have been suggesting, from a psychological point of view, we transition from one belief to another belief. When that transition is inferential, then that transition occurs according to some habit. When we look at this psychological process from a logical rather than psychological point of view, conclusions are taken to be informationally necessary consequences of their premises. Presumed in such inferencing are guiding principles. Guiding principles may be deductive logical guiding principles, such as the argument form modus ponens. Or guiding principles may be inductive guiding principles, such as that what is true regarding the magnetic properties of this copper sample is true of any other copper sample. The guiding principle will vary depending on the nature of the inference. Guiding principles express or presume ways of proceeding, for example, by taking random samples.

But logic is not exhausted by such psychological considerations. Even though we can look at the psychological process from a logical point of view, logic is not exhausted by the actual inferences people make. Rather, logic is concerned with all sorts of inferences, all the inferences that might possibly be made. Nevertheless, the psychological facts provide empirical fodder for broader logical considerations.

As noted earlier, the question of whether and to what extent Peirce's position may be described as psychologistic is somewhat vexed. My proposal is that on Peirce's view, logic can draw on empirical data about how persons in fact reason without acquiescing in the claim that logical norms are explained by facts of human psychology. We can look to the psychological facts for clues as to in what good reasoning consists. Having done so, we can abstract away from the psychological facts in order to start cataloguing varieties of guiding principles. From reflection on inferences as they are made, we can abstract some guiding principles as to what distinguishes good inferences from bad ones. In short, Peirce's psychological description gives us some empirical fodder for logical considerations. Peirce's view of logic is unpsychological in that the truth of guiding principles does not depend on the constitution of the human mind. But one can have an unpsychological view of logic while drawing on psychological considerations for logical insights.

One last point bears making before turning to Peirce's account of arguments. In a lecture from 1903, Peirce appears to criticize his account in "The Fixation of Belief" and "How to Make Our Ideas Clear" when he remarks:

> But *how do we know* that belief is nothing but the deliberate preparedness to act according to the formula believed?
>
> My original article ["How to Make"] carried this back to a psychological principle [in "Fixation"]. . . . I do not think it satisfactory to reduce such fundamental things to facts of psychology. For man could alter his nature, or his environment would alter it if he did not voluntary do so, if the impulse were not what was advantageous or fitting. Why has evolution made man's mind to be so constructed? That is the question we must nowadays ask, and all attempts to ground the fundamentals of logic on psychology are seen to be essentially shallow. (PPM 116)

It is easy to misread Peirce here. Peirce is not claiming that his early theory of logic was itself subject to the charge of psychologism. We have already seen that Peirce has an unpsychological view of logic as early as 1865. Neither is Peirce claiming that he ignored evolutionary considerations in his earlier work. That is patently false, for in "The Order of Nature" he explicitly appeals to evolutionary considerations. Peirce remarks, "the mind of man is strongly adapted to the comprehension of the world. . . . How are we to explain this adaptation? The great utility and indispensableness of the conceptions of time, space, and force, even to the lowest intelligence, are such as to suggest that they are the results of natural selection" (ILS 161, 1878). By 1903, it was generally recognized that evolutionary considerations are paramount, whereas in 1878 it was not. Peirce was not behind that curve but ahead of it.

What Peirce is claiming in the block-quoted passage is that his argument for the pragmatic maxim in "How to Make" rested on psychological claims in "Fixation." And so it does.[16] Peirce's argument in "How to Make" is that belief disposes us toward action. The contents of beliefs are, partly, ideas. Therefore, the meaning of an idea is, at least in part, how it disposes us to act.[17] Clearly, this argument is based in a psychological claim about what beliefs do, viz., dispose us toward action. This is why his early defense of the pragmatic maxim is psychological.[18]

The problem is that that argument for the pragmatic maxim fails to take into account the logical point of view. Ideas may have consequences which we fail to consider and so generate no dispositions in us. In his later work, Peirce is much more concerned to emphasize that ideas are independent of the vagaries of contingent human knowers. Recall that while psychological considerations may be fodder for logical inquiries, they do not exhaust logical considerations. We take a logical interest in the guiding principles actually used as well as those that might possibly be used. Similarly, an account of the meaning of ideas should not be restricted to how we are in fact disposed to act. It should include all the various ways in which an idea might possibly inculcate dispositions in an inquirer. That is why it is not satisfactory to rest a logical principle such as the pragmatic maxim on a psychological principle. The error consists not in using psychological principles as fodder for one's logical inquiries but in thinking one's logical inquiries can be completed or adequately supported by a study of psychological principles alone.

Argument as a Semiotic Category

Thus far, I have argued that whereas the logician will take an interest in inference as a psychological process, this does not threaten the project of taking an unpsychological view of logic. The logician will reflect on and criticize inferences that are actually made. But such critical scrutiny of inferences is for the purpose of gathering information about all the sorts of inferences that might be made. The logician's interests are not restricted to actual inferences but extend to all the sorts of inferences that might be made.

Moreover, I have drawn a distinction between the process of inferring and the contents of that inferring. When we reflect on an inference, we will endeavor to make all of the elements that go into the process of inferring explicit. These core elements are the premises, the habit or procedure used, and the conclusion. The habit will be described and functions as a guiding principle of the inference. Moreover, the person inferring will take it that the habit has some truth-producing virtue. She will intend to make a truth-preserving, truth-approximating, or truth-conducing inference.

Like many of us, Peirce will often use 'argument' and 'inference' interchangeably. We make arguments as much as we make inferences. In this vein, we find Peirce distinguishing between arguments and argumentations, where an argument "is any process of thought reasonably tending to produce a definite belief" and an argumentation is "an [a]rgument proceeding upon definitely formulated premisses" (EP 2:435, 1908). This distinction between arguments and argumentations tracks the distinction between inferences in which the guiding principles are only indistinctly apprehended and those in which the guiding principles are distinctly apprehended.

But Peirce also uses 'argument' in a distinctly technical and semiotic sense. In this sense, arguments are distinguished from propositions and terms. Briefly, a term is a common noun or phrase, such as *speaking monkey* or *glass*. Cognitively, a term is grasped in a simple apprehension; it cannot be asserted.

A proposition is a complex sign consisting of a symbol as well as an index. A symbol is a general term the signification of which is usually established by convention.[19] For example, the words *monkey* and *glass* are symbols. An index is a sign that stands in some "real, physical connection" with its object (W 2:225, 1868). Indexes may be demonstratives, definite descriptions, or quantifiers.[20] Accordingly, *monkeys speak* and *this is glass* are propositions. In contrast with terms, propositions can be asserted. Peirce states that a proposition is "a representation of which one part serves, directly or indirectly, as an index of its object, while the other part excites in the mind an image of that same object" (SWS 140, 1903). That is, the index indicates some thing to which the symbol is to be applied, or else the index supplies some rule or percept for finding some thing to which the symbol may be applied (see EP 2:168, 1903). The symbol functions to elicit an interpretant, which may literally be an image in the mind, or it may be a feeling, an action, or another proposition (see EP 2:430, 1907).

Importantly, though, a proposition does not show or state which interpretant is to be elicited.

In contrast with a proposition, an argument does state what the interpretant of the sign is. The conclusion of the argument is the interpretant of the premises, and the premises are the sign. From a set of premises, any number of conclusions might possibly be drawn. An argument, however, states which conclusion is to be drawn from the premises. Peirce remarks, "[a]n *Argument* is a sign which distinctly represents the Interpretant, called its *Conclusion*, which it is intended to determine" (CP 2.95, 1902). The conclusion is a logical interpretant of the premises.

Peirce's reference to an intention, however, is vexing. It is vexing because here again we seem to run up against worries of psychologism. How does the argument intend its conclusion? In other passages, Peirce states that the premises show the conclusion they aim at (SWS 99, 1901). How can an argument aim at a conclusion?

This peculiarity in the definition of an argument is not unique to Peirce. In the contemporary logic textbooks, we find one group of scholars who, in defining what an argument is, refer to intentions or claims of support. For instance, the *Blackwell Dictionary of Western Philosophy* states that an argument is "[t]he reasoning in which a sequence of statements or propositions (the premises) are intended to support a further statement or proposition (the conclusion)" (Bunnin and Yu 2008, 46). With its reference to reasoning, this definition risks conflating inference as a psychological process with argument as a semiotic category. Moreover, it appeals to an intention. Presumably, this is the intention of the reasoner. Also, Gregory Johnson defines an argument as "a set of statements, one or more of which are intended to support another in the set" (2018, 2). This definition of an argument doubtfully implies there are no arguments from an empty set of premises, and it requires that the premises be intended to support the conclusion. Again, though, we must ask how the intention is supplied. As a final example, the authors of *Logic and Philosophy* define an argument as a "a set of sentences, one of which (the conclusion) is claimed to be supported by the others (the premises)" (Hausman et al. 2021, 15). Presumably, the claim of support is made by the person who proffers the argument.

In contrast with this trend of defining arguments with reference to intentions and claims of support, other scholars make no such appeal when they define what an argument is. Perhaps most curiously, also in *Logic and Philosophy*, an argument is defined as "a set of statements, one of which is the conclusion (the thing argued for) and the others are the premises (reasons for accepting the conclusion)" (Hausman et al. 2021, 1). Notice that in this definition there is no mention of a claim of support; the authors of the book make no note of the difference.[21] In Merrie Bergman et al.'s *The Logic Book*, an argument is defined as "a set of two or more sentences, one of which is designated as the conclusion and the others as the premises" (2004, 9). Although we might wonder how the conclusion is designated, note that this definition does not state that premises are intended to support the conclusion. Volker Halbach, in *The Logic Manual*, states an argument "consists of a set of declarative sentences (the premisses) and a declarative sentence (the conclusion) marked as the concluded sentence" (2010, 17). Here again, we find no appeal to claims of support or to intentions. As a final example, Brian Skyrms defines an argument as "a list of *statements* [i.e., sentences that make a definite factual claim], one of which is designated as the conclusion and the rest of which are designated as premises" (2000, 13).

On one conception of an argument, an argument is but a list or set of declarative sentences, one of which is designated as the conclusion and the rest of which are the premises. Call this the list conception of an argument. On another conception of an argument, an argument is a list or set of declarative sentences, one of which is designated as the conclusion and the rest of which are the premises, together with some claim or intention regarding the support of the premises for the conclusion. Call this the list-plus-intention conception of an argument.

While this difference may seem to amount to little more than a quibble, it has significant implications for how we evaluate arguments. If we adopt the list conception of an argument, then we can ask what guiding principles might have been used to make the inference. If we adopt the list-plus-intention conception, we are supplied with information which suggests what principles were used to make the inference. For example, if the person intends for the argument to be truth-preserving, then the argument is deductive. If the person intends for the argument to be truth-approximating, it is inductive. The intention, together with the premises, is indicative of the habit or procedure used. We can then ask whether the argument in fact has the truth-producing virtue claimed for it.

The difference comes out most clearly when we consider two different but common ways of parsing how we classify and evaluate arguments. On one common way, arguments are either deductive or inductive. These arguments may be good or bad. With respect to logical evaluation, a good deductive argument is a valid argument; a bad deductive argument is an invalid one. In contrast, a good inductive argument is a strong one, and a bad inductive argument is a weak one. On this construal, arguments are either deductive or inductive, and each has its standards of evaluation. Such a view lends itself to the list-plus-intention conception of arguments. A person may present an argument which she intends to be truth-preserving. However, if her guiding principle is not truth-preserving—say, she affirms the consequent—then her deductive inference is invalid.

But on the list conception of an argument, there are not two kinds of arguments with two different standards of evaluation. Rather, there is a single scale, the top of which is that the argument is deductively valid and the bottom of which is that the argument is worthless. In between these two extremes are degrees of inductive strength and weakness. Such a view lends itself to the list conception of arguments. We are given a list of statements which includes at least a conclusion and may include some premises. We may then ask whether the argument is deductively valid. If it is not deductively valid, we may then ask whether and to what degree it is inductively strong. In so doing, we make no appeal to whether the person intended to make a deductive or inductive argument.

To illustrate, suppose a person makes the following inference (n.b., not argument!): *If it has rained, then the ground is wet. The ground is wet. Therefore, it has rained.* Now imagine that the logician reflects on this inference and wants to evaluate whether it was any good. The content of this inference is an argument. Suppose we adopt the list-plus-intention conception of an argument. Then the logician may ask the person who makes the inference whether she thinks the premises guarantee the truth of the conclusion. If she does, then the logician will politely inform her that the argument is deductively invalid. It is a case of affirming the consequent. If she instead states that she intended it to be an inductive argument, the logician will ask whether she considered other explanations for the wet grass, such as that it is a dewy morning or the lawn has been watered. These would be implicit commitments used in the inference which the logician ought to make explicit in the argument. Depending on whether these were considered and ruled out, the argument may be inductively strong or weak.

In contrast, suppose we adopt the list conception of an argument. We will first make sure that we have the full list of premises explicitly before us. Once we do, we will ask what sorts of procedures might make the argument good. Plainly, deductive procedures[22] will not be any good, for the truth of the conclusion is not guaranteed by the truth of the premises. We may then ask whether there is some inductive procedure the person making the inference might have followed. Depending on whether the person considered and eliminated other possible explanations, the argument (and by extension, the inference) will be strong or weak, or not inductively good at all.

Especially on the list conception of an argument, it is important to realize that evaluation of the argument from an inductive point of view will require positing or supposing some procedure that the person drawing the inference has followed. In such cases, the logician evaluating the argument will have to propose or conceive of procedures according to which the inference might be valid. For example, suppose an argument qua list of statements consists only of the premise *50% of the beans in this sample, which has been drawn from this bag, are white* and the conclusion *therefore, approximately 50% of the beans in this bag are white*. Plainly, such an inference is not deductively valid. But taken just at face value, it is not possible to evaluate whether this argument qua list of statements is inductively good or not. We must consider ways in which the sample might have been taken. Depending on the procedures of which we conceive, the argument may or may not be truth-approximating. If, for instance, the person sorted through the beans, found one white one and one non-white one, and declared *50% of the beans in this sample, which has been drawn from this bag, are white*, she could not validly infer that *approximately 50% of the beans in this bag are white* because such a procedure is not truth-approximating in the long run. A person may simply keep taking biased samples to get the conclusion she wants. In contrast, if a person mixes the beans thoroughly, draws several handfuls from different places in the bag, counts the beans in the handfuls she has drawn, and finds that *50% of the beans in this sample, which has been drawn from this bag, are white*, her inference is inductively valid. Although she may be wrong that *approximately 50% of the beans in this bag are white* just because it is possible her sample is biased, continued repetition of the procedure is truth-approximating in the long run.

The two accounts of evaluating arguments just canvassed are fully compatible with each other once we distinguish between the two different conceptions of an argument. If we conceive of an argument as a list of declarative sentences independent of any procedure the person making the inference used, then we may ask what sort of procedures could have been used to generate a good argument. But if we conceive of an argument as a list of declarative sentences plus some intention that the argument be truth-preserving, truth-approximating, or truth-conducing, the intention together with the premises is suggestive of some procedure which may or may not be truth-producing in the intended way. Once we are clear about the procedure, we can pronounce whether the argument is good or not.

This leads us back to Peirce's semiotic conception of an argument as well as his aims to develop an unpsychological view of logic. As noted earlier, Peirce defines an argument as a sign (the premises) which is intended to determine a conclusion, as well as a sign that shows or aims at a conclusion. One way to understand the notion of determining, intending, or aiming is in accordance with the list-plus-intention conception of an argument. Ordinarily, a logician who is reflecting on inferences as they have been made will first make the inference as explicit as possible. Doing so, she will not only list the premises involved in the inference but also the description of the procedure and its presumed truth-producing virtue. In this sense, the person who makes the inference aims at the conclusion by taking the procedure to have some truth-producing virtue relative to the premises and the conclusion drawn.

However, Peirce would also allow that we can abstract away from the inference as it has been made and consider what sorts of procedures might make the inference good. Such an account lends itself to the list conception of an argument. On this account, the sign shows its conclusion merely by designating it as the conclusion, that is, as the member of the set of declarative sentences which interprets the other declarative sentences, the premises. Once the premises and conclusion have been made explicit, we can then ask whether the truth of the premises guarantees the truth of the conclusion (deduction), whether there is some procedure that might have been used to approximate the truth over the long run (induction), or whether there is some procedure that might have been used to conduce toward the truth (abduction).

This second way of evaluating arguments is the more unpsychological view of logic, since it makes no reference to the procedures of the reasoner or her intentions to make truth-preserving, truth-approximating, or truth-conducing inferences. However, it is also more removed from the actual practices of inquiry. For oftentimes when we subject arguments to logical scrutiny, they are the arguments of inferences we have already made. In the process of inquiry, we wish to know whether and to what extent *that* inference, the very inference we have made, is good. Nonetheless, both manners of evaluation are appropriate to logic. It is quite possible, for instance, that we wrongly think the premises provide a deductive level of support for the conclusion when they do not. Upon reflection, we may reject the inference as deductively valid while continuing to endorse the inference as inductively strong.

Conclusion

Peirce aims to develop an unpsychological view of logic. He does not, however, deny the importance of psychological considerations to developing an unpsychological view of logic. Inference is a psychological process of drawing a conclusion from a set of premises by adhering to some procedure. The logician will reflect on these inferences. She will endeavor to make the premises and guiding principles used in the inference as explicit as possible. Yet the logician may also abstract from the inference as it has been made and consider the possible inferences that might be made. She may also abstract the premises and the conclusion of the argument from the procedure in fact used by the reasoner. In so doing, the logician may inquire into what sorts of procedures might make such an argument qua list of declarative sentences good. In these later cases, the logician reflects on and criticizes arguments at a higher level of abstraction but also at a level more removed from the actual practices of the inquirer. Both sorts of investigations—into inferences as they have been made and inferences as they might be made—are appropriate to logic.

2

Classifying Inferences

In the previous chapter, I distinguished between inferences and arguments. Inference is a psychological process in which one draws some conclusion from putatively asserted premises in accordance with some procedure. Every inference consists of a set of premises, a conclusion, and some procedure to which the person making the inference adheres, though she may only be dimly aware of the procedure she follows.

The logician may also abstract away from the inference as it has been made and consider what sorts of procedures a list of premises together with a conclusion may support. In such a case, the logician is not concerned with any actual inference that has been made but with inferences that could be made. In these cases, in order to have a complete argument, the logician must imaginatively reconstruct some proceeding and regard the list of declarative sentences to have been made in accordance with the proceeding. In cases of deduction, the reconstruction is fairly straightforward: We consider the logical form of the argument and the meanings of the words, and from these we can infer whether the conclusion is a logically necessary consequence of the premises given the meanings of those words. But in cases of induction, we can ordinarily conceive of circumstances wherein different procedures could have given us the same conclusion from the same premises. For example, we might have a sampling procedure that is biased (sorting to get the proportion that one wants) or one that is fair (random sampling), yet both could yield the conclusion that 50% of the beans in a given population are white.

The premises and conclusion alone cannot tell us what sort of inference has been made. As noted in the previous chapter, a person may make an inference from the premises *if it has rained, then the ground is wet* and *the ground is wet* to *it has rained*. However, the person may be making either an invalid deductive inference or an inductive inference in which she endeavors to arrive at what explains why the ground is wet. We can only distinguish between the two when we consider the procedure she intended to follow. Did she intend to make a truth-preserving inference by deducing the conclusion from the premises in accordance with some set of axioms and rules? In that

Peirce on Inference. Richard Kenneth Atkins, Oxford University Press. © Oxford University Press 2023.
DOI: 10.1093/oso/9780197689066.003.0003

case, her inference is an invalid deduction. Or did she intend to make a truth-approximating inference by ruling out various explanations for the ground being wet? In that case, her inference may be a strong or weak induction, depending on whether and which alternative explanations she ruled out.

It is different procedures which distinguish among the varieties of inference. On Peirce's mature classification of inferences, it is the truth-producing virtues of the diverse ways of proceeding which are determinative of the inference's *genus*. The genera of inference are, in turn, specified by the diverse ways of proceeding. For deductions, the truth-producing virtue is truth-preservation. For inductions, the truth-producing virtue is truth-approximation over the long run. For abductions, the truth-producing virtue is truth-conduciveness, provided there is any truth to be ascertained. These three genera of inference admit of further specification. For example, inductions may be either hypothetic (qualitative) inductions or statistical (quantitative) inductions. Similarly, deductions may be either necessary demonstrations or probable demonstrations.

Peirce's manner of classifying inferences as just articulated is his more mature account of inference, developed after 1900. The story with respect to Peirce's earlier work is more complicated and can easily obscure the subtler position just described. Some presentations of Peirce's classification of inferences make it seem as though Peirce distinguishes among deduction, induction, and abduction from his earliest writings to his last and that he did so on the same grounds throughout. This picture is deeply mistaken. In fact, Peirce's classification of inferences goes through several distinct phases and modifications. Even toward the end of his life, he had serious doubts about it. Unless one is attentive to these changes, Peirce's theory of inference will be misunderstood.

Perhaps one of the most pervasive errors concerns the relationship between induction and abduction. On Peirce's early theory of inference, inferences are either deductions or inductions. Inductions are of two varieties. They are either hypothetic inductions or statistical inductions. In his early work, what Peirce calls hypothesis is hypothetic induction and is not the same as what he later calls abduction. Moreover, what Peirce calls induction is more specifically statistical induction and not induction in general. In this chapter, I make the case for these claims more fully by examining Peirce's developing classification of inferences.

It will be helpful to orient the reader by distinguishing among three stages in Peirce's classification of inferences. In his early works of the 1860s and

1870s, Peirce's major distinction between the genera of inferences is between deduction and induction, or as he later calls them explicative and ampliative inferences. At these stages, there is no genus of abductive inference. The first stage is from the 1860s. Then, Peirce classifies inferences as follows:

I. Necessary (deductive)
II. Probable (inductive, broadly conceived)
  A. Hypothetic inductions[1] (hypothesis), or
  B. Statistical inductions (induction, narrowly conceived)

By the 1870s, the classification changes. At this time, he holds that inferences made in conformity with the probability calculus alone are deductive, even though they are probable. Accordingly, his classification is that inferences are as follows:

I. Explicative (deductive)
  A. Necessary (based on the propositional or predicate logic calculi), or
  B. Probable (based on the probability calculus)
II. Ampliative (inductive)
  A. Hypothetic inductions, or
  B. Statistical inductions

Peirce's classification continues to undergo a series of changes over the next decades. On his mature classification, which emerges in the early 1900s, inferences are as follows:

I. Deductive
  A. Logical analyses of ideas or hypotheses, or
  B. Demonstrations
    i. Necessary, or
    ii. Probable
      a. Axiomatic (based on the probability calculus), or
      b. Statistical deductions (inferences of probability-that statements from probability-of statements)
II. Inductive
  A. Crude inductions (induction by absence of counterexample), or
  B. Qualitative inductions (hypothetic inductions), or

C. Quantitative inductions (statistical inductions)
III. Abductive

In this chapter, I trace the emergence of Peirce's late classification by considering what motivates these shifts. First, I look at his distinction between deduction and induction in the 1860s and 1870s. Second, I differentiate between Peirce's late account of abduction and his early account of hypothetic induction. These first and second sections explore how Peirce identifies the three genera of inference, viz., deduction, induction, and abduction. In the third and final section of this chapter, I explain how Peirce enriches and deepens his classifications of deduction and induction. I defer a fuller discussion of what abduction involves to Chapter 7.

Deduction versus Induction in the 1860s and 1870s

In his early work, Peirce's primary distinction between genera of inference is between deduction and induction, broadly conceived. Peirce's distinction between these two genera of inference undergoes a series of changes. There are two basic phases or developments to sort out. The first phase, found in the *Journal of Speculative Philosophy* Series (1868–1869) and writings from that time period, regards the distinction between deduction and induction to be determined by the modality of a conditional in which the premises are the antecedent and the conclusion is the consequent. These are the modalized conditionals briefly mentioned in the previous chapter. The second phase, found in the *Illustrations of the Logic of Science* series from the 1870s, treats the distinction between deduction and induction as determined by whether or not the procedure consists in inference from axioms and definitions in accordance with rules.

Modalized Conditional Phase of the 1860s

The Account: The first phase is from the 1860s. In my judgment, Peirce's writings from the 1860s are confused but in deep and informative ways. This is largely because he is trying to navigate tensions between two different approaches to logic. The first is the traditional Aristotelian, syllogistic approach to logic. The second is the emerging mathematization of logic in the work of Boole, Jevons, and others. The result is a tangled skein of commitments not easy to untangle. Nonetheless, what is evident is that in the 1860s, Peirce regards all deductive inference to be necessary reasoning. Necessary reasoning contrasts with probable reasoning.

An inference is not simply a list of claims. Rather, the conclusion is taken to follow from the premises in some way. In Peirce's early works, there are only two truth-producing virtues of arguments, truth-preservation and truth-approximation in the long run. Either (a) it will be necessary that if such-and-such premises are true, then the conclusion is true or (b) it will be probable that if such-and-such premises are true, then the conclusion is true.

To illustrate, consider the following two inferences, the first of which is deductive and the second of which is inductive:

Inference A
  (1) All horses are mammals.
  (2) All mammals are animals.
  (3) Therefore, all horses are animals.

Inference B
  (4) The clouds are dark and heavy.
  (5) The atmospheric pressure has dropped.
  (6) Therefore, it will rain.

As 'therefore' indicates, the last statement in each argument is supposed to follow from the first two statements. It is not merely the case that all horses are mammals *and* all mammals are animals *and* all horses are animals. Rather, the third claim is taken to be a consequence of the first two. It is not merely the case that the clouds are dark and heavy *and* the atmospheric pressure has dropped *and* it will rain. Rather, the third claim is supposed to be a consequence of the first two. However, in the first case, the conclusion always follows from the premises. In the second case, the conclusion usually follows from the premises.

To the Peirce of the 1860s, these considerations suggest that every inference involves an implicit logical commitment. That implicit commitment is a conditional statement that has some modality. In the case of deductive inferences, the implicit commitment is that *necessarily, if such-and-such premises are true, then the conclusion is true*. In the case of probable inferences, the implicit commitment is that *probably, if such-and-such premises are true, then the conclusion is true.* Equivalently, the implicit commitment is that *probably, such-and-such conclusion is true on condition that such-and-such premises are true.* In Peirce's early works, he often calls these conditionals the leading principles of the inference. His later work, however, gives us a much more nuanced account of leading principles, as explained in the previous chapter. To avoid conflating Peirce's mature account of leading principles with this earlier account, I shall call these implicit logical commitments modalized conditionals rather than leading principles.

Importantly, the inference itself is not to be confused with any modalized conditional. The modalized conditional only states that *if such-and-such premises are true, then the conclusion is always or usually true.* It does not state that the premises are in fact true. Peirce writes, "[a]ll that is in the premises cannot however be thrown into the leading principle [i.e., modalized conditional], since there is no argument which states nothing. Nor is there an argument without a leading principle, for if nothing is implied the conclusion is already stated in the premises. But a mere statement is not an argument" (W 2:295, 1869). Every inference involves an implicit deductive or inductive modalized conditional, but that modalized conditional is not identical with the argument of the inference. The premises (antecedent) must also be stated or asserted.

Which sort of modalized conditional ought to be attributed to the person making the inference? If we are charitably reconstructing her inference, that will depend on the sort of information she uses to draw her conclusion. Peirce holds that "[a]n apodictic or deductive syllogism is one whose validity depends unconditionally upon the relation of the fact inferred to the facts posited in the premises" (W 2:215, 1868). That is, with respect to deductions, the information supplied by the premises has as a logically necessary consequence the conclusion. In contrast, "a syllogism whose validity depends partly upon the *non-existence* of some other knowledge, is a *probable* syllogism" (W 2:215, 1868). That is, for inductions, the truth of the conclusion is not guaranteed by the truth of the premises. There may be some other information of which we are unaware but were we to become aware of it, that information would vitiate the inference. Nonetheless, given the information we do have, the conclusion follows from the premises. Whether a conclusion is an informational consequence of some body of knowledge is always relativized to the premises used in drawing the conclusion. Plainly, this way of characterizing the difference allows for only two genera of inference, deductive and inductive.

The result is that when subjected to logical scrutiny, every inference can be stated as having a form akin to modus ponens. The premises will be (a) the premises on which the inference has been made and (b) a conditional with the conjunction of the premises as the antecedent and the conclusion as a consequent. Furthermore, for (b), the conditional will be modified with a modal operator, viz., that the conditional is either necessary or probable. Accordingly, to Inference A given earlier, we ought to add a premise (0^N) with a deductively modalized conditional. For Inference B, we ought to add a premise (0^P) with an inductively modalized conditional, like so:

Argument A
   (0^N) Necessarily, if all horses are mammals and all mammals are animals, then all horses are animals.
   (1) All horses are mammals.
   (2) All mammals are animals.
   (3) Therefore, all horses are animals.

Argument B
   (0^P) Probably, it will rain on condition the clouds are dark and heavy and the atmospheric pressure drops.
   (4) The clouds are dark and heavy.
   (5) The atmospheric pressure has dropped.
   (6) Therefore, it will rain.

We can further analyze these implicit modalized conditionals by abstracting away from what the premises state. If we do so for the inferences just given, then we are left with the following purely formal implicit modalized conditionals:

Argument A
   (0^N) Necessarily, if all S are M and all M are P, then all S are P.

Argument B
   (0^P) Probably, Z occurs on condition both X occurs and Y occurs.

More generally still, we can abstract away from the logical forms used in the inference and merely remark that for deductions the implicit logical principle is a necessity of the consequence where the colligated premises are the antecedent and the conclusion is the consequent. Accordingly, the implicit conditional of a deductive argument is □(P ⊃ C). This logical principle should not be confused with the quite different necessity of the consequent that P ⊃ □C. If the latter were the logical principle, then all the conclusions of deductive inferences would be necessarily true. In contrast with necessary reasoning (deduction), the logical principle of an induction is akin to a conditional probability. It is the probability that the conclusion is true on condition the premises (antecedents) are true, or Probably, C, on condition of A.[2]

According to Peirce, this deductive inference is valid:

(1) □(P ⊃ C)
(2) P
(3) C

However, this inference is invalid:

(1) □(P ⊃ C)
(2) P
(3) □C

One can see that the first is valid, whereas the second is invalid if P = I draw a square and C = I draw a figure with four sides. If the second inference were valid, then my drawing a figure with four sides would have been necessary. Presumably, though, I could have refrained from drawing any figure at all. Nevertheless, this third inference is valid (it is modal modus ponens):

(1) □(P ⊃ C)
(2) □P
(3) □C

Matters become more complicated when we turn to induction, but using a toy example, we can make an analogous point with respect to probable inference. Suppose A = all the balls in my sample are red and C = all the balls in the urn are red. Peirce holds that this inference is valid:

(1) Probably, C, on condition of A.
(2) A
(3) Therefore, C

Whereas this inference is invalid:

(1) Probably, C, on condition of A.
(2) A
(3) Therefore, probably, C

I may infer from my sample to the conclusion that all the balls are red, as in the first inference. However, it does not follow that it is probable all the balls in the urn are red, for that conclusion does not follow if my sample is very small, say just five balls from an urn containing a million. Note that Peirce would allow one to infer from a randomly drawn sample of nothing but five red balls drawn from an urn containing a million balls that all the balls are red. The inference is valid but weak. Because it is weak, Peirce would not allow one to infer from the limited sampling that it is probable all the balls are red.

We can make the second inductive inference valid by changing premise (2), as we did in the deductive case of modal modus ponens. However, it will not do to replace premise (2) with *Probably, A*, since either all the balls of the sample are red or they are not. There is no question of probability here. What will work is to replace the second premise with the claim that the sample is a good sample, that the sampling practices have been sufficiently strong to merit drawing the conclusion with confidence. In other words, this inductive inference, the analogue of modal modus ponens, is valid:

(1) Probably, C, on condition of A.
(2) A, as has been determined by thoroughly carrying out a strength-producing inductive procedure.[3]
(3) Confidently, C

Note here that it is confidence and not probability which attaches to the conclusion. Strictly speaking and as Peirce stresses, probability belongs exclusively to consequences (i.e., conditionals) and not to conclusions.

The upshot of these considerations is that every argument can be regarded as being made in accordance with some modalized conditional, where (i) the modality is either *necessarily* or *probably* and (ii) the conditional has as its antecedent the conjunction of the premises and as its consequent the conclusion. This provides us with grounds for distinguishing between deduction and induction based on the modality of the conditional. The modality of the conditional marks whether the inference is deductive or inductive.

Squaring the Account with Peirce's Aristotelian Framework: The foregoing is my reconstruction of how Peirce arrives at his primary distinction between deductive and inductive arguments in the 1860s. But there is one issue from Peirce's writings that we need to square with the preceding. The issue is that Peirce apparently arrives at his early classification of inference from within the context of Aristotle's syllogistic logic. In "A Natural Classification of Arguments," Peirce arrives at his natural classification by considering different arrangements of syllogisms in traditional logic.

On that syllogistic approach, every valid necessary inference may be put in the following form (where the parentheses taken together stand for alternate constructions):

(1) Any M is (is not) P.
(2) Any (Some) S is M.
(3) Therefore, Any (Some) S is (is not) P.

Through substitutions of the denial of a premise for the denial of the conclusion, we can derive two other figures. Here is the second figure:

(1) Any M is (is not) P.
(Denial of 3) Some (Any) S is not (is) P.
(Denial of 2) Some (Any) S is not M.

Here is the third figure:

(Denial of 3) Some (Any) S is not (is) P.
(2) Any (Some) S is M.
(Denial of 1) Some M is not (is) P.

These are the three basic figures of deductively valid syllogisms.

To derive the forms of inductive inference, Peirce first would have us insert collections for the predicate in the second figure and for the subject in the third figure. If we do so, from the second figure we get:

(1) Any M is P', P'', P''', etc.
(2) S is P', P'', P''', etc.

(3) Therefore, S is probably M. (see W 2:46, 1867)

P stands for a property, and P′, P″, P‴ stand for different properties. If S and M have the same properties, we may hypothesize that they are the same. Peirce gives us an example:

(1) Light is polarizable.
(2) Ether waves are polarizable.
(3) Therefore, light is ether waves. (see W 1:427, 1866)

This is Peirce's account of hypothetic induction. Today, we would put it this way: If light is ether waves, then light and ether waves should share their distinctive properties in common (viz., polarizability). They are both polarizable. So, probably, light is ether waves.

With respect to statistical induction, we will strip out the quantifiers and qualifiers of the third figure. Doing so, we get:

(1) S′, S″, S‴, etc. are P.
(2) S′, S″, S‴, etc. are taken at random as M's.
(3) Therefore, M is probably P. (see W 2:46, 1867)

S stands for a subject term, and S′, S″, S‴, etc. is a series of subjects. If the same class of subjects has some property in common, we may infer its genus extends over the property. Peirce illustrates this mode of inference as follows:

(1) Neat, swine, deer, and sheep are cloven-hoofed.
(2) Neat, swine, deer, and sheep are herbivora.
(3) Therefore, all cloven-hoofed animals are herbivora. (see W 1:427, 1866)

Today, we would put it this way: A random sampling of cloven-hoofed animals for whether they are herbivores shows that they are all herbivores. So, all cloven-hoofed animals are probably herbivores.[4]

These considerations also shed light on how Peirce distinguishes between hypothetic induction and statistical induction in his 1867 essay "Upon Logical Comprehension and Extension." Peirce states that hypothetic induction increases the comprehension of a term. In the example given earlier, the conclusion tells us something about the nature or essence of light: It is ether waves. In contrast, statistical induction tells us something about the extension of a term. We discover that being cloven-hoofed falls under the extension of being herbivorous. Although Peirce presents the distinction in this way in 1867, it does not appear in his later works. He later claims that being too caught up in considerations of logical extension and comprehension led him to confuse different kinds of reasoning (see CP 2.102, 1902).

It would be silly to deny that this is Peirce's approach to the distinction among deduction, hypothetic induction, and statistical induction in his early works. After all, Peirce is perfectly explicit about it. But if my earlier account is correct, then it appears the distinction between deduction and induction is not especially beholden to Aristotelian syllogistic logic. And this is correct. Underneath the Aristotelian veneer is a more contemporary and formal approach to classifying inference. For all Peirce is doing is rearranging a consequence (the rule), its antecedent (the case), and its consequent (the result). In fact, Peirce strongly implies that this is his approach. In "Some Consequences of Four Incapacities," he claims, "[e]very deductive syllogism may be put into the form If A, then B; But A; ∴B" (W 2:218–219, 1868). With respect to hypothetic induction, he writes that 'hypothesis' is used "[m]ost commonly in modern times, for the conclusion of an argument from consequence [i.e., if P, then Q] and consequent [Q] to antecedent [P]. This is my use of the term" (W 2:219, 1868). That is, all deductive reasoning can be put into the argument form of modus ponens. All hypothetic inductions are in the argument form of affirming the consequent. And although Peirce is not as explicit about statistical induction, his examples show that it infers from an antecedent and a consequent to a conditional. The antecedent is what we are sampling, such as the beans from a bag. The consequent is the nature of the sample, such as the beans being white. The conditional is the inferred rule, such as that all the beans in the bag are white.

In his later works, Peirce is explicit that this was indeed his earlier strategy. In a letter to the Italian philosopher Mario Calderoni, Peirce acknowledges that his classification of inferences is both a priori and a posteriori. He remarks, "these three [deduction, induction, and abduction] are the only elementary modes of reasoning there are. I am convinced of it both *a priori* and *a posteriori*. The *a priori* reasoning is contained in my paper in the Proceedings of the American Academy of Arts and Sciences for April 9, 1867" (CP 8.209, 1905). The article which sets forward Peirce's a priori reasoning is "A Natural Classification of Arguments." In 1910, Peirce clarifies that his a priori way of classifying arguments is based on the arrangements of a conditional, its antecedent, and its consequent: "We have, then, to consider, *Deduction*, which from a consequence and its antecedent infers the consequent; *Induction*, which from antecedents and consequents infers consequences; and *Retroduction* which from a consequent and a consequence infers an antecedent" (PSR 149, 1910). As already noted, this account is suggested in "Some Consequences of Four Incapacities." In my discussion of abduction, I will later raise some problems with this a priori classification precisely on a posteriori grounds. The key point for the moment is only that the Aristotelian syllogistic framework is but a veneer, a vestige of traditional logic. In the 1860s, Peirce is still beholden to the traditional Aristotelian logic, but he also has a vague sense of its inadequacies.[5] This is why we find the Aristotelian classification of inferences tangled up with his a priori classification based on the arrangements of consequence, antecedent, and consequent.[6]

Critique of the Account: By now the astute reader will sense that there is a problem, and in fact there is. Toying with arrangements of consequence, antecedent, and consequent does not provide grounds for grouping together hypothetic induction and statistical induction as varieties of induction, as Peirce does. Accordingly, while he does arrive at his threefold distinction among deduction, hypothetic induction, and statistical induction by arranging consequence, antecedent, and consequent, this does not explain why he groups hypothetic induction and statistical induction together as varieties of probable inference.

One way to try to explain this anomaly in Peirce's classification is to draw on the account I have given here. They are both inductions because they are both led or guided by the modalized conditional *probably, if such-and-such premises are true, then such-and-such conclusion is true.* The natural way to understand 'probably' for statistical inferences is as a conditional probability. But this would plainly be strange to do for cases of statistical induction. To see why, consider the argument about raining where it involves a statistical induction:

(1) On 90 out of 100 observed occasions when the clouds have been dark and heavy and the atmospheric pressure has dropped, it has rained.

(2) Therefore, there is a 90% probability it will rain when the clouds are dark and heavy and the atmospheric pressure has dropped.

On the modalized conditional account, the implicit commitment of the argument is:

(O$^{P^*}$) Probably, there is a 90% probability it will rain when the clouds are dark and heavy and the atmospheric pressure has dropped on condition that on 90 out of 100 observed occasions when the clouds have been dark and heavy and the atmospheric pressure has dropped, it has rained.

This is either trivial or it is confused. It is trivial if the conditional is just presuming how statistical inductions work. It is confused if the word 'probable' in its various grammatical forms is being used in different ways. For one way to regard probabilities is as numbers, but to make (O$^{P^*}$) non-trivial the first 'probably' must mean 'it is likely that.'

What gives Peirce trouble is that he wants to use 'probable' and its grammatical cognates in four different ways,[7] but in the 1860s he has still not clearly distinguished among them.[8] In the first way, he wants a probability statement to be the ratio of times when the premises and the conclusion are true to times when the premises are true. This is his numerical or statistical conception of probability based on our observations. On Peirce's account, the proposition *there is a 90% probability of rain on condition that the clouds are dark and heavy and the atmospheric pressure drops* means that when I infer *it will rain* from the premises *the clouds are dark and heavy* and *the atmospheric pressure has dropped*, I will be correct 90% of the time. Such a probability is ascertained by measuring how frequently it rains under those conditions.

In the second way, Peirce wants to use 'probability' to express the real frequency with which one event does or would follow another in nature, quite regardless of our observations of it. These are the real laws or uniformities in nature. I write "does or would" because Peirce wants 'probability' in this second manner of use to capture both uniformities in nature as well as propensities. In the "does" case, I may make the inference about the rain a finite number of times, but presumably there is some uniformity or regularity with which rain follows the clouds being dark and heavy and the atmospheric pressure dropping. That uniformity is quite independent of the finite number of times I make the inference. In the "would" case, there is some propensity of a die to show a six. The probability is the ratio of times it would show a six to the number of times it would show any number at all. Now in fact we can never exactly ascertain what the propensity at a given time is. The very act of rolling the die will cause it to change shape, and the propensities will change depending on the shape of the die. As Peirce states, we must suppose "the result of no throw [has] the slightest influence upon the result of any other throw" (ILS 125, 1910) and the die "has a certain habit or disposition of behaviour, in its present state of wear" (ILS 276, 1910).[9] What we can do, though, is approximate the propensity of the die to show a six by calculating the actual frequency with which it does on the supposition that the die is not deformed too much or, to the extent it is deformed, is deformed in ways that can be discovered.

In the third way, Peirce wishes 'probability' to be the next and stronger step beyond plausibility and verisimilitude in hypothetic inductions. In his later work, Peirce distinguishes among three different degrees "to which a theory ought to recommend itself" (ILS 275, 1910). The first is plausibility, wherein a theory recommends itself just because "instinct urg[es] us to regard it favorably" in spite of us having no other evidence in its favor (ILS 275, 1910). The second is verisimilitude, wherein a theory recommends itself just because though there is evidence for it, the evidence is "insufficient because there is not enough of it" (ILS 275, 1910). The third is probability, which seems to have given Peirce the most trouble in his later writings.[10] Probability, in this sense, is the stage at which the evidence is sufficient to confirm some hypothesis. This third way is the way in which Peirce wishes to regard probability as the next degree, beyond plausibility and verisimilitude, at which a hypothesis is confirmed. He remarks in a marginal note that he is not satisfied with using the word 'probability' but prefers the phrase "so it would appear" (ILS 274, 1910). Notice, however, that this third conception of probability is different from the first two. It is not clear that probability in this sense can be expressed as a ratio, a point about which I shall say more in Chapter 5.

In the fourth way, Peirce wants to use 'probability' to express the nonmonotonicity of a mode of inference. There are some modes of inference such that, as we get more information, we might be compelled to abandon conclusions we have drawn at earlier stages. For instance, if my first sample of beans from a bag consists of nothing but white beans, I may infer that all of the beans in the bag are white. But if in my next draw some of the beans are black, I will have to abandon the conclusion that all of the beans in the bag are white.

And now to drive the point home: We have seen that Peirce's major distinction between deduction and induction in the 1860s is based on whether the premises supply sufficient information to guarantee the truth of the conclusion. If they do not, we take the informational relation between the premises and the conclusion to be one of probability. In this sense, the 'probably' of the modalized conditional is used in the nonmonotonic sense. But for hypothetic inductions, we will also want the 'probably' to function in the sense of supplying confirmation for the hypothesis, the third sense of probability just listed. Finally, statistical inductions are not to be made in accordance with probabilities in the statistical sense but to establish probabilities. This draws on the first and second conceptions of probability, where we expect that over the long run, continued sampling will approximate the real uniformities in nature. But statistical inductions are not made in accordance with probabilities in the sense of providing a plausible explanation or confirming a hypothesis.

These considerations indicate a second problem with Peirce's modalized-conditional conception of the distinction between deduction and induction. In the 1870s, Peirce discovers that deductive logic can be axiomatized. However, the probability calculus is also mathematically expressible in rules. Accordingly, if what marks an inference as deductive is whether it is inferred from a set of axioms or rules because such axioms and rules supply sufficient information to guarantee the truth of the conclusion, then those inferences made on the basis of the probability calculus will also be deductive. Some probable reasonings will be deductive, such as this one:

(1) The probability that Lena is vaccinated against measles is .98.
(2) The probability that Jane is vaccinated against measles is .98.
(3) Therefore, the probability that Lena and Jane are vaccinated against measles is .96.

Here, we have an inference that concludes with a probability statement, as statistical inductions do. However, it is deductive. This brings us to the next phase of how Peirce distinguishes between deduction and induction.

The Axiomatic Phase of the 1870s to 1900

Beginning in the 1870s, Peirce distinguishes between deduction and induction on the grounds that the former inferences can be axiomatized, whereas the latter inferences cannot be axiomatized. As arguments based solely on the probability calculus have conclusions the truth of which are (mathematically) guaranteed by the premises, drawing the distinction between deduction and induction based on whether the inference is necessary or probable is no longer tenable. Peirce has abandoned the modalized-conditional conception of the distinction between deduction and induction in favor of an axiomatic or proof-theoretic conception of the distinction between them. This position is first hinted at in 1870's "Description of a Notation for the Logic of Relatives." It comes to full bloom fifteen years later in "On the Algebra of Logic."

It is easiest to grasp Peirce's distinction from a more contemporary perspective. One approach to propositional logic is axiomatic, a species of the proof-theoretic approach. An axiomatic system for propositional logic can be constructed from just one rule and three axioms. The one rule is the rule of modus ponens. The axioms may vary. One option uses three axioms:

- $(P \supset (Q \supset P))$
- $((P \supset (Q \supset R)) \supset ((P \supset Q) \supset (P \supset R)))$
- $(\neg P \supset \neg Q) \supset ((\neg P \supset Q) \supset P)$

P, Q, and R should be treated as metalinguistic variables for which any well-formed formula may be substituted. Whatever is provable from these axioms together with the rule of modus ponens is a theorem. Moreover, these three axioms along with modus ponens suffice for the whole of propositional logic.

A similar claim can made for the probability calculus. The probability calculus can be axiomatized. In contemporary presentations, the axioms are three:

- *Non-Negativity*: $P(A) \geq 0$, for every event A.
- *Additivity*: If events A and B are disjoint, the probability of their union satisfies: $P(A \lor B) = P(A) + P(B)$.
- *Normalization*: The probability of the entire sample space Ω is equal to 1, or $P(\Omega) = 1$.

Added to these axioms is a definition of conditional probability:

- $P(A|B) = P(A \land B)/P(B)$

From these axioms and the definition can be derived various rules. The most familiar are as follows:

- *Rule of Conjunction:* $P(A \wedge B) = P(A) \times P(B|A)$.
- *Rule of Disjunction:* $P(A \vee B) = P(A) + P(B) - P(A \wedge B)$.
- *Bayes's Rule:* $P(B|A) = P(B) \times P(A|B)/[P(B) \times P(A|B) + P(\neg B) \times P(A|\neg B)]$.

We owe the contemporary axiomatization of the probability calculus to Andrey Kolmogorov, whose work was published after Peirce died. Nonetheless, Peirce gives some related rules (see ILS 132, 1878).[11]

Whatever is provable from these axioms is a theorem. For example, Bayes's Rule, about which I shall say more in Chapter 5, is provable from these axioms. Peirce would regard inferences made on the basis of these axioms to be probable deductions and not statistical inductions. In 1911, he writes, "I use the expression '*probable deduction*' as a convenient abbreviation of 'deduction of a probability.' Probable deductions include all the logically sound parts of the doctrine of chances, otherwise called, the calculus of probabilities" (NEM 3:172).

Clearly, axiomatic approaches to propositional logic and axiomatic approaches to the probability calculus have much in common. These commonalities lead Peirce to change the way in which he classifies inferences. In the 1860s, he marks the distinction by a modalized conditional. This gets Peirce into trouble because of the different senses in which 'probability' is used. In the 1870s, he marks the distinction by whether the inference is explicative or ampliative. Inferences that are based on a set of axioms and rules are explicative insofar as the conclusions are provable from those axioms and rules. Whether the axioms be the axioms of propositional or predicate logic or the axioms of the probability calculus, all such inferences are deductive. Peirce's discovery that logic could be axiomatized, combined with his recognition that the probability calculus works in much the same way, is what compels him to abandon his modalized-conditional conception of the distinction between deduction and induction. Instead, he adopts an axiomatic conception of the distinction between them.

Peirce claimed in the 1860s that every deductive inference can be put into the form of modus ponens; however, he had not yet developed an axiomatic approach to logic. In "Description of a Notation for the Logic of Relatives" (henceforth "Description") from 1870, he develops a notation for predicate logic that draws on the algebraic work of Boole and Jevons. The axiomatization of algebra is a project that began in the nineteenth century and was brought to fruition only in the twentieth century. But even when Peirce was writing, it was quite clear that at least some collection of axioms should be sufficient to describe algebraic structures. "Description" explicitly begins by extending various axioms of algebra to logic. These axioms include that addition is associative and commutative, that addition is invertible, and that multiplication is doubly distributive, associative, and invertible, among others.[12]

The axioms Peirce lists very obviously parallel the ten axioms which are now regarded as sufficient to establish an algebraic field in mathematics.[13] Moreover, Peirce is perfectly explicit that he is drawing an analogy between algebra and logic, extending the use of algebraic symbols into the logical domain. He writes, "[i]n extending the use of old symbols to new subjects, we must of course be guided by certain principles of analogy, which, when formulated, become new and wider definitions of these symbols" (W 2:360, 1870). Although Peirce ends "Description" by remarking, "these axioms are mere substitutes for definitions of the universal logical relations, and so far as these can be defined, all axioms may be dispensed with" (W 2:429, 1870), he is not denying that deductive logic may be axiomatized. He is only affirming that the axioms merely express "definitions and divisions" (W 2:429, 1870) such that provided one understands the definitions, one will readily grasp the truth of the axioms. There is hardly any ground for doubt that Peirce is endeavoring to develop an axiomatic system of logic by drawing on the axioms of algebra. It is an effort that will bear full fruit in 1885's "On the Algebra of Logic." In that essay, he identifies five "icons" (see W 5:170–173). Those five icons suffice as a set of axioms for propositional logic.

A word of caution is in order. In later works, Peirce appears to relax his axiomatic conception of deductive reasoning. He claims:

Deductive reasoning consists in first constructing in imagination (or on paper) a diagram the relations of whose parts is completely determined by the premises, secondly in experimenting upon the effects of modifying this diagram, thirdly in observing in the results of those experiments certain relations between its parts over and above those which sufficed to determine its construction, and fourthly in satisfying ourselves by inductive reasoning that these new relations would always subsist where those stated in the premises existed. (NEM 3:41–42, 1895)

On this conception, the axioms employed in deductive inferences may be regarded more loosely as those which frame the universe of discourse such that the diagram is completely determined by the premises as well as the rules in accordance with which we are permitted to modify the diagram. One might, for example, reject the equivalency rule of double negation. Rejecting such a rule changes how the diagram can be modified. Notice, though, that this is still an axiomatic conception of deduction, broadly conceived. The difference is that Peirce allows nonclassical but axiomatized logics to also be deductive.

The foregoing may seem surprising since Peirce claims he is a "total disbeliever in axioms" (CP 7.622, 1903). However, Peirce does not object to axiomatic approaches to logic or mathematics. Rather, he objects to treating any claim or proposition as axiomatic in the sense of being beyond dispute. Peirce does object to treating any claim as beyond criticism. To do so is to violate the first rule of reason, that is, "in order to learn you must desire to learn and in so desiring not be satisfied with what you already incline to think" (RLT 178, 1898). This rule's corollary is that we ought not to block the road of inquiry. One surefire way of blocking the road of inquiry is to treat any claim as axiomatic in the sense of being indisputable. Peirce singles out the treatment of postulates as axiomatic, which blocked the development of both non-Euclidean geometries (see RLT 179, 1898) and mathematical work on infinities. Similarly, Peirce's critical commonsensism holds that while there are some indubitable propositions which seem perfectly evident, "we may and ought to think it likely that some one of them, if not more, is false" (CP 5.498, 1905). These indubitable propositions may be axiomatic in the Peircean sense of being "a self-evident truth" (EP 2:302, 1904). But their seeming perfectly self-evident nowise excuses them from any and all critical scrutiny. Moreover, we may find it worthwhile to experiment with rejecting some axioms. Peirce's development of trivalent logic is one such example of logical experimentation.[14]

In contrast with deduction, induction is ampliative. Ampliative inferences go beyond the information contained in the premises in the sense that the conclusion cannot be inferred from the premises using axioms and truth-preserving rules alone. As I shall explain in more detail later, the procedure for hypothetic induction is the elimination, corroboration, or confirmation of hypotheses based on testing their predictions. The procedure for statistical induction is the sampling of a population. Insofar as these procedures yield conclusions which go beyond what is stated in the premises, induction amplifies our knowledge. But there is another sense of 'ampliative' that is worth noting. Not only do ampliative inferences go beyond the information contained in the premises, axioms, and rules, their conclusions are also always subject to further probation and improvement. Inductive inferences are nonmonotonic. In a legal context, ampliation is the deferring of a final judgment pending further evidence. Note that deferring final judgment is fully consistent with making tentative or for-now judgments. Although we do draw conclusions inductively, we cannot with finality declare them to be true given the premises. Rather, we can only hold that the procedure used in the inference is such that, if consistently adhered to, it will indefinitely approximate the truth. Furthermore, we cannot be assured that at any given moment we have in fact ascertained the truth. As we acquire more information, we may be compelled to abandon the conclusions we have previously inferred from the limited evidence we earlier had.[15]

Induction versus Abduction from 1900 Onward

I have just been arguing that Peirce's distinction between deduction and induction from the 1870s to about 1900 is based on whether the inference merely explicates, by use of various axioms and rules, what is already stated in the premises or whether it ampliates what is stated in the premises. But beginning in the 1900s, Peirce recognizes a third genus of inference, abduction. Although some commentators conflate abduction with hypothetic induction,[16] this is a mistake. My task in this section is to show why. I shall argue that Peirce's early conception of hypothesis, such as we find it in the 1860s and 1870s, is not a precursor to his later theory of abduction. Rather, hypothesis is hypothetic induction, or what Peirce later calls qualitative induction. There is no obvious precursor to abduction as a mode of inference in Peirce's early classifications of inference.

Peirce's Admitted Confusion

I earlier noted that in a letter to Calderoni, Peirce mentions he is fully convinced there are three elementary modes of reasoning. He states he is convinced of this both a priori and a posteriori. His a priori argument is based on the arrangement of consequence, antecedent, and consequent. In deduction, a consequence and an antecedent imply a consequent. In induction, an antecedent and consequent imply a consequence. In retroduction, a term usually regarded as synonymous with abduction, a consequence and consequent imply an antecedent.

Peirce's stated conviction in his letter to Calderoni is at variance with what we find in his other writings. His other writings reveal serious doubts about his classification of inferences. These doubts most clearly emerge around the year 1900, and Peirce is perfectly explicit about it. In particular, Peirce admits he had confused abduction with hypothetic induction. In at least three autobiographical passages, Peirce states he was confused. First, in a draft for his 1903 Harvard *Lectures on Pragmatism*, he states that he had "confused Abduction with the Second Kind of Induction [i.e., hypothetic induction or qualitative induction], that is the induction of qualities" (PPM 277n3). The second passage is found in a 1910 letter to Paul Carus. Trying to rescue his early terminology by regarding hypothesis as aligned with abduction, he writes, "in almost everything I printed before this century [i.e., before 1900][17] I more or less mixed up Hypothesis [= abduction] and Induction [= statistical induction and hypothetic induction]" (ILS 279, 1910). The third passage is from 1902: "I was too much taken up in considering syllogistic forms and the doctrine of logical extension and comprehension. . . . As long as I held that opinion, my conceptions of Abduction [i.e., reasoning about hypotheses] necessarily confused two different kinds of reasoning. When after repeated attempts, I finally succeeded in clearing the matter up, the fact shone out that probability proper had nothing to do with the validity of Abduction, unless in a doubly indirect manner" (ILS 182, 1902).[18]

That Peirce is worried his readers might also confuse abduction and hypothetic induction is evident in oblique ways. In 1910, he remarks without explanation that "[r]etroduction is naturally liable to be mistaken for" qualitative induction, which is hypothetic induction (PSR 150, 1910). That Peirce cautions his reader in this way suggests he is cognizant of a similarity between abductive inferences and hypothetic inductions.

Peirce's comments give us a timeline for when we should find him clearing up this confusion, namely around the year 1900. And in fact, we do find him working his way through the issue in 1901. Peirce sometimes refers to "vague induction" as "*abductory induction*" (HP 2:897, 1901). His example is of being in a railway car and a fellow passenger asking him whether some other man might be a Catholic priest. Upon observing the other man's style of dress and movements, Peirce infers that the man is a Catholic priest. Note that Peirce's inference is not aimed at suggesting a hypothesis—a hypothesis was already suggested by the inquiring passenger—but confirming it. For that reason, the inference is inductive. However, Peirce calls the inference abductory because the "units" of what is observed cannot be "counted or measured" (HP 2:897, 1901). Peirce would come to replace "abductory" with "qualitative" and distinguish between qualitative and quantitative induction. In the latter, which includes statistical induction, the units can be counted or measured.

What we see in these passages is that Peirce has confused two different kinds of inference that involve reasoning about hypotheses. The first kind is hypothetic induction, that is, qualitative induction. This is reasoning aimed at eliminating, corroborating, or confirming hypotheses by putting them to the test. This is what Peirce calls hypothesis in his early works. The second kind is abduction. Abduction is the genus of inference (a) that suggests hypotheses which may be pursuit-worthy because they will give us a good leave (as in billiards) for future inquiries or (b) that suggests the best explanation given the evidence we have. It does not eliminate, corroborate, or confirm hypotheses by putting them to the test at all.

There is no correspondent to abduction in Peirce's early works. But this raises a question: In his early work, from whence does Peirce think hypotheses come? Peirce holds that observations or sensations introduce new elements of thought, not reasoning. He writes, "[t]he process of investigation itself consists necessarily of two parts, one by which a belief is generated from others, which is called *reasoning*; and another by which new elements of belief are brought into the mind, which is called *observation*" (W 3:55, 1872) and "[t]he process [of investigation] is therefore clearly one which introduces new elements of thought; and these are termed *sensations*" (W 3:105, 1873). These observations or sensations are not hypothetic inductions. Hypothetic inductions are determined by something previously in the mind, for they are syllogisms.

It is unclear what observations or sensations are on Peirce's early theory. They are not reasonings. Neither are observations judgments. Peirce claims, "[n]o two observers can make the same observation" (W 3:55, 1872), but patently two observers may make the same judgment.[19] Neither are observations bare percepts or uninterpreted sensations. If they were, it would be unclear how they could introduce new elements of belief. Moreover, as Peirce later admits, "[m]odern psychology shows us that there is no such thing as pure observation free from reasoning nor as pure reasoning without any observational element" (W 4:400–401, 1883), which helps to explain his waffling use of 'observation' and 'sensation'. Observations or sensations would seem to be something like interpreted percepts. In later works, Peirce calls such interpreted percepts *percipua* (see CP 7.642–681, 1903). As it is not clear what observations or sensations are and as it is beside the main aim of this book to investigate it, I shall set the matter aside.

Nonetheless, one might worry that I have overstated my claim that there is no correspondent to abduction in Peirce's early writings. After all, Peirce claims that he confuses abduction and hypothetic induction in his early works. If there is some confusion, then presumably we will find some correspondent to abduction in his earlier writings, only it will be mixed up with other considerations. And in fact, given what I have just stated, Peirce ought to have recognized that if there is no sharp distinction between pure observation and pure reasoning, then there may be some hypothesis-forming reasoning going on behind the scenes when observations suggest a hypothesis. This would be, on his more mature classification, abduction.

This riposte is well-taken. Peirce ought to have recognized this, but so far as I can tell, he did not. It is true Peirce sometimes suggests that hypothesis gives us conjectures. Here is an example: "Fossils are found; say, remains like those of fishes, but far in the interior of the country. To explain the phenomenon, we suppose the sea once washed over this land. This is another hypothesis" (ILS 171, 1878). But Peirce is not here regarding hypothetic induction as generating hypotheses or conjectures by inference. That he is not is evidenced from his later comment that "[h]ypotheses are sometimes regarded as provisional resorts, which in the progress of science are to be replaced by inductions. But this is a false view of the subject" (ILS 179, 1878). The hypothesis that a sea once washed over the land is not inferred from the observation but resorted to in order to explain what is observed. That we have rightly resorted to the hypothesis is confirmed in subsequent inquiries, and those subsequent inquiries are hypothetic inductions. In earlier works yet, Peirce criticizes those who regard "hypothesis, not as an inference, but as a device for stimulating and directing observation" (W 2:45, 1867). In Peirce's example of fossils, the observation of the fossils suggests a hypothesis, but it does not do so by inference. Rather, as Peirce would later obliquely criticize his earlier view, the hypothesis is suggested by a sort of "logical supernumerary" (HP 2:899, 1901), an observation that compels us to resort to a hypothesis but is not properly an inference. A hypothesis (i.e., a theory or conjecture) is to be distinguished from the mode of inference which Peirce calls hypothesis. Hypothesis as a species of inference (hypothetic induction) eliminates, corroborates, or confirms a hypothesis by putting it to the test. But hypotheses themselves need not be, and on Peirce's early theory are not, generated by inference. They are suggested by observation, or they are resorted to in order to explain what we observe and are confirmed or refuted by subsequent inference.

Five Arguments against Conflating Abduction with "Hypothesis"

The foregoing considerations give us five arguments for distinguishing abduction, in Peirce's later works, from hypothetic induction (i.e., hypothesis), in his earlier works. The first argument is that hypothetic inductions can confirm hypotheses, whereas abduction only suggests hypotheses. In his early work, Peirce contends that "certain premises will render an hypothesis probable, so that there is such a thing as legitimate hypothetic inference" (W 2:45, 1867). In his late work, Peirce avers that abduction "produc[es] no conclusion more definitive than a conjecture" (PSR 99, 1906).

The second argument is that the grounds for the validity of hypothetic induction differ from the grounds for the validity of abduction. I will explore the topic of validity more fully in future chapters. For now, it suffices to remark that in his early work, Peirce holds that the grounds for the validity of hypothetic induction and of statistical induction are the same: "All probable inference, whether induction or hypothesis, is inference from the parts to the whole. It is essentially the same, therefore, as statistical inference" (W 2:268, 1869). In his late work, Peirce's account of the validity of induction is in basic details the same as in his earlier work. However, his account of the validity of abduction consists in the fact that our minds have a natural bent toward discerning the order of nature itself and "one *must* trust one's instincts" (ILS 283, 1910). It has nothing to do with parts or wholes or continued testing and sampling.

The third argument is that in the 1860s and 1870s, Peirce clearly classifies inferences as, first, deductive or inductive and, second, inductive inferences as hypothetic inductions or statistical inductions. Hypothesis is not a genus of argument distinct from deduction and induction (broadly speaking) but a species of induction: "That synthetic inferences [i.e., inductions] may be divided into induction [i.e., statistical induction] and hypothesis in the manner here proposed, admits of no question" (ILS 179, 1878). Yet in the earlier writings, Peirce contends that inductive inferences are probable or ampliative inferences, whereas in his late writings he claims that abduction does not render hypotheses probable but only supplies conjectures. Accordingly, on his early classification of inference, hypothetic induction is not merely conjecture but probable, ampliative inference. In contrast, abduction only gives us hypotheses we might put to the test and does not increase our knowledge. This third argument differs from the first argument in that it is not based on his statements about hypothesis and abduction with respect to confirming hypotheses. Rather, it is based on his classification of the varieties of inferences.

The fourth argument is that in his early work, Peirce claims that the boundary between hypothetic induction and statistical induction is blurry. Moreover, he regards hypothetic induction as a part of the experimental verification of a hypothesis. What differentiates hypothetic induction and statistical induction are different techniques for investigating nature (see ILS 181, 1878). Hypothetic induction is part of the project of inquiring into the truth of some hypothesis. This makes it unlike abduction. Abduction generates and ranks hypotheses which will then be put to the test. It ranks hypotheses as pursuit-worthy or as being the best explanation given the evidence we have. Abduction does not eliminate, corroborate, or confirm those hypotheses.

The fifth and final argument for distinguishing abduction from hypothetic induction concerns how known information relates to the hypothesis under consideration. Abduction consists in reasoning to a general explanation of the facts at hand. Abduction is a "method of forming a general prediction" (CP 2.270, 1903). In contrast, hypothetic induction ascertains whether such predictions are borne out. Hypothetic induction proceeds by arguing that if such-and-such hypothesis is true, then we should observe such-and-such phenomena. When we do observe those phenomena, it supplies some corroboration, perhaps even confirmation, of the hypothesis such that were we to accept the hypothesis, then we would take it that continued investigation of the predictions of the hypothesis will not overturn the hypothesis. Abductions proceed by arguing that if such-and-such hypothesis is true, then it would explain such-and-such phenomena. We have observed those phenomena. Therefore, there is reason to suspect pursuing the hypothesis will be truth-conducive or that it is the best explanation given the evidence we have.

Peirce's A Priori versus A Posteriori Classification of Inferences

We are now in position to clarify how the distinction between abduction and hypothetic induction relates to Peirce's a priori classification of inferences. As it turns out, both hypothetic inductions and abductions have similar logical forms, and the similarities of their forms can cause us to confuse them. This is evident in the last argument I have just given against conflating abduction with Peirce's early theory of hypothetic induction. It is also evident in Peirce's explicit accounts of the matter. Consider, first, abduction. In 1903, Peirce states that an abduction has the following form:

> The surprising fact, C, is observed.
> But if A were true, C would be a matter of course.
> Hence, there is reason to suspect that A is true. (EP 2:231)

If we abstract away from the modalities of surprise and suspicion, it is plain that this is an argument from a consequence and its consequent to an antecedent.

Consider, second, hypothetic induction. Hypothetic inductions (when they succeed in corroborating or confirming the hypothesis) have the following form:

(1) If H, then $P_1, P_2, P_3 \ldots P_n$.
(2) $P_1, P_2, P_3 \ldots P_n$.
(3) Therefore, H.

where H is a hypothesis and P_n is a prediction from the hypothesis to be put to the test.

The prediction itself is a conditional the antecedent of which is a description of some experimental conditions and the consequent of which is an expected observation. Peirce holds that hypotheses are (a) distinguished from one another by their predictions and (b) demonstrable as true or false from their predictions. Let hypothesis H_1 have a total set of predictions $P_I = \{P_1,$

$P_2, \ldots P_n\}$ and H_2 have a total set of predictions $P_{II} = \{P_1, P_2, \ldots P_n\}$. If $P_I = P_{II}$, then H_1 and H_2 are equivalent hypotheses. Moreover, if H_1 is false, there will be some member of set P_I that, if put to the test, will compel us to eliminate H_1, where to eliminate H_1 is to table it (eliminate it from present consideration), to refine it (eliminate it in its current form), or to reject it (eliminate it entirely). If some prediction P_n of the set P_I is found to obtain, then H_1 is corroborated. If all of the hypotheses we have to explain some phenomena have been eliminated except one about which we have no real doubts, then the remaining hypothesis is confirmed. (I shall add some nuance to this account in Chapter 5.)

Although this position may raise the hackles of contemporary theorists from a philosophical point of view, it is much less clear that the practicing scientist who aims to confirm some theory over others would disagree with Peirce so far as it concerns her specific researches. Ordinarily, she will have some finite set of hypotheses which have suggested themselves. She will then begin the labor of sorting through them so as to put them to the test. If she has no way to experimentally differentiate her hypotheses, then she has no way to confirm or disconfirm one hypothesis with respect to some others.[20] I shall take up some of these philosophical issues later on. For now, the key point is that patently, a corroborating hypothetic induction is an argument from a consequence and its consequent to an antecedent. Hence, it has the same form as an abductive inference.

But now it is clear that while abductions and hypothetic inductions have the same logical form, they are used in quite distinct ways. Once we no longer ignore the operators of surprise, suspicion, confirmation, and the like, abductions only tell us which hypotheses might be good explanations or pursuit-worthy. In contrast, hypothetic inductions eliminate, corroborate, or confirm hypotheses by testing their predictions.

The result is that we need not abandon Peirce's a priori classification for *arguments*. Instead, we need to supplement it with a posteriori considerations regarding *inferences*. On Peirce's a priori classification of arguments, there are deductive argument forms (from consequence and antecedent to consequent), inductive argument forms (from antecedent and consequent to consequence), and retroductive argument forms (from consequence and consequent to antecedent). Although his classification of argument forms into deduction, induction, and retroduction is a priori, we must also attend to the a posteriori facts on the ground about how such argument forms are used in inferences. Once we do that, we find that the retroductive argument form (arguing from consequence and consequent to antecedent) is used in two quite different kinds of inference. In the abductive kind of inference, it is used to generate and rank conjectures. In the inductive kind of inference, it is used to eliminate, corroborate, and confirm hypotheses. Because the same retroductive form is used in two ways, Peirce claims, "Retroduction [i.e., abduction as a mode of inference rather than a form of argument] and Induction face opposite ways" (CP 2.755, 1905). It is this later conception on which we reason toward the elimination, corroboration, or confirmation of a hypothesis which Peirce's early conception of hypothesis represents. There is nothing in his early work that clearly corresponds to Peirce's later conception of abduction.

This distinction between the a priori classification of argument forms and the a posteriori classification of inferences as employing argument forms also sheds light on another question that has vexed Peirce scholars. Peirce sometimes seems to treat arguments from analogy as inductions and at other times as abductions. But in fact, such argument forms can be used in both ways in inferences. I may notice that two phenomena are alike in some way G_1, G_2, and so on and conjecture that they are alike in some additional respect F. Such an inference is an abductive analogical argument. But I may also hypothesize that if two things are alike in some respect F, then they will be alike in some other respects G_1, G_2, and so on. Finding they are alike in those other respects G_1, G_2, and so on, I may infer that they are alike in respect F. Such an inference is a hypothetic induction.

Peirce's Mature Classification of Inference's Genera

With the preceding in place, we are now in a position to understand how Peirce's mature classification of the genera of inferences emerges. He cannot classify inferences by their argument forms, for the same argument form is used in both hypothetic induction and in abduction. Instead, Peirce classifies inferences based on their truth-producing virtues. Peirce continues to endorse his position from the 1870s that deductive inferences are explicative. They explicate what is contained in the premises together with a set of axioms and rules. Insofar as they do so, such inferences are truth-preserving.

In contrast, inductive inferences aim at discerning which hypotheses are true or at approximating some real frequency in nature. These sorts of inferences are truth-approximating in the long run. Hypothetic induction is truth-approximating in the long run because we will gradually sort through the false hypotheses that have been proposed for testing and embrace the true ones through a process of testing those hypotheses, provided inquiry continues indefinitely into the future and the true hypothesis is among those proposed for testing. Statistical induction is truth-approximating in the long run because increased sampling will tend toward the theoretical mean over the long run. Although we may never make a final judgment that we have reached the theoretical mean, continued sampling will tend toward it.

Finally, abductive inferences are truth-conducing, provided there is any truth to be attained. Abduction generates hypotheses. We select which ones are the best for us to pursue. We also entertain which ones are the best given the evidence we presently have. If our guessing at hypotheses is good, and if our way of pursuing them is efficient, then abduction will be truth-conducive. It is the task of induction, however, to eliminate, corroborate, or confirm those hypotheses. Such are the grounds for Peirce's mature classification of inferences. A more detailed account of the nature of these inferences will be forthcoming in future chapters.

Enriching Peirce's Classification of Inferences

Thus far I have been focused on examining the genera of inferences in Peirce's classification of inferences. We have seen that the genus of the inference is determined by its putative truth-producing virtue. These genera, however, further divide into species of inference. The species of inference are determined by the general procedure followed in making the inference. It behooves me here to provide a brief overview of these species of inference by way of introducing what will come in future chapters of the book.

In 1878's "Deduction, Induction, and Hypothesis," Peirce notes that one of the reasons for distinguishing between hypothetic induction and statistical induction is that "different kinds of scientific men" may be distinguished by "differences of their *techniques*" (ILS 181, 1878). What is it that scientific inquirers do; what are their diverse techniques? Answering that question enables us to distinguish among the different species of inference by differences in procedures, or techniques of investigation. Here, my comments on those procedures shall be brief, as I take a more in-depth look at them in future chapters. I first look at the general procedures used in induction.

Following that, I consider the general procedures used in deduction. Peirce's comments on abduction, however, are vexed. As such, I shall defer a fuller examination of the species of abduction (if there be any; Peirce is of different minds on the matter) to Chapter 7.

Inductive Procedures

Procedure One: Hypothetic Induction: Beginning with the species of induction, in his early work Peirce identifies two different techniques. The first is to probe a hypothesis by putting it to the test. One generates some predictions from the hypothesis. This is a conditional with the hypothesis as the antecedent and the prediction as its consequent. These predictions are themselves conditionals. The antecedent of the prediction describes some observational or experimental conditions. The consequent of the prediction describes some expected observation. These statements have the form:

$$H \supset (C \supset O)$$

where H is the hypothesis and $C \supset O$ is the prediction, consisting of a description of some observational or experimental conditions (C) and an expected observation under those conditions (O).

Ordinarily, the inquirer will have a set of hypotheses to put to the test. Letting P abbreviate the prediction $C \supset O$, first, the inquirer will have a list of predictions from all the suggested hypotheses[21] that are to be considered as explaining the phenomena, for example:

(1a) If H_1, then P_1, P_2, P_3, P_4.
(1b) If H_2, then $P_1, P_2, P_3, \neg P_4$.
(1c) If H_3, then $P_1, P_2, \neg P_3, \neg P_4$.
(1c) If H_4, then $P_1, \neg P_2, \neg P_3, \neg P_4$.

If she has done her preparatory work well and it is economical to do so, she will likely start by testing for P_3, since doing so will enable her to halve her number of hypotheses. Recall that a prediction is a conditional in which the antecedent of the conditional describes some experimental conditions and the consequent an expected observation. Let us suppose she sets up the experimental conditions and observes what she expected. In that case, it follows:

(2) P_3.

Because she has asserted P_3 (for which reason we are treating it as a premise), we suppose she has no doubts about the quality of her observation or the satisfactoriness of her testing conditions. If she were to have such doubts, then she ought to repeat the test or reconsider her testing strategy. Moreover, because she has found P_3, she will eliminate from consideration both H_3 and H_4. Her remaining premises are (1a) and (1b).

Whereas H_3 and H_4 have been eliminated because P_3 has been observed, H_1 and H_2 have been corroborated by observation of P_3. It would be wrong to claim, however, that either has been confirmed. For plainly one of them must be false as they have inconsistent predictions. At this stage, it would be useless to test for any of P_1 through P_3, since both hypotheses share these predictions. Instead, our inquirer will test for P_4. Suppose she sets up the experiment described in the prediction but does not observe what she expected. In that case, she has found that:

(3) $\neg P_4$.

Then she will conclude:

(4) $\therefore H_2$.

Now she has confirmed H_2 relative to the other hypotheses under consideration. We might stop short, however, of making the unqualified claim that the hypothesis has been confirmed. We may also want to test whether H_2 passes tests P_1 and P_2. Moreover, we should not ever claim that any hypothesis has been indefeasibly confirmed (recall the sense of 'ampliative' in which final judgment is deferred), since we will not ordinarily be able to test all the predictions of a hypothesis. These caveats made, note that Peirce will allow that hypothetic inductions confirm hypotheses and do not merely corroborate them, unlike the Popperian view. Of course, any given hypothesis that has been confirmed may be refuted down the line. However, if we have done well our jobs of sorting through hypotheses and putting them to the test, then if only one of the hypotheses remains after testing, the hypothesis is confirmed, at least until another hypothesis comes along or observations inconsistent with the previously confirmed hypothesis compel us to eliminate it. Once again, making for-now judgments is compatible with deferring final judgment.

So far as I have described hypothetic induction, I have described only the general procedure. The devil will be in the details. The person's testing may be vitiated for all sorts of reasons, including flaws in the instruments, poor observation conditions, fatigue, carelessness, and so on. And she may be wrong about the predictions she has derived from the hypotheses. Moreover, in our short lifespans, we will be able to put only a limited number of those predictions to the test. The inference will be probative in degrees dependent on the thoroughness and severity of the testing. By thoroughness, I mean the number of tests it has passed. Complete thoroughness for hypothetic inductions will consist in testing every prediction. By severity, I mean passing a test that is unlikely to be passed if the hypothesis is false.[22] A hypothesis might pass one hundred easy tests and no severe ones. That will not speak much in favor of the hypothesis. Alternatively, it may pass one severe test and be untested for any others. That will speak well for the hypothesis. Newton's theory of gravity passed lots of tests. Einstein's theory of relativity has passed several very severe ones. Whether the hypothesis follows as a matter of plausibility, of verisimilitude, or of probability (in the third sense mentioned earlier) will depend on how many and which sorts of tests the hypothesis has passed.

One question these considerations raise is whether we should consider the preparatory steps for a hypothetic induction (gathering and ranking hypotheses and deducing their consequences) to be part of the hypothetic induction. Peirce is quite clear that these preparatory steps are rather abductions and deductions. Yet he also refers to these as stages in the process of inquiry. Here, it is helpful to distinguish between hypothetic induction as a mode of inquiry and hypothetic induction as a species of inference. As a mode of inquiry, hypothetic induction does include the preparatory steps. It is the so-called hypothetico-deductive method. As a species of inference, it does not include these steps. Yet when we evaluate any given hypothetic induction as a species of inference, some of the premises will have been derived from the preparatory stages of the inquiry. If we want to know whether an inference is good, we will want to know whether the premises are true. To answer that question, we will need to turn to the preparatory stages as parts of hypothetic induction qua mode of inquiry.

Procedure Two: Statistical Induction: A different technique is used when one wants to know the relative frequency of some event given some other events. Here, we have some unknown quantity, and we aim to estimate what that quantity is. For example, one might want to know the relative frequency with which it rains when the clouds are dark and heavy and the atmospheric pressure drops. Or one might want to know by what percentage some candidate is favored win the next election. Or one might want to know the relative frequency with which some die rolls a six. These are statistical inductions, not hypothetic inductions. For statistical induction, the procedure will consist in sampling so as to ascertain these ratios. A completely thorough sampling will consist of a sampling of every instance. That is typically impossible; a die might be tossed an infinite number of times, for example. To make our statistical inductions strong in the short run, we will need to predesignate for what we are sampling, take our samples randomly, sample as much as we can, and the like.

The example of a die suggests a problem: Hypothetic induction and statistical induction blend into each other. I might suspect some die is not fair. I want to put this hypothesis to the test. This will require hypothetic induction: If I roll the die thirty times, I should find that about fifteen times, the number is even. Now I will perform a statistical induction: I roll the die thirty times. I find that on fifteen of the thirty rolls, the die shows an even number. I now have calculated a relative frequency. I conclude the die is fair. But what was my inference? The inference involves both hypothetic induction and statistical induction.

In 1878, Peirce concedes that the distinction between hypothetic induction and statistical induction is blurry. He also claims that the distinction between deduction (explicative inference) and induction (ampliative inference) is blurry. "Even in regard to the great distinction between explicative and ampliative inferences," he writes, "examples could be found which seem to lie upon the border between the two classes, and to partake in some respects of the characters of either. The same thing is true of the distinction between induction [i.e., statistical induction] and hypothesis [i.e., hypothetic induction]" (ILS 175–176, 1878).

The die example is a case where the distinction between statistical induction and hypothetic induction is blurry. In order to clear matters up, it is helpful to distinguish between induction as a stage of inquiry and induction as a genus of inference which has species. Induction as a stage of inquiry may involve both hypothetic induction and statistical induction.

Statistical inductions will often help us to eliminate, corroborate, or confirm hypotheses as we put those hypotheses to the test. Nonetheless, within that stage, we can distinguish between the hypothetic induction and the statistical induction. These are two different species of inductive inference, both of which are used in the inductive stage of inquiry.

A case that blurs the line between inductive and deductive reasoning is when one samples every member of a population. A statistical induction that involves a complete enumeration of the population to be sampled has inductive and deductive characteristics. If I draw every ball out of an urn and conclude that 12.5% are red and large, the inference is inductive since it involves sampling, here the extreme case of sampling every member of the population. If we classify an inference by the procedure, then such an inference ought to be classified as inductive. However, the truth of the conclusion is guaranteed by the premises. In such a case, the truth-producing virtue of the procedure is stronger than mere truth-approximation in the long run. Peirce's considered view is that such apparent statistical inductions by complete enumeration are not really inductions at all. Rather, he calls such an inference a "logistic deduction" (CP 2.757, 1908), for its truth-producing virtue is truth-preservation.

Procedure Three: Crude Induction: In Peirce's late work, he adds a third kind of inductive procedure to the two just mentioned. He calls this crude or rudimentary induction. Crude inductions are inferences wherein a conclusion is inferred just because to think otherwise would be "utterly at variance with all past experience" (CP 2.756, 1905). Peirce gives the example of someone inferring from the fact that no one has ever established an instance of a person having clairvoyant powers that no one has such powers. Induction by absence of a counterexample is a crude induction. Notice that crude induction does not involve in fact putting hypotheses to the test; it only "pooh-poohs" hypotheses based on what is already known (CP 7.111, 1903).

Deductive Procedures

Procedure One: Logical Analysis: One deductive procedure is what Peirce calls logical analysis. Logical analysis should not be confused with subjecting an inference to logical scrutiny and criticism. By logical analysis, Peirce means both framing a definition of an idea and tracing out the practical consequences of an idea or a hypothesis. This conception of logical analysis is of a piece with his pragmatism. According to Peirce, ideas may have three grades of clearness, viz., clarity, distinctness, or pragmatic adequacy. First, an idea has clarity to the extent that we are able to use it with facility. As an example, our idea of water is clarified to the extent that we do not confuse water for vodka or bleach. Second, an idea is distinct to the extent that we are able to provide a real definition of it. Our idea of water is distinct in that we know it to be H_2O. Third, an idea is pragmatically adequate to the extent that we can trace out the consequences or effects of something being it as opposed to something else. We pragmatically develop our ideas by tracing out their consequences, and those ideas are pragmatically adequate when they are developed sufficiently for our purposes. Our idea of water is pragmatically developed in that we know at sea level, water boils at one hundred degrees Celsius and freezes at zero degrees Celsius.

Logical analysis as a variety of deduction takes an idea or hypothesis and analyzes it with respect to each of these three grades. It should be evident that logical analysis will typically be preparatory to hypothetic induction, as I indicated earlier. By logical analysis qua deduction, we will trace out predictions of our hypotheses. For example, suppose I hypothesize that some substance is water. I predict that if I am at sea level and the temperature is below zero degrees Celsius, it will begin to freeze. That prediction is a deductive consequence derived from my idea of water. Hypothetic induction is the procedure of putting those predictions to the test.

Procedure Two: Necessary and Probable Demonstration: A second deductive procedure is demonstration. Demonstrative inferences do not analyze ideas and hypotheses but infer conclusions from rules and sets of premises or axioms, as in proofs. Peirce distinguishes between species of demonstrative deductions in two separate ways. The first way is based on whether the demonstration is necessary or probable. I have already explained the distinction between necessary and probable demonstrations as dependent on whether they employ (1) the propositional of predicate logic calculi or (2) the probability calculus. The former are necessary demonstrations; the latter are probable demonstrations.

One question Peirce's distinction between logical analysis and necessary demonstration raises is how we should understand such inferences as *John is a bachelor; therefore, John is unmarried.* Such an inference could be regarded as a logical analysis of the term 'bachelor.' But it could also be regarded as a demonstration from the premises *All bachelors are unmarried* and *John is a bachelor.* This is another case in which we should distinguish between stages of inference and species of a genera of inference. A stage of deductive inference may involve both logical analysis and demonstration. In the example just given, there is one deductive stage of inference. That deductive stage first logically analyzes the concept of a bachelor. Second, the deductive stage demonstrates from the logical analysis and the premise *John is a bachelor* that *John is unmarried.* The one deductive stage of inference involves two species of deductive inference.

The second way Peirce identifies species of demonstrative inference is based on whether the demonstration is corollarial or theorematic. Generally speaking, a theorem is any statement provable from some set of axioms. However, a distinction is sometimes drawn between major theorems and corollaries. Major theorems are the most important results, whereas corollaries are straightforward extensions of the major theorems. Major theorems must often be proven by introducing a supporting claim— a lemma—which serves as a stepping stone to the major theorem from which the corollaries are drawn. When demonstrative deductions introduce this additional lemmatic element, they are theorematic deductions. Peirce explains the distinction:

> Deductions are of two kinds, which I call *corollarial* and *theorematic.* The corollarial are those reasonings by which all corollaries and the majority of what are called theorems are deduced; the theorematic are those by which the major theorems are deduced. If you take the thesis of a corollary, i.e., the proposition to be proved, and carefully analyze its meaning, by substituting for each term its definition, you will find that its truth follows, in a straightforward manner, from previous propositions similarly analyzed. But when it comes to proving a major theorem, you will very often find you have need of a *lemma,* which is a demonstrable proposition about something outside the subject of inquiry; and even if a lemma does not have to be demonstrated, it is necessary to introduce the definition of something which the *thesis* of the theorem does not contemplate. (EP 2:96, 1901)

Peirce denies that this distinction tracks the distinction between explicative and ampliative inferences from the 1870s. Even though theorematic deductions do seem to increase our knowledge relative to the "thesis of the theorem," the lemmata are still provable from the axioms: "It now appears that there are two kinds of deductive reasoning, which might, perhaps, be called *explicatory* and *ampliative.* However, the latter term might be misunderstood; for no mathematical reasoning is what would be commonly understood by *ampliative*, although much of it is not what is commonly understood as *explicative*" (NEM 4:1, 1901). The theorematic/corollarial distinction is different from the explicative/ampliative distinction that marks the division between deduction and induction.

Why does theorematic deduction appear akin to ampliative reasoning? First, major theorems once proven are valuable insofar as they increase the power of a calculus and our knowledge of what is provable from the axioms. For example, derived rules in axiomatic logic, such as hypothetical syllogism, enable us to complete proofs with much greater efficiency. Second, theorematic deduction sometimes requires introducing new objects, such as when one infers from *every horse is a mammal* to *horses are a subset of mammals*.[23] In these respects, theorematic deduction seems ampliative and not what is commonly understood as explicative. But the major theorems—not to mention the corollaries and lemmata—are demonstrable from the axioms nonetheless. Accordingly, theorematic deductions are explicative.

This development does not call into question Peirce's axiomatic way of drawing the distinction between deduction and induction. It only calls for clarification regarding Peirce's earlier terminology of explicative versus ampliative reasoning. Whether theorematic or corollarial, deductive inference proceeds on the basis of a set of axioms and rules, together with the meanings of words. Induction does not.

Procedure Three: Statistical Deduction: A third procedure in deduction is really only a special case of probable demonstration. However, it deserves special attention because it resolves a problem with Peirce's theory of probability. Peirce is typically presented[24] as holding that probabilities are properly understood to be frequencies expressed as arguments or conditional probabilities—for instance, there is a 90% probability it will rain on condition the clouds are dark and heavy and the atmospheric pressure drops. This raises a question, however: We sometimes assign probabilities to categorical and not merely conditional probability statements. How can we do so?

The answer lies in statistical deduction. We make a statistical deduction when we infer a categorical probability from a conditional probability. Following Peirce, I shall call this a statistical deduction. For example, suppose I know that 98% of the population of Goessel, Kansas, is vaccinated. Knowing that Lena is a resident of Goessel, Kansas, I infer that there is a 98% probability she is vaccinated against measles. This is a statistical deduction.

Technically, Peirce draws a subtle distinction between probable deduction in a narrow sense and statistical deduction. He does so in "A Theory of Probable Inference," written in 1883. Probable deductions in the narrow sense and statistical deductions have very similar forms. A probable deduction in the narrow sense has this structure:

(1) The proportion N of Ms are Ps.
(2) S is an M.
(3) Therefore, with a probability N S is P.

Peirce gives us this example:

About two per cent of persons wounded in the liver recover;
This man has been wounded in the liver:
Therefore, there are two chances out of a hundred that he will recover. (W 4:408, 1883)

Statistical deductions differ from probable deductions in the narrow sense only in that the set S is taken at random from the collection of Ms. Here is an example Peirce gives of such a statistical deduction:

A little more than half of all human births are males:
Hence, probably a little over half of all the births in New York during any one year are males. (W 4:415, 1883)

The suppressed second premise is that the births in New York are a random sample of all human births, as would be the births in Lisbon or Gaborone or Bangkok. As there is a "known proportion of exceptions" (W 4:409, 1883), the conclusion must be stated as a probability to capture these known exceptions.

Although Peirce does draw this technical distinction, I shall use 'statistical deduction' to refer to any inference of a probability-that statement from a probability-of statement. Statistical deductions differ from statistical inductions. In statistical deduction, one applies the statistic to "many cases, and, while it may be false in any one, it will probably and approximately be true in the long run" (CN 1:148, 1892). In contrast, the conclusion of a statistical induction "may be false—only, if so, the further pursuit of the same method will in the long run probably and approximately *correct* it" (CN 1:148, 1892).

Peirce discusses the distinction initially in his review of John Venn's *The Logic of Chance*. He notes, "there are two principal phrases in which the word *probability* occurs; for, first, we may speak of the probability of an event or proposition, and then we express ourselves incompletely, inasmuch as we refer to the frequency of true conclusions in the genus of arguments by which the event or proposition in question may have been inferred . . . and, secondly, we may speak of the probability that any individual of a certain class has a character" (W 2:101–102, 1870.)[25] The distinction is drawn in the context of discussing a problem that arises with regard to statistical deduction. Suppose it is known both (a) that 90% of those with European ancestry who go to Tahiti get drunk and (b) that 10% of the devoutly religious people who go to Tahiti get drunk. Now suppose that Carmen is a devoutly religious person of European ancestry who is going to Tahiti. What is the probability that she will get drunk? By the first statistic, we ought to be able to statistically deduce that the probability is 90% that Carmen will get drunk. That is, $P(C) = .9$. By the second, we ought to be able to statistically deduce that the probability is 10% that Carmen will get drunk. That is, $P(C) = .1$. But plainly $P(C)$ cannot equal both .9 and .1.

Four points are in order. The first is that statistical deductions are always made relative to the information contained in the statistic. As Peirce states, probabilities-that are about individuals of a certain class and character. Relative to the information that Carmen is of European ancestry, the probability is 90%. Relative to the information that Carmen is religiously devout, the probability is 10%. Both inferences are valid, so long as the background information against which the statistical deduction is made is borne in mind.[26] What we need to be clearer about is the proposition which bears the probability assignment. In the first case, we should be more specific and let C_1 = Carmen qua person of European ancestry will get drunk, so that $P(C_1) = .9$. In the second case, we should be more specific and let C_2 = Carmen qua religiously devout person will get drunk, so that $P(C_2) = .1$. These are two different propositions with two different probabilities.

The second is that the question whether the inference is valid in each case is different from the question of what is the probability that religiously devout persons of European ancestry get drunk when they go to Tahiti. In the example, we are supposing that we do not have that statistic. Since we do not have it, we cannot assign a probability to the statement C_3 = Carmen qua religiously devout person of European ancestry will get drunk when she goes to Tahiti. We cannot assign a probability to C_3 because we have no primary statistic as to how probable it is that one gets drunk when one goes to Tahiti on condition one is a religiously devout person of European ancestry. Peirce states it is examples such as these that highlight the need of "ceasing to speak of probability, and of speaking only of the relative frequency of this event to that" (W 2:23, 1867). We infer from "certain indications" to the conclusion (W 2:22, 1867). This is the "untechnical sense" in which we assign probabilities to events (W 2:22, 1867). It is in this way one may claim *it will probably rain today*, even though in fact it either will rain or will not rain. Having observed the clouds are dark and heavy and the atmospheric pressure has dropped, and knowing that under those conditions there is a 90% probability of rain, one infers that it will probably rain today. 'Probably' here is not used in a technical way but has the sense of being confidently asserted. Note, however, that in drawing this statistical deduction one will have already ascertained by statistical induction the frequency with which rain follows the clouds being dark and heavy and the atmospheric pressure dropping.

Nevertheless and third, the mere fact we do not now have information as to the frequency with which religiously devout persons of European ancestry get drunk when they go to Tahiti in no way implies we cannot acquire such information. Once we have that statistic, we will be able to statistically deduce the probability of C_3. We cannot, however, obtain the probability of C_3 simply by averaging C_1 and C_2. Neither can we do it by conjoining or disjoining the probabilities. The strength of inductions for C_1 and C_2 may be wildly different, giving very inaccurate probabilities if we were to calculate them in any of these ways.

Fourth and finally, however, we may sometimes have a pressing need to assign a probability to C_3. This pressing need might arise if an insurance company must price a policy for Carmen's trip to Tahiti. Suppose all the information the company has for pricing such a policy is $P(C_1)$ and $P(C_2)$. In that case, $P(C_1)$ and $P(C_2)$ may serve as parameters for assigning a probability to $P(C_3)$. The company, however, will have to draw on various heuristics in order to price the policy, and these heuristics will be drawing on data beyond the specific question of what probability may be assigned to C_3. While $P(C_1)$ and $P(C_2)$ might be parameters for making a judgment about $P(C_3)$ when there is a pressing need, neither may be close to the facts of the matter. It may be the case that no religiously devout persons of European ancestry get drunk when they go to Tahiti.

Conclusion

In this chapter, I have argued for three major theses. The first major thesis I have argued for is that in his earlier works up until about 1900, Peirce's primary distinction is between deductive and inductive inferences. Both hypothetic induction (hypothesis) and statistical induction (induction) are species of induction. Although Peirce initially draws the distinction between deduction and induction based on the modality (necessary or probable) of a conditional in which the conjunction of the premises is the antecedent and the conclusion is the consequent, he later abandons this account. Instead, he draws the distinction between deduction and induction based on whether or not the conclusion is a consequence of the premises together with a set of axioms or rules and the meanings of words.

Second, scholars have wrongly conflated Peirce's later account of abduction with his earlier account of hypothesis. This is a mistake because hypothesis is rather hypothetic induction, or what Peirce later calls qualitative induction. Abduction does not eliminate, corroborate, or confirm hypotheses. Rather, abduction and hypothetic induction have different truth-producing virtues. Abductions (when they are good) are truth-conducive. Hypothetic inductions (when they are good) are truth-approximating in the long run.

The third major thesis is that in 1900, Peirce adds abduction as a third genus of inference distinct from deduction and induction. This change leads Peirce to distinguish among the three genera of inference based on their truth-producing virtues. Peirce's account of deduction is not changed by this addition; deductions are truth-preserving because they make inferences based on the premises and the relevant calculi. However, Peirce now distinguishes between inferences that are truth-approximating in the long run (induction) and those which are truth-conducive provided there is any truth to be ascertained (abduction).

In arguing for these theses, I have also shown how Peirce continues to deepen and enrich his classification of inferences. Deduction comes to embrace not only necessary and probable deduction but the logical analysis of ideas and hypotheses, a distinction between corollarial and theorematic deduction, and probable deductions as either statistical deductions of categorical probabilities from conditional probabilities or probable demonstrations based on the probability calculus. Induction is enriched to include crude induction along with qualitative induction (hypothetic induction) and quantitative induction (statistical induction). I shall examine abduction in more detail in Chapter 7.

3

Validity

In the previous chapter, I examined Peirce's classification of inferences. Although in his early works Peirce recognizes only two genera of inference, deduction and induction, after 1900 he adds a third genus of abductive inference. Furthermore, we have seen that inferences are first classified by the truth-producing virtues of their procedures and then by the general nature of their procedures. In cases of deduction, the procedures are all truth-preserving. Those procedures are to clarify the meanings of words and to draw conclusions from the premises in accordance with axioms and rules. In his earlier works, Peirce recognizes two general kinds of inductive procedure. The first is the procedure involved in hypothetic induction, viz., having traced out the predictions of a hypothesis, testing whether they obtain. Peirce later calls this qualitative induction. The second is the procedure involved in statistical induction, viz., sampling a population. Peirce later calls this quantitative induction. In cases of induction, the procedures have the virtue of being truth-approximating in the long run. In his later work, Peirce adds to this list a third kind of induction, crude induction, whereby one's procedure is merely to consult the mass of past experience and determine whether a claim is consistent with that experience or not. Lastly, for abductive inference, the procedure is to ascertain those hypotheses which are likely to be true given the evidence one has as well as those which are the best to pursue. In core cases, that procedure will be truth-conducing, provided there is any truth to be ascertained. I shall have more to say about abduction in Chapter 7.

In Chapter 1, I also drew a distinction between inferences as they have been made and inferences that could be made. If we think of an argument as no more than a list of declarative sentences, one of which is designated the conclusion and the others of which (if there be any others) are designated as the premises, then we may ask what sorts of procedures might yield such a list and thereby be an inference. In such a case, the list of claims cannot be classified as deductive, inductive, or abductive until we consider what sorts of procedures might be involved in making such an inference. Once again, if the conclusion cannot be inferred from the meanings of the premises' words

Peirce on Inference. Richard Kenneth Atkins, Oxford University Press. © Oxford University Press 2023.
DOI: 10.1093/oso/9780197689066.003.0004

or a set of truth-preserving axioms and rules, then the argument could not possibly be deductive and a fortiori could not be a good deductive inference. If the conclusion could not be inferred from the premises by following a truth-approximating procedure, then the argument could not possibly be inductive and a fortiori could not be a good inductive inference. If the conclusion could not be inferred from the premises by following a truth-conducing procedure, then the argument could not possibly be abductive and a fortiori could not be a good abductive inference.

I have just now referred to good deductions, inductions, and abductions. Peirce holds that inferences and arguments can be evaluated according to two different standards. The first standard is validity. According to Peirce, not only are deductive inferences valid or invalid, but inductions and abductions are also valid or invalid. Plainly, then, Peirce uses 'validity' more broadly that we do today. On the contemporary account, all inductions are invalid because to be valid an argument must be such that there is no interpretation on which all the premises are true but the conclusion is false.[1] But it is patent that in cases of induction a conclusion may be false even if the premises are true. Even so, the induction might be good, such that further testing or sampling would correct the error. This is one sense in which inductive inferences are ampliative: They defer final judgment (but not tentative, for-now judgment). While Peirce does accept the contemporary account of validity for necessary demonstrative deductive inferences, he thinks of validity more broadly than just deductive validity. The aim of the present chapter is to cash out Peirce's broader account of validity.

When we begin digging into what Peirce has to say about validity, we find that there are at least four different accounts of validity in his writings. The first aligns with our contemporary view of validity, but for Peirce that is to be restricted to deductive demonstration alone. The second is that an inference is valid just in case its leading principle is true. We find this account most frequently in Peirce's early writings, but I shall argue that it is confused. The third account of validity we find in Peirce's writings is that an inference is valid just in case the conclusion is always or usually true granted the premises are true. I shall argue that this account faces significant problems and, moreover, Peirce must abandon it when he recognizes abduction as a distinct genus of inference. The fourth and final account of validity is that validity is a property of the genera of inferences together with the procedure used in drawing the inference. On Peirce's mature account, an inference is valid if and only if (1) the procedure followed in fact has the truth-producing virtue claimed for it (truth-preservation, truth-approximation, or truth-conduciveness) and (2) the person making the inference has in fact followed the procedure she professes to follow. "In order to be valid," Peirce claims, "the argument or inference must really pursue the method it professes to pursue, and furthermore, that method must have the kind of truth-producing virtue which it is supposed to have" (CP 2.780, 1901). In what follows, I examine each of these four accounts of validity in turn. Before doing so, however, some preparatory comments on inference, arguments, and semiosis are in order.

Inference, Argument, and Semiosis

As I noted in Chapter 1, one of the major tasks of the logician is to scrutinize inferences as they have been made and to evaluate how far those inferences are good. When the logician is engaged in this task, she is reflecting on actual inferences that have been made. Inference is a psychological process. The logician will consider that psychological process and endeavor to make the commitments of the person drawing the inference as explicit as possible. Moreover, the content of that process of inference is an argument. For Peirce, 'argument' is a semiotic category. An argument is a sign, the conjunction of the premises, that separately shows or aims at some interpretant, the conclusion. A complete argument will include a description of the procedure used. This implies that the process of inference itself, in addition to being a psychological process, is a semiotic process. It is the semiotic process of interpreting some set of declarative sentences to imply (deductively, inductively, or abductively) some conclusion.

Improvements to his logic and semiotics aside, Peirce never strays from this central insight on inference and the task of the logician. In his early works, Peirce regards logic as objective symbolistic and thinks of an argument or inference as one sign (the conclusion) substituting for other signs (the premises) of the same object. Logic is not restricted to formal logic. Rather, Peirce conceives of logic as semiotics. It is the study of signs and how to use them so as to ascertain the truth. Francesco Bellucci explains Peirce's early view on the matter nicely:

Logic, in its turn, studies symbols only in their relations to their objects, and is thus to be defined as "objective symbolistic." This amounts to saying that logic is the study of the validity of arguments. An argument is valid if it leads from true premises to a true conclusion, that is, if the object of the premises is also an object of the conclusion. In an argument, therefore, a symbol is substituted for another that has *the same object* (i.e., which is true when the former is true). (2018, 16)

For example, consider the premises *all horses are mammals* and *all mammals are animals*. We interpret these two claims with the conclusion *all horses are animals*. When we do so, we have made a deductive inference. Note, though, that the conclusion *all horses are animals* is not the only way we could have interpreted the relation between these premises. Alternatively, we could have inferred that *both all horses are mammals and all mammals are animals* by conjunction introduction, for example.

Peirce holds that the conclusion interprets the premises of the inference, and so it does. But that is not the only interpretative relation that occurs in inference, for the procedure also interprets the relation between the premises and the conclusion. In his early work, Peirce's example of an interpretant is that *b is like p by way of turning on a horizontal axis*: "Suppose we wish to compare the letters p and b. We may imagine one of them to be turned over on the line of writing as an axis, then laid upon the other, and finally to become transparent so that the other can be seen through it" (W 2:53, 1867). The phrase "by way of turning on a horizontal axis" tells us how to the interpret the relation of resemblance between b and p. The same point can be made about procedures used in inference. The premises *if p, then q,* and *p* entail *q* by way of modus ponens. Modus ponens tells us how to interpret the relation of entailment between the premises and the conclusion. Premises substitute for their conclusions because procedures permit such substitutions to be made. A good procedure will permit such substitutions in ways that have truth-producing virtues.

Evaluations of Inferences as Concerning Sign Substitution

In the quotation from Bellucci provided earlier, we have a statement as to in what validity consists. When we interpret a set of premises, we substitute the conclusion qua interpretant for the premises qua sign. That substitution is permitted by or made in accordance with some procedure. In the example of horses and animals, we may always substitute *all horses are animals* for the premises *all horses are mammals* and *all mammals are animals* without loss of truth. Similarly, it is usually (not always) truth-preserving to substitute the conclusion *it will rain* for the premises *the clouds are dark and heavy* and *the atmospheric pressure has dropped*. In a lecture from 1865, Peirce explains validity in this way:

> Validity is legal tender. Greenbacks are not true cash but they will buy proportionally to cash because they are valid. In the same way colour is not true of objects because it is an affection of the mind and cannot be in matter but its modifications are true because they correspond to modifications of things. That therefore is valid whose modifications are true. A representation is subjectively valid which is consistent. . . . A representation is objectively valid whose modifications are true of objects; or synthetically consistent. (W 1:248)

American dollar bills (greenbacks) are mere paper, but they substitute for gold by virtue of a convention. (At the time Peirce was writing, America was on a gold standard such that the dollar stood for a fixed quantity of gold.) Colors substitute for real features of objects, perhaps microphysical properties, by virtue of physical and psychological laws. Similarly, conclusions of inferences substitute for their premises by virtue of various procedures. Minimally, our commitments must be consistent. Premises also substitute for some object of cognition. He uses the word 'object' broadly. In a late letter, he states, "I use the term 'object' in the sense in which obiectum was first made a substantive early in the XIIIth century; . . . I mean anything that comes before thought or the mind in any usual sense" (SS 69, 1908). Peirce does not restrict the meaning of 'object' to material objects but includes abstract objects, sets, facts, and laws. An inference is valid just in case in fact the conclusion may always or usually substitute for the premises and still be true of the same object the premises are about. This rough and ready idea, based on semiotic considerations, is at the heart of Peirce's early account of validity. Ultimately, though, he is compelled to modify it for reasons I shall explain later.

For the moment, I want to stress that although it may seem peculiar to claim that inductions can be valid, Peirce's account of validity is in lockstep with theories of logical consequence in the medieval philosophers who were his primary influence.[2] As Gyula Klima explains, medieval philosophers had widely varying conceptions of logical consequence, but they did not restrict the validity of inference to deduction alone:

[T]he result is a cluster of several theories covering consequences from conditional propositions of various strengths to argument forms of various strengths, ranging from what we would recognize as formal validity to mere probability. Yet this cluster of theories all relate to the focal idea that a consequence is valid just in case the denial of the consequent is in some way "repugnant" to the antecedent. This idea of "repugnancy" was spelled out in several ways with regard to different forms of consequences. (2016, 339)

In his early writings, Peirce spells out this idea of repugnancy in terms of his objective symbolistic. Conclusions interpret premises as substitutes for the objects those premises are about, and they do so by virtue of some procedure. When we make good inferences, our conclusions do a good job of substituting for those premises. Of course, we will need an account of what goodness consists in. That will be a theory of validity.

These considerations indicate that Peirce's conception of deductive validity roughly aligns with our broadly contemporary account of validity as truth-preservation. A valid deductive inference is truth-preserving since the conclusion will always be true provided the premises are true. But Peirce holds that the premises and conclusion must be true of the same object. Requiring that the objects of the premises and conclusion be the same implies that inferences constructed such that their premises and conclusion are unrelated are invalid. For example, this inference would be invalid because the conclusion has nothing to do with the premise:

(1) All horses are animals and some horses are not animals.
(2) Therefore, cats purr.

Nevertheless, the argument is valid on our standard contemporary accounts of logical validity since there never is a time when the premises are all true and the conclusion is false (the premise is always false). Peirce would have to admit, though, that the following argument is valid,[3] since the conclusion is about the same object, viz., horses:

(1) All horses are animals and some horses are not animals.
(2) Therefore, all horses are animals.

Similar points may be made about inferences where the conclusions are logically true. This inference would be invalid since the premise and conclusion are not about the same object:

(1) Cats purr.
(2) Therefore, either all horses are animals or some horses are not animals.

Though this one would be valid:

(1) All horses are animals.
(2) Therefore, either all horses are animals or some horses are not animals.

Peirce does hold that validity should be restricted along these lines: "we certainly never can be warranted in drawing any conclusion about S from a premise, or set of premises, which does not relate in any way to S" (RLT 131, 1898). If this is correct, then Peirce's conception of validity would appear to lean in the direction of contemporary relevance logic, in the broadest sense that the conclusion needs somehow to be relevant to the premises. However, Peirce would not spell out the relevance condition formally but semiotically, based on whether or not the conclusion and premises are signs of the same object. A detailed but incomplete development of logic as semiotic emerges in Peirce's later work.[4]

There is another consideration that can be brought to bear here. As noted earlier, logicians will often be interested in evaluating inferences as they have actually been made. The logician will endeavor to make the procedures and premises used in the inference as explicit as possible. In so doing, she will ignore background beliefs of the person making the inference that do not bear on the inference under consideration. As Peirce states, to be a premise of an inference, a proposition must "act" as a premise (BD 2:1, 1901). Accordingly, beliefs and other statements extraneous to what is needed to draw the conclusion will never be listed among the premises in such cases.

Peirce distinguishes between formal logic, which is a part of mathematics, and logic as semiotics. From a purely formal and logical point of view, the inferences just canvassed are valid. The conclusion is formally entailed by the premises, even though they are about different objects. *Cats purr* does formally entail that *either all horses are animals or some horses are not animals*. But for Peirce, validity is not a purely formal and mathematical notion but a logical and semiotic notion. That is, we should distinguish between

(a) formal (or mathematical) validity and (b) semiotic (or logical) validity. Although an inference from P to $\neg(Q \land \neg Q)$ is formally, mathematically valid, it is not semiotically, logically valid. That these two conceptions of validity come apart is evidenced by the requirement that the conclusion and premises be of the same object for an inference to be valid.

Both hypothetic induction and statistical induction must also satisfy the constraint that premises and conclusions be of the same object. This constraint poses no special problems for hypothetic induction, since the conclusion will typically be contained in the antecedent of the conditional which states the prediction of the hypothesis. At first glance, though, the requirement may appear to pose a problem for statistical induction; that is, *the clouds are dark and heavy and the atmospheric pressure has dropped*, on the one hand, and the fact *it will rain*, on the other, do not seem to be about the same object. The premises state nothing about rain. Yet we must bear in mind that statistical inductions are not about instances of such inferences but about the frequency with which they lead to the truth. Suppose that on one hundred occasions, a person has noted whether rain follows the clouds being dark and heavy and the atmospheric pressure dropping. She finds that on ninety of those occasions, it has rained. These individual cases are the premises. The conclusion is that 90% of the time, when the clouds are dark and heavy and the atmospheric pressure has dropped, it will rain. This conclusion is clearly about the same object as the premises, viz., about it raining under some conditions on various days and the frequency with which that occurs. The conclusion of that statistical induction is then used in a statistical deduction in order to infer that *it will rain*. In both cases, the conclusion is clearly about the same object as the premises.

Peirce's Four Accounts of Validity

With the foregoing semiotic caveats in mind, we are now in a position to examine Peirce's various accounts of validity. I should state at the outset that in distinguishing among four different accounts of validity that we find in Peirce's writings, it is not always clear that Peirce himself recognizes that he is providing four different accounts of validity. In fact, he appears to think they are equivalent, sometimes expressing two different accounts in one and the same series of essays. Moreover, as we shall see, these accounts of validity are closely related. Nonetheless, they are different accounts, and only by prizing them apart and noting their differences can we get Peirce's mature account of validity in view.

Deductive Validity

There is significant contemporary debate as to how we should understand logical consequence, or validity, as it pertains to logic. The most common account is the model-theoretic account of validity. In propositional logic, an argument is valid just in case there is no interpretation (i.e., assignment of truth-values) on which the premises are true but the conclusion is false. This rough account, satisfactory for propositional logic, requires more detailed and rigorous mathematical statement for other logics, such as quantified predicate logic and modal logic. Nonetheless, the underlying idea is the same: An argument is valid just in case there is no model on which the premises are true and the conclusion is false.

There can be no question that Peirce would endorse such an account, but he would also hold it is overly narrow in one way and overly broad in another. It is overly narrow because, as I have already mentioned in Chapter 1, Peirce also thinks that the inference from *John is a bachelor* to *John is unmarried* ought to be regarded as deductive. Peirce admits it as valid because the logician also considers the meanings of words when evaluating whether an argument is valid. Furthermore, Peirce would claim that the contemporary account is overly broad in another way. As just explained, Peirce thinks that inferences are such that the conclusion can substitute for the premises. There must be some semiotic connection among the premises, the conclusion, and the object. Accordingly, an inference from *cats purr* to *all horses are animals or some horses are not animals* is not valid on Peirce's view, even though it is valid on the contemporary account of deductive validity. Nonetheless, Peirce can concede that the argument form used in making the inference is mathematically valid, though the inference itself is not semiotically valid.

These two caveats noted, Peirce's account of deductive validity is roughly aligned with the contemporary view. However, he restricts its applicability to deduction while affirming that inductions and abductions can also be valid. Peirce remarks, "Kant gave the principle of [deduction]. In the premises the conclusion is, in substance, asserted" (ILS 283, 1909). He also states that it is "only in Deduction that there is no difference between a *valid* argument and a *strong* one" (EP 2:232, 1903). Inductions and abductions can be valid and yet weak, whereas valid deductions are always also strong.

The Leading-Principle Account of Validity

Mainly in his earlier works, but also in some of his later ones, Peirce claims that an inference is valid just in case its leading principle is true. In this vein, he writes:

- "A *valid* argument is one whose leading principle is true" (W 2:23, 1867).
- "A *valid* argument . . . is one whose leading principle is true" (W 2:294, 1869).
- "the validity of the argument depends on the truth of a general principle called the *consequence*" (W 2:432, 1870).
- "An inference whose leading principle is true is said to be a valid inference" (W 4:246, 1881).

We may call this the leading-principle conception of validity: An inference is valid if and only if its leading principle is true.

Notice that in the final quotation from 1881, Peirce states that the inference is valid just in case its leading principle is true, whereas in the earlier quotations he states that an argument is valid just in case its leading principle is true. As I mentioned in Chapter 1, Peirce sometimes distinguishes between inference and argument, but at other times he seems to regard them as synonymous. Attending carefully to the distinction will help us to tease out a problem with the leading-principle account of validity.

For part of the question we face here is: What is the leading principle of an inference? In Chapter 1, I distinguished among four different accounts of leading principles in Peirce's writings. The first account is that leading principles are modalized conditionals, either *necessarily, if such-and-such premises are true, then such-and-such conclusion is true* or *probably, if such-and-such premises are true, then such-and-such conclusion is true*. On the second and third accounts, a leading principle is a guiding principle, a commitment used in drawing the inference, such as that *what is true of the magnetic properties of this sample of copper will be true of the magnetic properties of all samples of copper*, which can be analyzed into logical and factual guiding principles. The procedure one uses in drawing the inference can be described and treated as a logical guiding principle of the inference. For instance, if one draws a conclusion using modus ponens, then *modus ponens is truth-preserving* is a guiding principle of the inference. Also, if one is sampling a population, then a description of the procedure one follows when sampling the population is a guiding principle of the inference. These described procedures are logical guiding principles. Fourth, a leading principle may be the entire effect of inferring, which is a conditional in which the conjunction of the premises and a description of the procedure used is the antecedent and the conclusion is the consequent.

In Chapter 2, I argued that in his works from the 1860s, Peirce thinks of leading principles in the first way just mentioned. However, in the 1870s and 1880s, Peirce thinks of the leading principles in the second and third ways. In the quotations provided earlier, we have the leading-principle account of validity expressed in both time periods. Consequently, we need to consider what the leading-principle account of validity might come to on both the first account of a leading principle and on the second and third accounts of a leading principle.

On the first account of leading principles as modalized conditionals, the leading-principle account of validity states that (i) an argument is deductively valid just in case provided the premises are true the conclusion is always true, whereas (ii) an argument is inductively valid just in case provided the premises are true the conclusion is usually true. This suggests a specification of the leading-principle account of validity, viz., that an argument is valid just in case it always or usually leads from true premises to a true conclusion. In fact, in Peirce's early works we find the leading-principle account of validity appearing in close connection with this other account of validity, which I shall call the always-or-usually-true account of validity. Peirce writes at roughly the same time as the first three quotations above that "valid inference . . . proceeds from its premise, A, to its conclusion, B, only if, as a matter of fact, such a proposition as B is always or usually true when such a proposition as A is true" (W 2:214, 1868). As I shall treat of the always-or-usually-true account of validity in the next section, I pass it over for the moment.

Let us, then, turn to the other account of a leading principle as a guiding principle. The guiding principle of an inference may involve several commitments, both logical and factual. On this account of leading, or guiding, principles, we face problems if we claim an inference is valid just in case its guiding principle is true. First, it is not clear what the guiding principle of any given inference is. Second, it is not clear which sort of guiding principles are relevant to the validity of the inference. These claims stand in need of elaboration.

Let us start with the second point. Consider again the example from Chapter 1 about copper disks. Suppose a person notices that a copper disk stops spinning when placed between the poles of a magnet. She infers that this will be true of all copper disks. One way to understand the guiding principle of this inference is that *if this disk of copper is a representative sample of copper, then what is true of this disk of copper will be true of any disk of copper.* Such a guiding principle may be analyzed into two sub-commitments. The first is the logical guiding principle that *if S is a representative sample of the individuals comprising population P, then the properties of S are the properties of the individuals comprising P.* The second is the factual guiding principle that *this sample of copper is a representative sample of all coppers.*

One question we may ask at this juncture is which of the commitments just listed must be false for the argument to be invalid: the guiding principle as a whole, the logical guiding principle, or the factual guiding principle? It must not be the last of these. Suppose we are only mistaken about whether this sample of copper is a representative sample. It has, let us say, some impurities, whereas we had been under the impression that the sample was of pure copper. If the factual guiding principle sufficed to make the inference invalid, then we should have to claim the inference just made when the sample is impure is invalid. But this is wrong. The inference is not invalid; rather, it proceeds upon a false premise. Just as for deductive inferences false premises make the arguments unsound not invalid, so for inductive arguments false premises do not imply a logical error has been made, only that the model used is not representative of reality.[5]

These considerations indicate that what matters to the validity of an inference is not the factual guiding principle of an inference but its logical guiding principle, for the guiding principle as a whole is constituted of the factual and logical components. This brings us to the first point, that it is not clear what the logical guiding principle of any given inference is. Peirce admits that we may think we have drawn an inference according to some guiding principle when in fact we have drawn the inference in accordance with some other guiding principle. Accordingly, we ought not to evaluate an inference as valid except derivatively as its argument being valid, for we may not even know which guiding principle we have used in making the inference. We can formulate each argument using different guiding principles, the guiding principle we think we have used in the inference and the guiding principle we might have used in the inference. We can then evaluate whether the argument is valid based on each guiding principle.

To illustrate, suppose a person draws the following inference: *if it is raining, then the ground is wet. The ground is wet. Therefore, it is raining.* What is the guiding principle of this inference? The guiding principle might be the deductive fallacy of affirming the consequent. In that case, the inference is invalid. Alternatively, the guiding principle may have the structure of being a hypothetic induction. In that case, depending on whether other explanations have also been eliminated, the inference may be inductively valid and strong. Accordingly, what we must consider is the argument in relation to various guiding principles first, and only then may we ascertain whether the inference has been valid or invalid.

But now Peirce is in a pickle, as we can see if we consider the case of affirming the consequent. On the one hand, Peirce can claim that the inference from *if it is raining, then the ground is wet; the ground is wet* to *therefore, it is raining* is invalid because it involves the procedure of using affirming the consequent as a deductively valid argument form when it is in fact invalid. On the other hand, he concedes that one could draw such an inference while being explicitly committed to the principle that *"if p, then q, and q, therefore p" is a deductively valid argument form.* Now that will be a false premise, no doubt. But one might endorse this false premise about affirming the consequent and also endorse the procedure of using modus ponens as a valid argument form. But then inference from *if p, then q, and q, therefore p; if it is raining, then the ground is wet; the ground is wet* to *therefore, it is raining* is valid! Evidently, all one must do to make her invalid inferences valid is to ensure her implicit logical guiding principles are explicit.[6]

There is a clause that is doing some important work in the previous example. It is the clause that the person accepts the guiding principle as a deductively valid inference form. This clause hints at what I take to be the right account of validity in Peirce's mature writings. One could think a procedure has some truth-producing virtue when it in fact does not. What is requisite for an inference to be valid is that the procedure one follows in fact has the truth-producing virtue claimed for it. That procedure can be described and treated as a premise. It will be a guiding principle of the inference. If the leading-principle account of validity is meant to be about the logical guiding principles of the inference, then (as I shall argue momentarily) the leading-principle account of validity is roughly correct, provided one does not also treat logical guiding principles as additional premises constituting an inference that itself is governed by yet a second logical guiding principle which makes the entire argument valid, in the vein of Lewis Carroll's "What the Tortoise Said to Achilles." However, when Peirce articulates his leading-principle account of validity in the 1860s, this is not his theory of leading principles. Rather, his theory of leading principles in the 1860s is the modalized-conditional theory described earlier. That theory implies the always-or-usually-true conception of validity, to which I turn now.

The Always-or-Usually-True Account of Validity

I turn now to a third account of validity we find in Peirce's writings. He claims that an inference is valid just in case its conclusion is always or usually true provided the premises are true. As I noted earlier, we oftentimes find this account of validity intertwined with Peirce's leading-principle account of validity. This is because we may think of every inference being led by one of two principles, that *necessarily, if such-and-such premises are true, then such-and-such conclusion is true* or *probably, if such-and-such premises are true, then such-and-such conclusion is true*. This is Peirce's modalized-conditional conception of a leading principle. With respect to deduction, Peirce takes it that an inference is valid just in case it always leads from true premises to a true conclusion. With respect to induction, Peirce claims that a valid induction will usually lead from true premises to a true conclusion.[7]

The always-or-usually-true account of validity faces two problems. The first problem is that it makes specific or tokened arguments valid just in case they always or usually lead from true premises to true conclusions. Suppose that, entirely unbeknownst to me, it never happens that when I am hungry, I am also thirsty. If so, then the inference *I am hungry or I am thirsty; I am hungry; therefore, I am not thirsty* will be valid, for ex hypothesi whenever the premises are true, the conclusion is true. But such an argument is not valid. Inclusive disjunctions allow for both premises to be true, and none of the premises is that being thirsty is exclusive to being hungry. It just happens to be the case that whenever I am hungry, I am not thirsty. The mistake is that we have taken the always-or-usually true conception to apply to specific, tokened inferences rather than to kinds or genera of inferences.

This could easily be resolved by noting that validity pertains to a genus of inference rather than its tokens, as Peirce usually does note. However, there is a second, glaring problem with the always-or-usually-true account of validity. Once Peirce identifies a third genus of inference—abduction— he can longer hold that an inference is valid just in case it always or usually leads from true premises to a true conclusion. That is because abductions are valid, but they neither always nor usually lead from true premises to a true conclusion. Rather, abductions generate hypotheses, and abductions rank hypotheses by pursuit-worthiness and by whether the hypotheses are the best explanation given the evidence we have. But pursuit-worthy hypotheses need not be true, even if pursuing them may be conducive to ascertaining the truth. In fact, if we validly generate several hypotheses only one of which can be true, then usually an abduction will not lead from true premises to a true conclusion. Usually, the conclusion will be false because most of the hypotheses will be false. So the always-or-usually-true account of validity is insufficient as a theory of validity: It does not account for the validity of abductive inferences.

The Genus-of-Inference Account of Validity

Finally, we turn to Peirce's mature account of the validity of inference. This is the genus-of-inference account of validity. We find the genus-of-inference account of validity even in Peirce's earliest works, as when he writes, "A genus of argument is valid when from true premises it will yield a true conclusion,—invariably if demonstrative, generally if probable" (W 2:99, 1867). Here, consistent with the account given in the previous chapter, we see that Peirce recognizes only two genera of inference, deduction and induction. He will need to expand the account to include abduction. Notice, though, that this statement differs from the two immediately previous accounts examined. First, it makes validity apply to the genus of the inference and not to its leading principle or guiding principle. Second, it differs from the claim that an inference is valid just in case the premises always or usually lead from true premises to a true conclusion. This is because it makes validity a property of the genus of the argument rather than any specific or tokened argument. Accordingly, the inference from *I am hungry or I am thirsty; I am hungry; therefore, I am not thirsty* will be deductively invalid even if it happens to be the case I am never both hungry and thirsty. It will be deductively invalid because as a member of the genus of deductive inferences it is not valid, for it is logically possible that I am both hungry and thirsty (even if in fact I never am).

Although we do find the genus-of-inference account of validity in Peirce's earlier writings, it comes out most clearly in his later writings. In his entry for Baldwin's *Dictionary of Philosophy and Psychology*, Peirce states, "[e]very argument or inference professes to conform to a general method or type of reasoning, which method, it is held, has one kind of virtue or another in producing truth" (CP 2.780, 1901). These methods are procedures of inferring. Notice that in this passage, Peirce refers to the argument or the inference. Whether we are evaluating an argument or an inference, we will need to look to the procedure used and its putative truth-producing virtues in order to determine whether it is valid. As I have already noted, these virtues are whether the inference is truth-preserving, truth-approximating, or truth-conducing. A person who draws an inference takes it that the truth of the conclusion is guaranteed by the truth of the premises, or that the procedure followed is truth-approximating in the long run if consistently adhered to, or that pursuing or accepting on probation the conclusion is truth-conducive, provided there is any truth to be ascertained.

One may think an inference is truth-preserving, truth-approximating, or truth-conducing but be wrong. Peirce puts two conditions on validity. The first is that the inference "must really pursue the method it professes to pursue" (CP 2.780, 1901). For instance, I may profess to pursue the proper procedures for making statistical inductions but improperly ignore data inconsistent with the result I want. In such a case, I have not really pursued the "method" I profess to pursue. Second, the "method must have the kind of truth-producing virtue which it is supposed to have" (CP 2.780, 1901). I may think that the right procedure for making statistical inductions about how people in a state will vote is to survey my friends. That is a mistake, as my friends may not be a representative sample of the state's population. The procedure is not truth-approximating in the long run.

These examples from induction show that validity does not have to do with the form of inference in particular. Although the form of inference will oftentimes bear on whether or not a deduction is valid,[8] it will not bear on the validity of inductive or abductive inferences. To the contrary, to evaluate the validity of inductions and abductions, we will need to examine the procedures used in drawing the inference.

This has consequences for how we evaluate inductive and abductive inferences. On the one hand, the logician will often be interested in evaluating inferences as they have actually been made. In such cases, the premises, procedures, and conclusions will indicate the genus of the inference. The logician will then ask whether the person making the inference has in fact followed the procedure she professes to follow and whether that procedure is in fact truth-approximating or truth-conducing.

On the other hand, the logician will sometimes be interested in considering inferences that might be made. An argument may be merely a list of declarative sentences, one of which is designated as the conclusion and the others of which (if there be others) are the premises. In these cases, the logician may first evaluate whether the argument is deductively valid. This will be determined by the form of the inference together with the meanings of the words. But if the argument is not deductively valid, she may still wonder whether the argument could be inductively or abductively valid. In such cases, she will have to propose or conceive of procedures according to which the argument might conceivably be valid.

For example, suppose an argument qua list of statements consists only of the premise *50% of the beans in this sample, which has been drawn from this bag, are white* and the conclusion *therefore, approximately 50% of the beans in this bag are white*. That the premise is a statistic and the conclusion is not a specification of the statistic applied to an individual suggests that the inference is supposed to be a statistical induction.[9] Taken just at face value, though, it is not possible to evaluate whether this argument qua list of statements is inductively valid or not. We must consider ways in which the sample might have been taken, and depending on the procedures we conceive of, the argument may or may not be valid. If, for instance, the person sorted through the beans, found one white one and one non-white one, and declared *50% of the beans in this sample, which has been drawn from this bag, are white*, she could not validly infer that *approximately 50% of the beans in this bag are white* because such a procedure is not truth-approximating in the long run. A person may simply keep taking biased samples to get the conclusion she wants. In contrast, if a person mixes the beans thoroughly, draws several handfuls from different places in the bag, counts the beans in the handfuls she has drawn, and finds that *50% of the beans in this sample, which has been drawn from this bag, are white*, her inference is inductively valid. Although she may be wrong that *approximately 50% of the beans in this bag are white* just because it is possible her sample is biased, continued repetition of the procedure is truth-approximating in the long run.

The distinction between evaluating inferences as they have been made and evaluating inferences as they could have been made puts us in view of another important question. Thus far, it may seem as though it is relatively easy to determine the genus of an inference. If we know the procedure followed, that is a good indication of the genus of the inference. Alternatively, if we know whether the person making the inference takes the procedure to be truth-preserving, truth-approximating, or truth-conducing, that will be a good indication of the genus of the inference.

There are, however, liminal cases. Consider a case of complete enumeration. Suppose Sally wishes to know what proportion of her marbles are blue. She spreads them out, separates the blue ones from the non-blue ones, counts them up, and finds that 33% of them are blue. The procedure, insofar as it involves sampling a population (in this case, the extreme case of sampling the entire population), is a statistical inductive procedure. Note, though that provided the premises are true, Sally is guaranteed that in fact 33% of her marbles are blue. Her inference seems to be truth-preserving and so deductive. Here we have a case when an inference partakes of characteristics of both inductive and deductive inferences.

Complete enumerations are liminal cases of sampling, and Peirce holds that ordinarily in statistical induction we cannot make a complete enumeration: "[t]hat induction is only through complete enumeration; or, at least, that there is such an induction, is the doctrine of all logicians. I object to it *in toto*, nevertheless" (W 1:263, 1865). Peirce, in fact, denies that complete enumerations count as statistical inductions, writing, "what is called 'complete induction' is not inductive reasoning, but is logistic deduction" (CP 2.757, 1908). This is consistent with the position I have been defending: The genus of the inference is determined by its truth-producing virtue, not by its procedure. Species of inference's genera are determined by their procedures. Statistical induction is used when the members of a population cannot be completely enumerated. One way to regard Sally's procedure is not as sampling but as sorting. In that case, her procedure is not to sample the population as is done in a statistical induction but a straightforward case of applying a bit of set theory to a population of marbles: the blue marbles are a subset of all the marbles in the jar.

Furthermore, it is important not to conflate stages of inquiry with genera of inferences. Suppose I wish to know whether some die is fair. I hypothesize that it is not. I deduce that if I roll it thirty times, it will not come up even approximately half of those times. I roll it thirty times and find that it shows even only six times. I infer my hypothesis is corroborated. Such a process of inference involves all three stages of abduction, deduction, and induction. In the abductive stage, I generate the hypothesis that the die is not fair and commit myself to testing that hypothesis. The deductive stage will involve logical analyses of the idea of being fair as well as probable demonstrations about the minimum number of times I need to perform the experiment of rolling the die in order to make a good inference. These are two different species of the genera of deduction: logical analysis and probable demonstration. Furthermore, the inductive stage will involve both hypothetic induction and statistical induction.

Conclusion

Peirce holds that an inference is valid if and only if (1) one's procedure in fact has the truth-producing virtue claimed for it and (2) one in fact follows the procedure one professes to follow. In a late review, Peirce remarks that there are different schools of logic: "One appeals chiefly to mathematics, another to metaphysics, a third to the general notion of a sign, a fourth and fifth to this and that branch of psychology, a sixth to linguistics, a seventh to the history of science; and still the list is incomplete" (CN 3:279, 1906). Peirce draws largely on the first (mathematical), third (semiotic), and seventh (scientific) schools of logic in his own work.

The mathematical and semiotic schools inform Peirce's account of deductive validity. An inference will be deductively valid just in case the truth of the conclusion is guaranteed by the truth of the premises, either because of the meanings of the words or because there is no model or interpretation on which the premises are true but the conclusion is false. Peirce adds a semiotic caveat: The premises and conclusion must also be of the same object, where 'object' is understood broadly to include ordinary physical objects, facts, laws, and the like. Such a view goes hand in hand with a proof-theoretic account of validity, on which a deductive inference will be valid just in case the conclusion is provable from the premises together with a set of axioms and rules. This shall be relevant in the next chapter.

The scientific school informs Peirce's account of inductive and abductive validity. Whether an inference is valid will be determined by whether (1) one's procedure is in fact truth-approximating or truth-conducing and (2) one has in fact adhered to those procedures. As we have seen in Chapter 2, induction is further divided into species by procedures used in drawing the induction. Foremost among these species are hypothetic (qualitative) induction and statistical (quantitative) induction. But in any specific case of inquiry, it is not possible to state in advance the details as to how these procedures should be implemented. Studying ice cores is different from studying human subjects. Each may require different techniques and different controls. Although we can give general recommendations, different sorts of investigations may entail some procedures aptly used in some domains of scientific investigation are inapt in other domains. This is a thesis on which I shall elaborate in Chapter 5.

4

Deduction

Peirce holds, as we do today, that deductive inferences are to be evaluated in two respects: validity and soundness. An inference is sound if and only if it is valid and its premises are true.[1] In our classical logic, an inference is valid if and only if there is no model on which the premises are all true but the conclusion is false, or the conclusion is provable from the premises together with the axioms and rules of the system. As explained in Chapter 2, Peirce's approach to deductive logic in most of his writings is proof-theoretic and axiomatic. For propositional logic, the axioms he uses are equivalent to ones found in standard presentations of axiomatic systems today. Nonetheless, Peirce recognizes that one might also have a nonclassical logic—he was a discoverer of nonclassical logics—and that one will then need to define validity in different ways. Moreover, regardless of how the semantics or proof theory for the deductive system may be designed, when endeavoring to ascertain whether the premises have as a deductive logical consequence the conclusion, Peirce maintains that we will attend not only to the formal structure of the inference, as we do in model-theoretic and proof-theoretic constructions of our logical calculi, but also to the meanings of words.

   Peirce credits Kant with giving the right account of deductive validity: "Kant gave the principle of [deduction]. In the premises the conclusion is, in substance, asserted" (ILS 283, 1909). It is for this reason that deductive inference is essentially explicative: It explicates what is contained virtually in the premises, axioms, and rules of the system. Elsewhere, he characterizes deductive reasoning as "any reasoning which will render its conclusion as certain as its Premisses, however certain these may be" (LoF 1:549, 1911). The clause 'however certain these may be' is added since when we use the probability calculus, the premises will typically be but probable. Peirce provides an example: "That any homogenous cubical die will, at any given throw, turn up six is a chance against which the odds are 5 to 1 decidedly improbable. Therefore, it is very improbable that a pair of quite disconnected good dice should turn up sixes at any given throw without referring to any calculation of Probabilities" (LoF 1:577, 1911). Notice that Peirce waffles

Peirce on Inference. Richard Kenneth Atkins, Oxford University Press. © Oxford University Press 2023.
DOI: 10.1093/oso/9780197689066.003.0005

between constructing the argument as stating the exact probabilities and as making no specific reference to probabilities. Both are deductive.

In this chapter, I shall take a closer look at Peirce's conception of deduction. In the first section, I examine how Peirce conceives of deduction for each of the species he describes, viz., logical analysis, demonstration (whether necessary or probable), and statistical deduction. (I should note that Peirce regards statistical deduction to be part of probable demonstration, but I shall treat it separately here.) Peirce is not merely concerned with giving an account of deductive inference but also to reply to objections to its validity. In the second section, I survey Peirce's replies to various objections raised against the validity of deductive inference.

Peirce on Deductive Inference

Peirce describes deductive inference as involving four different steps. The first step consists of "constructing in imagination (or on paper) a diagram the relations of whose parts is completely determined by the premises" (NEM 3:41–42, 1895). This step in fact involves three different elements. The first element is settling on notational conventions and defining what a well-formed expression is. This will be the grammar. The second element is specifying rules or permissions for transforming expressions. The third element is stating the premises of the reasoning. Among these premises may be a priori truths known from the definitions of terms, or they may be practical consequences (i.e., pragmatic elucidations) of those terms. I shall have more to say about this later under the heading of logical analysis.

Notice that these elements apply both to ordinary mathematical expressions as well as logical expressions. Just as $\wedge PQ\neg$ is not well-formed, neither is $= 123x$. Just as if we negate the left side of $\neg P \wedge \neg Q \Longleftrightarrow \neg(P \vee Q)$ to get $\neg(\neg P \wedge \neg Q)$, we must negate the right side to get $\neg\neg\,(P \vee Q)$, so, too, if we divide the left side of $3x = 12$ by three, we must also divide the right side by three. And just as we can infer from the premise $\neg(\neg P \wedge \neg Q)$ to the conclusion $(P \vee Q)$, so, too, we can infer from $3x = 12$ that $x = 4$.

As the preceding comments suggest, Peirce's conception of a diagram involves more than just an image or a picture, though it may involve these. Peirce defines a diagram as "a representamen which is predominantly an icon of relations and is aided to be so by conventions. Indices are also more or less used. It [inference] should be carried out upon a perfectly consistent system of representation, founded upon a simple and easily intelligible basic idea" (CP 4.418, 1903). This definition is helpful only once we realize that an icon typically represents its object (i.e., the thing it professes to represent) by standing in structural isomorphism with it. That isomorphism will identify relations among features of the object represented. For example, a map of the New York City subway system is a diagram of the subway system itself because the train stations as represented on the map in fact stand in the relations depicted on the map, even though the map distorts the shapes of Manhattan and the surrounding boroughs. Similarly, the Newtonian law $F = ma$ (i.e., force equals mass times acceleration) is a diagram of the relations among force, mass, and acceleration because these in fact stand in the relations depicted or expressed in the formula. Peirce states that indexes will also be used, such as letters used to name features of diagrams. To take another example, the Pythagorean theorem is the diagram that for a right triangle, $a^2 + b^2 = c^2$. It also uses indexes because a and b indicate the legs of a right triangle, whereas c indicates the hypotenuse. Lastly, inferences on these formulae are carried out upon a perfectly consistent system of representation. The rules we use to make transformations of the formulae are consistent. From $F = ma$ we may infer that $F/m = a$ and from $a^2 + b^2 = c^2$ we may infer $c = \sqrt{(a^2 + b^2)}$ and doing so will not lead us to infer falsehoods provided the premises are true.

I have already hinted at in what the second part of deduction consists. When we make deductions, once we have our completely determined diagram, we transform the diagram in accordance with the rules or permissions of the system. Peirce describes this step as "experimenting upon the effects of modifying this diagram" (NEM 3:41–42, 1895). Such transformations are experiments only in a broad sense of the term. We are not performing physical experiments, such as laboratory experiments. Nonetheless, the process of transforming diagrams may be experimental in two respects. First, we may be engaged in the mental occupation of trying to figure out what sorts of conclusions could be drawn from the premises we have. Second, having already suspected a conclusion can be drawn from the premises, we may try to ascertain whether it can be drawn. This second way of experimenting upon diagrams will be familiar to those who have struggled through complex proofs in logic textbooks: One is given the conclusion in advance, and one must try to prove it.

The third and fourth steps of deduction are more straightforward. Peirce states that the third step in deduction is to observe in "the results of those experiments certain relations between its parts over and above those which sufficed to determine its construction" (NEM 3:41–42, 1895). That is, having completed the transformation, we will observe that the result is different from that with which we started and note which new relations are expressed. Fourth, we will double-check our work. We might double-check our work in various ways. First, we might simply review the transformations we made to assure ourselves we made no errors in doing so. Second, we might endeavor to complete the proof in some different way within the system of representation we have used. Third, we might endeavor to draw the conclusion in accordance with some different method, for example, using truth trees instead of truth tables, or using Peirce's unique logical notation the Existential Graphs rather than our standard propositional logic.[2] Peirce describes this fourth step as "satisfying ourselves by inductive reasoning that these new relations would always subsist where those stated in the premises existed" (NEM 3:42, 1895). 'Inductive reasoning' here should not be taken to mean that we perform some statistical induction to ascertain that the conclusion follows. Rather, Peirce means, as already indicated, that we assure or satisfy ourselves that we were right by double-checking our work.[3]

With the account of deductive inference just given in mind, let us take a closer look at the species of deductive inference Peirce recognizes. These are logical analysis, demonstration (both necessary and probable), and statistical deduction.

Logical Analysis

Peirce regards logical analysis as a species of deductive inference. Logical analysis, in this context, does not mean the scrutiny and criticism of inferences themselves but the analysis of the meanings of words. As I have noted, when the logician takes an interest in deduction, she considers not only the logical structure of various inferences but also takes account of the "meanings of words, and their consequences" (SWS 28, 1895). There is much to be said under this head, so much so that it would require a book of its own. Here, my main aim is to guard against an overly narrow conception of what logical analysis involves. I can only make some brief comments on how broad it really is.

Peirce has a rich account of the meanings of words. Words, or complexes of words, express ideas, loosely speaking. Most familiarly, words are used to express concepts, as do the words 'gravity' and 'circularity.' But words also refer to individuals (e.g., proper names such as *John*), express logical relations (e.g., *and, or*), express other relations (e.g., *the book is on the desk*), and modify other words or ideas (e.g., *he ran quickly*), among serving other functions. The logician will take it that we understand the meanings of words used in an inference, but in so doing she will not restrict herself to considering only the meanings of words that express concepts.

To illustrate, the meanings of words will often include the definitions of terms. Consider, for example, the inference from *John is a bachelor* to *John is unmarried*. Such an inference follows from the definition of the word 'bachelor.' But it would be a mistake to limit the meanings of words to definitions. The person who infers from *John is a bachelor* and *John is a resident of New York City* to *John is a bachelor and a resident of New York City* is understood to refer to the same person, John, in the premises and the conclusion. But this meaning of 'John' is not determined by a definition but by the referential function of the proper name.

Furthermore, the meanings of words will not be restricted to the meaning of a single word but may take into account the meaning of a complex phrase. This will be obvious in cases where some words modify other words, as in *Sally writes humorous poetry*. The phrase 'humorous poetry' is a complex phrase, and we understand that from *Sally writes humorous poetry*, we may infer that *Sally writes poetry* but not that *Sally is humorous*, for 'humorous' modifies the nature of her poetry and does not refer to Sally's personality traits. Sally may in fact be quite dour except when she is writing poetry, which happens to be rarely. By the same token, to use Peter Geach's terminology, we will take ourselves to know when adjectives are used predicatively as opposed to attributively. From *Lena is a good cyclist*, we may infer that she is a cyclist but not that she is good. In short, the meanings of words include how those words function in complex expressions.

This brings us to a further point: The logician will take herself to know not only the meanings of words and complex phrases but also of entire propositions and their consequences. Here, we will be well-served to take stock of Peirce's account of how we make our ideas clear, as he titles his well-known essay discussing the matter. For Peirce holds that ideas may have clarity in at least three respects. First, an idea is clear so far as we know how to use it. Peirce usually restricts his examples to common nouns and adjectives; for example, my idea of hardness is clear so far as I do not mistake hard things for soft things. But ideas need not be conceived so restrictively. I also have an idea of the location of my house, for instance, or an idea of whom another speaks. In the example the inference of John being a bachelor and living in New York City, given earlier, my idea of John is clear so far as I know how to identify him, not mistaking him for Sally or Jack or others.

Second, an idea is distinct when I know how to define it. Peirce recognizes a distinction between nominal definitions and real definitions. The former are useful for getting an initial grasp on a concept; the latter state the essence of something or carve nature at its joints, so to speak. A nominal definition is one given to a person who is unfamiliar with the idea. For example, water is nominally defined as the clear liquid that runs from faucets and is found in rivers and lakes. A real definition analyzes the idea or identifies the essence of its object, such as that water is H_2O. Peirce states, a "*Definition* is either *Nominal* or *Real*. A nominal definition merely explains the meaning of a term which is adopted for convenience. ... A *Real Definition* analyzes a conception" (NEM 4:285, 1903). The inference from *John is a bachelor* to *John is unmarried* is based on the real definition of 'bachelor.'

Yet ideas may have a third grade of clarity. This is the pragmatic adequacy of an idea, or its pragmatic grade of clarity. An idea is pragmatically adequate to the extent that we can trace out the consequences or effects of some thing being it as opposed to something else. We pragmatically develop our ideas by tracing out their consequences, and those ideas are pragmatically adequate when they are developed sufficiently for our purposes. Our idea of water is pragmatically developed in that we know at sea level, water boils at one hundred degrees Celsius and freezes at zero degrees Celsius. With regard to pragmatic adequacy, Peirce characterizes the third grade of clearness both (a) as identifying procedures for isolating some substance and (b) as identifying certain marks or indicators. As an example of (a), he pragmatically develops the idea of lithium by providing a procedure to chemically isolate it (see EP 2:286, 1903). As an example of (b), he claims that a hard thing "will not be scratched by many other substances" (ILS 91, 1878).

Importantly, not every idea can be clarified along all of these dimensions. For instance, some ideas may have no real definition. Wittgenstein's example of our idea of a game may be a case wherein we cannot provide a real definition of 'game' even though we can identify games, provide a nominal definition of them, and even supply certain marks or indicators of them. Moreover, even when an idea can be clarified along all of these dimensions, we may not yet have made it clear along all of these dimensions. We may have a clear idea and even a pragmatically adequate idea without having a real definition of the idea. Some ideas have a life of their own in that we can discover facts about them well after we have used them, such as we discovered that water is H_2O long after we had the idea of water.

When we understand the meaning of a word, we may not merely understand its definition (if we do at all) but also its conceivable practical consequences. For instance, when I claim that *this diamond is hard*, built into the meaning of 'hard' is the experimental consequence that it will not be scratched by many other substances. Hence, the words *this diamond is hard* also mean that *this diamond will not be scratched by many other substances*, for that is part of the meaning of the word 'hard.' But my idea of a diamond may also involve several other components. Peirce notes, "[f]rom some of these properties [e.g., being carbon, being a transparent crystal, having a certain chemical structure] hardness is believed to be inseparable. For like it they bespeak the high polymerization of the molecule" (EP 2:356, 1905). The idea expressed in *the diamond is hard* is not merely something about the nature of hardness but about the chemical composition and structure of diamonds. One way to conceive of Peirce's point is that the meanings of words do not merely include their definitions and grammatical functions but a mental model that is built up from the meanings of the ideas the words express.[4] The mental model is an icon (not an image) that is, in some important ways, structurally isomorphic with the object represented (e.g., the very diamonds which are hard).

The upshot of these considerations is that deduction involves not merely stating a priori definitions of terms but also tracing out the conceivable practical consequences of a mental model. Such consequences may be entailed by a complex set of commitments. Those commitments are sometimes expressed by stating the experimental conditions under which an observation should be had. The Rutherford Gold Foil experiments illustrate such a case. The theory that atoms consist of negatively charged subatomic particles (plums) suspended in a positively charged volume (pudding) has as a consequence that a beam of alpha particles should show minimal scattering when directed through a thin sheet of gold foil. This is a logical, deductive consequence of the model given the meanings of the words, where those meanings are the conceivable practical consequences of the mental model which has been constructed. Notice that the predicted observation (minimal scattering) is in fact a consequence of a complex set of commitments, viz., that atoms have a structure akin to plum pudding, that alpha particles pass through gold foil, that negatively and positively charged particles will repel each other, and so on.

We are clearly now moving up against some thorny problems. In particular, the last considerations seem to suggest that our ideas will face the tribunal of experience as a whole, and not just piecemeal. The problem of underdetermination rears its head. To raise these worries at this juncture, however, is to put the cart before the horse. For the topic under consideration here is not whether experience can confirm or disconfirm any given theory but to trace out the logical consequences of those theories. No doubt, in doing so we will draw on a vast set of background beliefs. (The *whole* of them is surely an exaggeration.) The point of the present considerations is that tracing out those consequences is a deductive project. That deductive project consists in the logical analysis of our ideas. These ideas are expressed in words and phrases. When the logician is engaged in the task of logical analysis, she takes herself to understand the meanings of those words and their correlative ideas, together with the consequences of them. Those ideas may be complex, amounting to a certain model or structure describing experimental conditions by which a hypothesis can be put to the test. No doubt there is a very complex story to tell about how such models get elaborated and developed. It is not my task here to give an account of *that* process, only to point out that doing so is a species of what Peirce regards as deduction, falling broadly under the fact that the logician takes account of the meanings of words and their consequences. Tracing out the consequences of our hypotheses is a deductive project, albeit one that will inform other inquiries.

It should be clear that when deduction is logical analysis, the deduction will be truth-preserving just in case the meanings of the words claimed really are the meanings of the words. To infer from *John is a bachelor* to *John is unmarried* is valid, but to infer from *John is a bachelor* to *John is married* is invalid. In the former case, being unmarried is part of the meaning of the word 'bachelor.' For that same reason, the latter inference is invalid, as being married is inconsistent with being unmarried. Similarly, to infer from *diamonds are hard* to *diamonds will not be easily scratched* is valid, but to infer from the same premise that *diamonds will be easily scratched* is invalid.

Peirce's Axiomatic Conception of Demonstration

In addition to logical analysis, Peirce identifies demonstration as a species of deduction. Demonstration is of two varieties, necessary and probable demonstration. The former is based on the axioms of logical calculi. The latter is based on the axioms of the probability calculus. I have already mentioned how these calculi can be axiomatized in Chapter 2.

Peirce rests the validity of necessary and probable demonstrations on the manifest truth of the axioms and rules of the calculus in question. Since the truth of the axioms and rules is manifest, there is no need to supply a separate argument in favor of them. In his defense of the validity of deduction such as we find it in "Grounds of the Validity of the Laws of Logic" (henceforth, "Grounds"), Peirce does not endeavor to demonstrate the validity of deduction. Rather, he focuses on responding to objections that call the validity of deduction into question. Andrew Howat describes this as Peirce's quietist conception of grounding, according to which we may "remain *quiet* on the question of their [the laws of logic] validity, unless/until a genuine doubt comes along" (2014, 483).

Deductive Calculi: In "Grounds," Peirce's defense of deduction's validity amounts to little more than an explanation of the claim that "if one sign denotes generally everything denoted by a second, and this second denotes generally everything denoted by a third, then the first denotes generally everything denoted by the third" (W 2:243, 1869). This is the doctrine of the *nota notae.* It gives expression to the argument form Barbara, that:

(1) All M are P.
(2) All S are M.
(3) Therefore, all S are P.

Peirce remarks that this is "not doubted by anybody who distinctly apprehends the meaning of these words" (W 2:243, 1869). All of the other deductive syllogistic forms may be reduced to Barbara or are of its same general form (see, e.g., W 2:29–38, 1867). Accordingly, there is no need for a defense of the validity of deduction—or at least syllogistic logic—beyond grasping the class relations signified in Barbara. Other than that, all the logician must do is respond to objections to such forms of reasoning.

Furthermore, on his axiomatic conception of logic such as we find it in incipient form in "Description of a Notation for the Logic of Relatives" of

1870,[5] Peirce feels no particular need to demonstrate the truth of his algebra-analogous axioms. The axioms of algebra are sufficiently familiar so that Peirce thinks their logical analogues need no further support than a brief explanation. In his late work, he insists that the mathematician need not rely on anything other than the fact that the mathematical statement is evident: "The mathematician does not 'rely' upon anything. He simply states what is *evident*, and notes the circumstances which make it evident. When a fact is evident, and nobody does or can doubt it, what could 'reliance' upon anything effect?" (HP 2:845, 1904)

As one would rightly suspect, Peirce's insistence on the thesis that the axioms and rules of the deductive calculi need not be demonstrated is intended to defuse the objection that any attempted proof of the axioms would employ the very deductive calculi one purports to demonstrate. In his attempt to ground the validity of the laws of logic, Peirce acknowledges the "sweeping objection to my whole undertaking" that "my deduction of logical principles, being itself an argument, depends for its whole virtue upon the truth of the very principles in question" (W 2:243, 1869). Peirce replies that his aim is not to convince "absolute sceptics nor men in any state of fictitious doubt" (W 2:243, 1869). He only asks his reader to be candid and "if he becomes convinced of a conclusion, to admit it" (W 2:243, 1869).

In Peirce's late writings, he is explicit that we can in fact perceive necessary truths, such as the truth of the axioms of a deductive calculus. 'Perceive' here is not to be restricted to sense perception but to a sort of cognitive compulsion which is analogous to sense perception. The truth of an axiom (in the cases under present consideration) is forced upon one; it is impossible to doubt it with one's present cognitive resources or within the context of the determinate diagram one has constructed in the mathematical imagination. In these contexts, we can perceive mathematical truths in the diagrams we have constructed. Peirce gives as an example our ability to perceive that the sum of the angles produced by one line abutting another will always be 180 degrees by drawing a perpendicular line to the point of abutment. Peirce remarks, "[t]his perpendicular must lie in the one angle or the other. The pupil is supposed to *see* that. He sees it only in a special case, but he is supposed to perceive that it will be so in any case. . . . [A]ll reasoning, so far as it depends on . . . mathematical reasoning, turns upon the perception of generality and continuity at every step" (EP 2:207, 1903).

Two points are in order. The first is that mathematical and logical statements have practical consequences insofar as we can experiment on them in the mathematical imagination through the manipulation of diagrams. Such experimentation will be familiar to anyone who has wondered whether accepting some proposition is permitted given (i.e., is consistent with) one's other commitments. Indeed, such experimentation is at the heart of all reductio ad impossibile arguments. Cheryl Misak puts the matter nicely when she writes of mathematical and logical truths that "[t]hey have practical consequences in thought experiments or experiments on diagrams" (1995, 115). The second point is that just as we make efforts to correct hallucinatory or illusory experiences when we are subject to them but admit the veridicality of our percepts when we cannot alter them by voluntary, direct, and immediate mental effort,[6] so with necessary truths we admit their truth when attempts to refute or construct arguments without some axiom are unsuccessful. Of course, in any given case, we may be wrong. For this reason, Peirce admits that axioms are not beyond the scope of all doubt. Nonetheless, until and unless we are satisfied such axioms are wrong or needless, we are compelled to candidly concede their truth.[7]

Peirce holds that the axioms of our deductive calculi are evident and in need of no proof, and similarly for the rules used. Importantly, this is so not only for the axioms and rules of necessary reasoning but also for the axioms and rules of probable reasoning. We can see the truth of an axiom such as $P \supset (Q \supset P)$ since the consequent of $Q \supset P$ is already posited in the antecedent of the entire statement. To make the point in a semantic way, if P is false, then the antecedent of the whole conditional is false and so the whole conditional is true, whereas if P is true, then the consequent of the whole conditional is true and so the whole conditional is true. Similar points can be made for the axioms of the probability calculus. For instance, the truth of the rule of additivity that if events A and B are disjoint, the probability of their union satisfies: $P(A \vee B) = P(A) + P(B)$ can be perceived by thinking of the total probability space as a one-by-one square and the probabilities of A and B as shapes occupying some area of that square. As A and B will be disjoint, their areas must not overlap. Therefore, the probability of the disjunction $A \vee B$ must be the sum of each area. Such geometric ways of thinking about the axioms of the probability calculus are familiar from introductory courses on probability.

The Emergence of a Semantic Conception of Deduction: In my previous comments, I made recourse to a semantic way of looking at the axioms of propositional logic in order to make the truth of the axiom $P \supset (Q \supset P)$ evident. In addition to his axiomatic conception of demonstration, Peirce has a semantic conception of validity for propositional logic. The conception emerges from the work of his student Christine Ladd-Franklin. Her essay "On the Algebra of Logic" develops a semantic approach to logical validity. It was published in 1883 as part of the volume on logic that Peirce edited.[8] In "Studies in Logical Algebra," written in 1885, Peirce resorts to such a semantic conception because the "fifth icon" or axiom—that $((P \supset Q) \supset P) \supset P$—is "hardly axiomatical" (W 5:173, 1885). Peirce means that the truth of the fifth icon is not evident, unlike the first icon (for example) which is that $P \supset P$. He argues for the fifth axiom by remarking:

> That it is true appears as follows. It can only be false by the final consequent [P] being false while its antecedent $[((P \supset Q) \supset P)]$ is true. If this is true, either its consequent, [P], is true, when the whole formula would be true, or its antecedent $[(P \supset Q)]$ is false. But in the last case the antecedent of $[(P \supset Q)]$, that is [P], must be true. (W 5:173, 1885)

In 1893, Peirce provides a truth-table for the conditional (R 946:4[9]). In 1902, he again uses something like a truth table. The table he constructs is equivalent to the biconditional $P \equiv Q$ (see CP 4.262). In 1909, Peirce introduces a third truth value—L, for the limit between truth and falsity—and presents truth tables for a trivalent logic.[10]

In accordance with the semantic conception of validity, Peirce can regard valid deductive arguments as truth-preserving as well, since there will be no interpretation on which the premises are true and yet the conclusion is false. One will rightly wonder whether Peirce recognized the distinction between his axiomatic conception of validity and his semantic conception of validity, and if he did whether he endeavored to prove soundness or completeness. Atwell Turquette (1964) shows that Peirce could have proven completeness from his five icons. Pietarinen (2020, 272) has found that Peirce does recognize the problem and attempts a demonstration of both soundness and completeness. However, I shall not delve into the minutiae of these proofs. Peirce's arguments are superseded by contemporary work on the topic.[11]

Statistical Deduction

The other species of inference Peirce classes as deduction is statistical deduction. Statistical deduction is the application of a conditional probability to a particular case. As Isaac Levi explains, Peirce appeals to the weak law of large numbers and the central limit theorem to defend the validity of statistical deduction.[12] Simply stated, the law tells us that as our number of samples increases, the sample mean will tend toward the population mean. The theorem tells us that given a large enough sample, the sample mean will tend to have a normal sampling distribution, regardless of the underlying population. As Deborah Mayo points out (1996, 439), Peirce was well aware that sample sizes of thirty or greater are usually sufficient for the theorem to hold.

Taken together, the law and the theorem tell us that as we sample more, we will tend in the long run to get closer to the real uniformity of nature, with thirty samples usually being sufficient to draw the conclusion that probably and approximately such-and-such a proportion of S's are P, even though we may be wrong in any given case and at any given time the frequency we have ascertained may not represent the real probability. Nevertheless, provided that our statistical induction has been sufficiently strong, we may infer from a probability-of statement to a probability-that statement since the inference will hold good at approximately the ratio established. For example, we may infer from the probability of someone being vaccinated against measles on condition the person lives in Goessel, Kansas, to the probability that Lena qua resident of Goessel, Kansas, is vaccinated against measles. As Levi states, "the inquirer who knows nothing about the outcome of the random sample to be taken other, perhaps, than it will be taken, can use statistical deduction and the calculus of probabilities to conclude that the outcome of random sampling will be representative with very high probability concerning the proportion of P's among the M's" (2004, 270–271). Since the outcome is representative of the population, we may apply it to individual cases and we will be probably and approximately correct at the ratio ascertained by statistical induction. Having made these probable or statistical deductions, we can then use the probability calculus to calculate other probabilities. To use the example from Chapter 2, if 98% of the persons in Goessel, Kansas, are vaccinated against measles, then the probability that any two people (Lena and Jane) from Goessel, Kansas, are vaccinated against measles is the product of .98 and .98, which is .96.

It is important, though, not to confuse the strength of the statistical induction with the validity of the statistical deduction. A statistical deduction assumes that the ratio ascertained by statistical induction is correct. If that ratio is incorrect, then the inference proceeds on a false premise but is not invalid. Rather, in the simplest sorts of cases,[13] statistical deduction works as follows. First, one has a statistic, for example, that there is a 98% probability of S being vaccinated against measles provided S is from Goessel, Kansas. Next, one substitutes the name of a person for S, for example, Lena. It follows that there is a 98% probability of Lena being vaccinated against measles provided Lena is from Goessel, Kansas. Next, we know that Lena is from Goessel, Kansas. That is, the probability that Lena is from Goessel, Kansas, equals one. It follows straightforwardly from the probability calculus that the probability Lena is vaccinated against measles is the same as the probability of S being vaccinated against measles provided S is from Goessel, Kansas.[14] As the conclusion follows from the probability calculus, it is a deductive inference. This account of the validity of statistical deduction is in lockstep with Peirce's axiomatic conception of deduction. The inferences are made on the basis of the axioms of the probability calculus and the known exceptions to the general rule.

Objections to the Validity of Deduction and Replies

As already noted, Peirce's main task in defending the validity of deduction is to respond to objections against the validity of deduction. He undertakes that task in "Grounds." Many of the objections he considers are lamentable catches. Those that are not lamentable catches are still lively debated. Peirce's "Grounds" was published in 1869, and in reviewing the objections he considers, it is important to bear in mind his views of mental action and validity in those earlier works. In particular, in these early works, Peirce maintains (1) that mental action is of the nature of a valid inference and (2) that logic is objective symbolistic. The latter assumes Peirce's semiotic conception of logical validity, discussed in Chapter 3. The conclusion of the argument must be substitutable for the premises of the argument and be true of the object the premises are about. This account presupposes that the premises are about objects at all, such that the conclusion can also be true of those objects. Consequently, objections which purport to show that subsequent cognitions (conclusions) are not determined (i.e., made to be as they are) by their objects mediately via the premises are also objections to the validity of deduction.

The Regress Objection

Objection: The first objection Peirce considers is that "no proof can be of any value, because it rests on premises which themselves equally require proof" (W 2:247, 1869). Peirce concedes the objection shows proving anything beyond the possibility of doubt is impossible, but he claims the thesis usually taken to be even stronger: "the objection is intended to go much further than this, and to show (as it certainly seems to do) that inference not only cannot produce *infallible* cognition, but that it cannot *produce* cognition at all" (W 2:247, 1869). Why Peirce thinks this stronger thesis is implied is evident from considerations just mentioned: If all mental action is of the nature of an inference, then there would have to be an infinite series of cognitions. Since mental action is of the nature of an inference (or so Peirce holds in 1869) and there cannot be such an infinite series, there must be no mental action.

Reply: Peirce replies that just "because there has been no first in a series" it does not follow that the "series has had no beginning in time" (W 2:247, 1869). It here becomes important to distinguish among three ways in which a cognition may be a first cognition. In the first way, it may be a temporally first cognition or set of cognitions. Patently, there must be a temporally first cognition or set of cognitions if only because Peirce accepts Darwin's theory of evolution and there was a time when there were no cognizing agents at all. In the second way, a cognition may be a logically first cognition. Peirce thinks that for any given inquiry there are logically first cognitions. These are the cognitions with which we begin our inquiries, those prejudices or observations from which all inquiry starts out. In the third way, a cognition may be a mentally first cognition. Suppose that there is some sharp dividing line between mind and world. A mentally first cognition would be the first cognition at the limit and on the mental side of that divide.

When Peirce denies that there are any first cognitions, he means to deny that there are any mentally first cognitions. In "Questions Concerning Certain Faculties Claimed for Man," he had illustrated this idea. Take a triangle and dip the apex in a glass of water. The water line is the current cognition of the object. There is an infinite series of lines between the water line across the triangle and the apex. Each of those lines is a prior cognition. The apex is the object itself: "The apex of the triangle represents the object external to the mind which determines both these cognitions [i.e., the present cognition and any previous cognition]" (W 2:210, 1868). Peirce denies the supposition that there is some sharp dividing line between mind and world. Our minds are in the world and continuous with the world, just as there is a continuum between the apex of the triangle and any given water line across the triangle.

Why does this objection call into question the validity of deduction? It calls deduction's validity into question because if there can be no first cognition of reality, then we have lost our cognitive grip on the world. That is, our cognitions may not be determined by their objects and so be substitutable in a series for those objects. In fact, this objection calls into question not only the validity of deduction but of all inference. As briefly explained in Chapter 1, Peirce is trying to rise to the challenge set forward by his rejection of Cartesianism: If mental action is thoroughly inferential, then why is our cognitive grip on the world not compromised by the possibility that we have some cognitive defect or an evil demon deceives us? If there is no sharp dividing line between mind and world, there is no cause for alarm. The mind and the world shade into each other. One way to understand such a seemingly mystical claim is to recognize that the statement *horses are animals* is both a fact when considered from the side of the world and the object of belief when considered from the side of the mind. When those statements we believe are also facts, our beliefs are true.[15]

The Petitio Principii Objection

Objection: The second objection Peirce considers is that every deduction is a petitio principii because the conclusion is implicit in the premises. He attributes this objection to "Locke and others" (W 2:247, 1869). The objection is that since the conclusions of deductions are mere explications of their premises, whatever the conclusion states is contained virtually in the premises. Consequently, the conclusion is assumed in the premises, which is the logical fallacy of petitio principii.

Reply: Peirce replies that the logician aims to account for "what forms of facts involve what others" (W2:248, 1869). In contrast, to commit the fallacy of a petitio principii, one must reason from the "unknown to the unknown" (W2:248, 1869). Deduction does not require one to reason from the unknown to the unknown but from putative truths to other truths that follow from them given the formal relations implicit in the putative truths. It is not a petitio principii to claim that the putative truths (1) All men are mortal and (2) Socrates is a man entail that (3) Socrates is mortal. With respect to the validity of deduction, the question is not whether (1) and (2) are known but whether granted (1) and (2), (3) is a logical consequence of them.

The Reasoning Machine Objection

Objection: The third objection, which Peirce illustrates by appeal to a passage in Jonathan Swift's "Voyage to Laputa," is that necessary reasoning is mechanical. However, mental action, inference, requires a soul. Consequently, necessary reasoning must not be inference at all, let alone valid inference: "A much more interesting objection is that a syllogism is a purely mechanical process.... This being so (and it is so), it is argued that this cannot be *thought*; that there is no life in it" (W 2:248, 1869). Aristotelian syllogistic logic is decidable. Peirce knew machines could be designed to carry it out. Blaise Pascal had designed a calculator in the 1640s. Charles Babbage had designed computing machines in the 1820s.[16] Peirce's contemporary William Stanley Jevons had worked on designing reasoning machines and presented his results to the Royal Society in 1870.

Reply: Peirce concedes, based on these researches, that necessary reasoning can be mechanical. He denies that this implies necessary reasoning cannot be valid. Just because the syllogistic process is not the whole of mental action, it does not follow that the syllogistic process fails to represent it or a part of it. Analogously, just because a surveyor's map does not represent every feature of a plot of land, it does not follow that the map fails to represent the plot of land. Thought may proceed syllogistically, even though the syllogistic representation of it does not represent every feature of thought. To put it another and more poetic way that parallels the objection, syllogistic logic may represent the mechanism of thought without representing the soul of thought.

Another way to think of this reply is that the objection confuses an argument with an inference. Arguments are (at least) lists of declarative sentences, one of which is the conclusion and the others of which, if there be others, are the premises. Although machines can be designed to draw conclusions from such sets of premises, these are not inferences. Inferencing requires a habit, a logic in use on the part of the person making the inference. Typically, the logician will be concerned with inferences as they have been made, and these inferences can in fact be valid or invalid. Arguments, in contrast, may be valid or invalid derivatively, as inferences that could have been made.

The One-Sidedness Objection

Objection: Peirce attributes the next objection to Hegel. The objection is that necessary inference is "one-sided" because it teaches us "a part only of all that is true of an object" (W 2:250, 1869). Although it was not Hegel's way of making the point, the objection may be grasped by considering the fact that the thesis gives us only one side of the truth. The other side is the antithesis. While these may be mediated in a synthesis, that synthesis will itself be a thesis with an antithesis in need of mediation through a synthesis, and so on ad infinitum. Peirce claims that the objection comes down to an inconsistency among the claims that there is nothing incognizable, that by inferential mental action everything is cognizable, and yet if inference is one-sided, something is incognizable.

Reply: There are two replies to be made to this objection. First, Peirce responds that the argument "proves too much" because it would make "all finite knowledge to be worthless" (W 2:250, 1869). Inference that is one-sided leaves something uncognized. But it does not follow from the fact that something is uncognized that the one-sided knowledge we do have compromises our cognitive grip on the world. We can concede that not everything is cognized or even cognizable by us now without conceding that the knowledge we do have is somehow compromised by what we do not know. This point is especially pertinent to Peirce's appeal to the long run of inquiry, which plays a key role in his account of the validity of induction. At any given phase of inquiry, there will be more to be known. The long run of inquiry is not something attained. Nonetheless, what would be believed further down the line can be approximated at the present stage of inquiry. The gap between the present state of inquiry and the long run will never be bridged, but that does not entail a plank in a bridge spanning the chasm cannot be laid. The consequence is that Peirce is a fallibilist in that we cannot attain absolute precision or absolute certainty. As Elizabeth Cooke rightly remarks, this sort of "fallibilism is not a curable condition" (2004, 159). However, by following standards of inquiry that make our inductive inferences strong, we can manage the symptoms.

Second, Peirce does hold (1) that there is nothing which is incognizable,[17] and (2) that mental action will ultimately root out error and alight on the truth, provided the questions we ask have any definite meaning. It does not follow from this, however, that there is some time at which everything is known. Hegel's demand that inference teach us the whole of the object is misplaced. Even if our knowledge is not limited and so everything is cognizable, it does not follow that there is some assignable time at which everything is known or cognized. We can hold that everything is knowable at some time without conceding that there is some time at which everything is known.[18]

The Subjectivity Objection

Objection: The fifth objection is that a syllogism's "'therefore' is merely subjective; that, because a certain conclusion syllogistically follows from a premise, it does not follow that the fact denoted by the conclusion really depends upon the fact denoted by the premise, so that the syllogism does not represent things as they really are" (W 2:251, 1869). The objection gains its force from the fact that we can infer a cause from an effect. For example, from smoke we can infer a fire. In the syllogism, the effect is prior to the cause since the effect (smoke) is stated in the premises, whereas the cause (fire) is stated in the conclusion. In reality, however, the cause is prior to the effect.

Reply: This objection confuses cause-and-effect relations with the relation of logical entailment. Granted that if there is smoke, then there is a fire, the fact that there is smoke logically entails that there is a fire. However, this in no way implies that fire is causally posterior to smoke. Peirce admits that "there is often an appearance of reasoning deductively from effects to causes" (W 2:251, 1869). However, the reason we can make such inferences is "because there really are such facts as that 'If there is smoke, there has been fire,' in which the following event is the antecedent" (W 2:251, 1869).

The Truth-to-Falsity Objection

Objection: Hegel's work is also the source of the sixth objection. The objection is that sometimes a conclusion can be false even though the premises are true and the syllogistic form is correct. An example is this: Whatever is painted blue is blue. Some wall has been painted blue. It follows the wall is blue. However, if the wall has also received a coat of yellow paint, then it will be green and not blue.

Reply: Peirce replies that 'painted blue' is equivocal between being "painted with blue paint, or painted so as to be blue" (W 2:253, 1869). If the former, the premise that *whatever is painted blue is blue* is false. If the latter, then the wall is blue and not green.

The Continuity Objections

Objection: Objection seven marks a turn away from these rather spurious objections to deduction's validity and toward some of the stronger and more classical objections to it. Peirce credits these objections to "the Eleatics and Sophists" (W 2:254, 1869). He identifies the objections as falling into three general classes: those pertaining to continuity, those pertaining to arguments by reductio ad absurdum, and those pertaining to semantic paradox. The seventh objection concerns continuity.

Zeno's paradoxes of motion are together the seventh objection. To take just one example, suppose Achilles is racing toward a finish line. He must pass half the distance to the finish line, half the distance to the halfway point, half the distance to the quarter-way point, and so on. However, it is impossible to traverse an infinite number of distances in a finite time. Consequently, the runner's completion of the race is only apparent.

This objection is clearly related to the first. If mind and world shade into each other continuously but are distinguishable nevertheless (as red and orange shade into each other but are distinguishable nevertheless), is the world really ever able to reach or make contact with the mind? The continuity objection calls into question our cognitive grip on the world. To be metaphorical, if we are incapable of traversing the distance from our own minds to the world or if the world is incapable of traversing the distance to our minds, then the conclusion of an inference cannot represent in a truth-preserving way the same object the premises are about.

Reply: Peirce's response to Zeno's paradoxes is that the sum of an infinite series of finite polynomials may be finite. For example, the sum of a half, a quarter, an eighth, a sixteenth, and so on is one, not infinity. As Peirce puts it, "nothing more can be concluded than that he [the runner] passes over a distance greater than the sum of any finite number of the above series of terms. But because a quantity is greater than any quantity of a certain series [i.e., 1 is greater than ½, ¼, etc.], it does not follow that it is greater than any quantity [i.e., it is infinite]" (W 2:255, 1869). In short, one can traverse an infinite series of distances (½, ¼, etc.) without traversing an infinite distance.

The Reductio Objection

Objection: The eighth objection pertains to arguments wherein two true propositions seem to be inconsistent with one another. If there are two true propositions which are inconsistent with each other, but both assume the existence of their subjects, that would imply reductio ad absurdum is not a valid inference form. Peirce, though, accepts reductio ad absurdum, as indeed all classical logics do. His example of two (apparently) true but inconsistent propositions is *If he were deliberately to upset his inkstand, no ink would be spilt* and *If he were to deliberately upset his inkstand, the ink would be spilt.* The first is true because no one with an inkstand with ink in it would deliberately upset it. The second is true on the supposition there is ink in the inkstand. But if we assume the existence of the man deliberately upsetting his inkstand, it seems to follow both that ink is and is not spilt, which is absurd.

Reply: Peirce contends that these sorts of cases arise from two general errors. The first general error concerns the problem of existential import. For example, from *All of John Smith's family are ill,* we would ordinarily be able to infer that each individual member (Jane, Julia, Jeremiah, etc.) is ill. However, if there are no members of John's family, then it would be false that each individual member of John's family is ill. On the Aristotelian view, inference on the universally quantified proposition presupposes the existence of members of the class. But Peirce also recognizes that the class may be empty and so the inference "would be sophistical if the universal were one of those propositions which does not assert the existence of its subject" (W 2:259, 1869). *All of John Smith's family are ill* is false if the universally quantified proposition has existential import, since there are no members of John's family. Supposing there are no members of John's family, the universally quantified proposition is true provided we do not assume existential import. Since the apparently contradictory truth-value depends on existential import, the propositions are not in fact contradictory.

The second general error is failing to account for the modality of the proposition. He notes that reductio ad absurdum "also occasions deceptions in another way, owing to the fact that we have many words, such as *can, may, must*

&c., which imply more or less vaguely an otherwise unexpressed condition" (W 2:260, 1869). In the inkstand case, it is significant that the apparently contradictory propositions are counterfactuals. Accordingly, these propositions about the inkstand are not inconsistent: "These propositions are both true, and the law of contradiction is not violated which asserts only that nothing has contradictory predicates: only, it follows from these propositions that the man will not deliberately overturn his inkstand" (W 2:258, 1869).

The Liar Paradox

Objection: The final objection to the validity of deduction Peirce discusses is the liar paradox. The liar paradox is that these three commitments are inconsistent:

(1) Every proposition is either true or false.
(2) No proposition is both true and false.
(3) "This proposition is false" is a proposition.

The paradox arises when we ask whether *this proposition is false* (i.e., the liar proposition) is true or false. If the former, then it must also be false, since that is what it states. If the latter, then it must also be true, for it states it is false. But per (1) it must be one or the other, and yet per (2) it cannot be both. However, per (3) it is a proposition.

Reply: Peirce's reply to the liar paradox develops over the course of his life.[19] In a passage from 1901, Peirce identifies two general replies to the problem. The first holds that every proposition implicitly asserts its own truth. As the liar proposition also explicitly asserts its own falsity, the proposition is contradictory and so false. This is Peirce's position in "Grounds," viz., that *this proposition is false* is self-contradictory. Peirce attributes this manner of solving the paradox to Paulus Venetus, who "undertakes to show that every proposition virtually asserts its own truth" (CP 2.618, 1901, and see W 2:263, fn.7, 1869).

In 1903, Peirce disavows his solution of 1869 and claims his argument for the thesis that every proposition implicitly asserts its own truth is question-begging, a "huge *petitio principii*" (EP 2:167, 1903). The earlier argument is question-begging because Peirce's argument for the claim that every proposition asserts its own truth assumes that the liar proposition is paradoxical. Since it is paradoxical, there must be something which distinguishes it from other propositions. What distinguishes it from other propositions is the fact it asserts its own falsity. The paradox arises, therefore, because propositions must also assert their own truth. This argument is question-begging because whether the liar proposition is paradoxical is precisely what is at issue.

At this same time, Peirce turns to adopting the second general solution to the liar paradox. It is to hold that the liar proposition is no proposition at all. Peirce attributes this manner of solution to William of Ockham, who "admits the validity of the argumentation and its consequence, which is that there can be no such proposition, and attempts to show by other arguments that no proposition can assert anything of itself" (CP 2.618, 1901). Among those who opt for this solution, Peirce notes that some think no argument at all needs to be given in order to fortify the conclusion that the liar proposition is in fact no proposition, whereas others think some supporting arguments for the conclusion ought to be given.

Peirce falls into the latter camp. He gives three arguments purporting to show that there can be no proposition which asserts its own truth, or its own falsity: "if no proposition asserts its own truth, none asserts its own falsity" (EP 2:167, 1903).[20] The first argument is that if propositions could assert their own truth or falsity, then their precise denials qua assertions should also be equivalent. But they are not. The precise denial of *it rains* is *it does not rain*, whereas the precise denial of *it is asserted it rains* is that *it is not asserted that it rains*. Evidently, Peirce is thinking that to predicate truth of a proposition is to assert the truth of the proposition. One might be suspicious of this claim, though. If every proposition has one of two contradictory predicates regardless of what we assert about it, then *this proposition is false* will be true or not true regardless of what we assert about it.

Peirce's second argument is a special case of the postcard version of the liar paradox. On the postcard version, we are to imagine a postcard with sides labeled A and B. Side A states *what is written on side B is true*. Side B states *what is written on side A is false*. In Peirce's example, these sides of a postcard are rather witnesses, one of whom testifies that whatever the other witness says is true and the other of whom testifies that whatever that first witness has said is false. Peirce claims, "[c]ommon sense would certainly declare that nothing whatever was testified to" (EP 2:167, 1903). Once again, however, if every proposition has one of two contradictory predicates regardless of what we declare about it, then each item of testimony will be true or not true regardless of what we declare about it.

Peirce's third and final argument is based on his analysis of the proposition. He maintains that every proposition minimally consists of two parts. The first is an index, which indicates the object the proposition is about. The second is a symbol, which states something about the object indicated. For example, the proposition *that man is tall* indicates some thing (perhaps by ancillary signs, such as pointing) and ascribes to it the properties of being a man and being tall. The problem with a proposition such as *this proposition is false* is that the object—the very proposition—does not exist at the time the words 'this proposition' are uttered or read. Consequently, the proposition cannot indicate its object, since its object does not exist. But there are two worries about this claim. The first is that it implies perfectly sensible propositions, such as *this proposition is in English*, are not propositions. Second, it confuses tokens and types. Although any given tokened proposition will not exist until uttered, the type of proposition is either true or false as well.

I have been casting some doubts on Peirce's arguments for the thesis that there can be no proposition which asserts its own truth or falsity. But there is an implicit commitment in Peirce's arguments that needs to be drawn out. Once we draw it out, we will see that Peirce has another reply to the liar paradox. Peirce notes that Aristotle "defines a proposition as a symbol that is either true or false" (EP 2:168, 1903). Although Peirce has doubts about the propriety of the definition, he does not deny it.[21] Granted Aristotle's definition, a string of words is either true or false or else it is not a proposition.

Moreover, propositions are true or false in virtue of functioning in a representational way. They purport to represent some fact, some way the world is. As Peirce writes:

> Truth belongs exclusively to propositions. A proposition has a subject (or set of subjects) and a predicate. The subject is a sign [viz., an index]; the predicate is a sign [viz., a symbol]; and the proposition is a sign that the predicate is a sign of that of which the subject is a sign [i.e., the proposition is a unified but complex sign constituted by interpreting the index and symbol as signs of the same object and then using it in a representational way]. If it [i.e., the object indicated by the index] be so [i.e., has the property represented by the symbol], it [i.e., the proposition] is true. (EP 2:379, 1906)

In short, for any string of words to be a proposition, that string must be assertible, that is, capable of being used to represent some fact. I suggest that this is why Peirce's first two arguments appeal to the notion of assertion. They appeal to assertion because the liar proposition can be a proposition only if it is assertible.

At this juncture, it is important to stress that on Peirce's account, propositions are a semiotic category. They are not abstract objects, let alone abstract objects that supervene on states of affairs. Neither are propositions mental events or states. Rather, propositions are sign substitutes for (pro-posits) facts or states of affairs. A person may wake from a dream and spontaneously declare *someone somewhere is eating breakfast*. No doubt the utterance is perfectly grammatical. Moreover, it is likely a fact that a person somewhere is eating breakfast. But insofar as the utterance does not represent any person eating breakfast, it is not a proposition and is not true. It is not a proposition because it is not functioning to represent any fact that a person is eating breakfast, even though it corresponds with that fact. Truth is not mere correspondence of an utterance with a fact but the representation of that fact by the utterance.[22] For a string of words to function representationally, the object of the fact must be indicated, and something must be said about that object. Peirce remarks, "[i]n driving along the road with a compagnon, he points at a house and remarks, 'That is a pretty house.' It is a proposition. But I make a voyage to a distant island and repeat the remark, 'That is a pretty house.' 'What house,' asks my interlocutor. 'Oh, a house far away.' He will be right in telling me that I do not enunciate anything at all, because I do not indicate what I am talking about" (LoF 1:335, 1898). A string of words is a proposition only if that string is used to represent some fact or state of affairs. Not all grammatically well-formed declarative sentences are propositions. Sometimes a string of words is, as Peirce claims, an "imitation proposition" (HP 2:734, 1901).

The problem is that the liar proposition cannot function to represent a fact. It cannot function to represent a fact because were it to so function, it would imply its own contradictory. This is evident from the paradox itself: We tie ourselves up in logical knots when we try to use it to represent some fact. The liar proposition is not assertible. One may think she can assert the liar proposition, and she may defend it with all her heart. But as it is one thing to think you can fly and quite a different thing to be able to fly, so also it is one thing to think you can assert a claim and quite a different thing to be able to assert it. As jumping out the window will quickly prove you wrong in the first case, the knots you logically tie yourself in will quickly prove you wrong in the second. The liar proposition is obviously assertable (one can say it, even with conviction), but it is not assertible (one cannot use it to represent a fact).

Here is the argument in sum: A proposition is defined as a sign that is true or false. For anything, either it is true or it is false, or else it is not a proposition. A string of words is capable of being true or false only if it is capable of being used to represent some fact or state of affairs. The liar proposition, however, is not capable of functioning to represent some fact or state of affairs, for any attempt to use it commits one to a contradiction. Therefore, the liar proposition is no proposition at all. The Ockhamist solution that there can be no such proposition is vindicated.[23]

At this juncture, one might argue that if the liar proposition is not true, then the paradox rears its head again. In fact, I have argued that something is a proposition only if it is assertible, and the liar proposition is not assertible. It follows it is not a proposition, and so like chairs and tables it is not true. But then, using the strengthened liar, *"this proposition is not true" is not true*, and we are off to the races again.

This is a mistake. To get the strengthened liar off the ground, we must treat the strengthened liar proposition as a premise and so a proposition. To infer from *this proposition is not true* to *"this proposition is not true" is true* by the t-schema *"p" is true iff p* is to treat *this proposition is not true* as a proposition. Similarly, to infer from *this proposition is not true* to *"this proposition is not true" is not true* by substituting the very proposition for the subject of the premise is to use it as a premise. But for the same reasons the liar proposition cannot be a proposition, neither can the strengthened liar.

This brings us to another point. The argument just developed seems plainly incompatible with Peirce's claim that the liar proposition "is certainly a proposition" (EP 2:167, 1903). But I have argued that it is not a proposition. Even if the solution I have just given is accurate, it cannot be Peirce's. Two replies are in order. The first is that Peirce makes that claim in the context of explaining his earlier solution to the liar paradox. Consequently, in context, it may not express his mature view on the matter but only be a report of his earlier position. The second is that I have not claimed the solution on offer here to be Peirce's solution, only a Peircean solution. Peirce may not have realized that there are some important distinctions for which the solution just developed calls. If he did realize it, he is not as explicit about it as one would hope.

Conclusion

Deductive inferences, when they are valid, have the virtue of being truth-preserving. Peirce distinguishes among logical analysis, demonstration (whether necessary or probable), and statistical deduction. Logical analysis explicates the meanings of words, and so far as those meanings explicated are in fact the meanings of the words, the deduction is truth-preserving. Demonstration proves other propositions from premises, axioms, and rules. The axioms are truths which are in no need of demonstration because their truth is evident. Provided the axioms are true and the rules are truth-preserving, deductions on the basis of those axioms and rules will be truth-preserving, too. Statistical deductions apply conditional probabilities to individual members of the population the conditional probability is about. Such inferences are truth-preserving because provided the conditional probability is accurate, it is a fact that there is the same probability any given member has the property or characteristic in question. Peirce also responds to various objections purporting to show that deductive inference is not truth-preserving. When these objections are not lamentable catches, they are deep problems with which philosophers and logicians continue to grapple. Foremost among the latter sort of objection is the liar paradox, which occupied Peirce's attention throughout his life.

5

Induction

Validity and Strength

Peirce distinguishes among three genera of inference. These genera are deduction, induction, and abduction. In addition to premises and conclusions,
all inferences involve some procedure the person making the inference more
or less dimly apprehends using. The procedures of each of these genera of
inference are evaluated relative to whether they have some truth-producing
virtue. Deductions are to be truth-preserving. Inductions are to be truth-
approximating in the long run if the procedure used is consistently adhered
to. Abductions are to be truth-conducing, provided there is any truth to be
ascertained. An inference is valid if and only if the procedure followed in
making the inference in fact has the truth-producing virtue claimed for it
and the person making the inference in fact followed the procedure in question. Inductions and abductions can also be strong or weak, depending on
the procedure used relative to its domain of application. Deduction is the
only genus of inference for which there is no distinction between validity and
strength (see EP 2:232, 1903).

In considering induction, then, we must distinguish between its validity
and its strength, as well as consider objections and replies to the validity of
induction. These are large topics, and so I have divided my discussion of
induction into two chapters. This chapter will consist of two main sections
examining what Peirce states about the validity and strength of induction.
In the first section, I will explain the general procedures of induction and
why they are truth-approximating. This discussion will be general because
the procedures used in actual inductive inquiries cannot be specified in advance. Particular strategies need to be devised for more specific fields of inquiry. Different controls and considerations will need to be considered for
different objects of study and the conditions under which they are studied.
We can easily mix a bag of beans to get a random sample, but we cannot
do the same for boulders distributed over a large region. Humans are

Peirce on Inference. Richard Kenneth Atkins, Oxford University Press. © Oxford University Press 2023.
DOI: 10.1093/oso/9780197689066.003.0006

subject to self-deceptions; amoeba are not. Ice cores melt at room temperature. Phosphorus ignites in open air. Fish die if taken out of water for long periods of time. Dolphins die if left underwater for long periods of time. The conditions under which we can study things change depending on the object of study. As T. L. Short aptly notes, right methods are discovered and improved upon over time. Scientists in pursuing their "vaguely defined aims . . . come, by degrees, to an ever more exact conception of that aim and correspondingly to a better idea of what their methods should be" (1998, 296). For these reasons, we can give only a general account of the procedures used in inductive inquiry and explain why in general those procedures will be truth-approximating in the long run. But as the devil is in the details, so, too, for any given inquiry the question of whether the procedure used is in fact truth-approximating in its specific application can be answered only relative to the actual investigation.

In the second section, I shall cull from Peirce's writings some advice on how we can make our inductive inferences strong. Peirce was a practicing scientist, and his scientific writings evidence techniques he employed for making strong inductive inferences. His writings on the history of science as well as his engagements with scientists of the day also reveal his views on practices inquirers are well-advised to follow if they wish to make their inductive inferences strong. In the next chapter, I will examine various objections to the validity of induction.

The Validity of Induction

In his mature writings, Peirce distinguishes among three species of inductive inference. The first is crude induction. When one makes a crude induction, one consults the mass of past experience and determines whether some claim is supported by the mass of past experience or not. The second is qualitative induction, which in his earlier writings Peirce calls hypothesis. It is hypothetic induction, the procedure by which one eliminates, corroborates, or confirms hypotheses. Quantitative induction is the third species of inductive inference. It is paradigmatically statistical induction, but in fact includes all sorts of inductive inference which involve measurement. Let us consider each in turn, and following that, I shall give a broader account of the validity of inductive inference in general.

Crude Induction

The first species of induction Peirce identifies is crude induction, which he also calls rudimentary induction. When Peirce presents crude inductions, he does so with examples based on the mass of past experience. In 1903, he characterizes such inference as inferring from the fact that one "has no evidence of the existence of any fact of a given description and concludes that there never was, is not, and never will be any such thing" (CP 7.111, 1903). In 1905, he remarks, "[t]he first and weakest kind of inductive reasoning is that which goes on the presumption that future experience as to the matter in hand will not be utterly at variance with all past experience. *Example:* 'No instance of a genuine power of clairvoyance has ever been established: So I presume there is no such thing.' I promise to call such reasoning *crude induction*" (CP 2.756, 1905). Peirce claims that the argument proceeds on the basis of "the *absence* of instances to the contrary" (CP 2.756, 1905). Moreover, he maintains that it is the only sort of inductive inference that can conclude to a universal proposition. The absence of a counterexample leads us to conclude that no such thing ever occurs.

Inference based on the absence of counterexamples can also involve affirmed propositions. In a given context of inquiry, *It has never been established that p* may be regarded as equivalent to *so far as is known, not: p*.[1] *It has never been established that stones fall from the sky* may be regarded as equivalent to *so far as is known, stones don't fall from the sky*. Likewise, *it has never been established that humans have not had any internal organs* is equivalent to *so far as is known, humans have always had some internal organs*. By crude induction, we can conclude that *humans always have some internal organs*.

Importantly, in consulting the mass of past experience, Peirce does not claim that the experiences have to be the experiences of the person making the inference. The inference does not proceed from the premise that *I have never established that p* but from the premise that *it has never been established that p*. Peirce allows that the testimony of others should figure into claims about past experience. It is not the mass of my past experience but the mass of experience as a whole that is employed in making crude inductions. This is particularly noteworthy in special cases in which a person has heard reports of something utterly at variance with past experience. Is a mere report that there are unicorns sufficient to overturn a crude induction? The equivalence of *it has not been established that p* with *so far as is known, not: p* is helpful here. It is epistemically indefensible that *I know that John knows that p* and yet *I don't know that p*.[2] Accordingly, if one thinks the report is credible in such a way that it rises to testimony-based knowledge, then a report at variance with the mass of other past experience does suffice to overturn a crude induction. Peirce remarks that Sir Humphrey Davy was given the opportunity to study a purported miracle—the liquification of the blood of St. Januarius—and was "granted every facility for the thorough investigation of it" (CP 7.112, 1903). Davy was not able to explain the miracle by scientific means. His report is thorough, and so we are "not justified in pooh-poohing such observations," even though the miraculous event is at variance with past experience (CP 7.112, 1903). But the thoroughness and "circumstantial details" of Humphrey's report are important here (CP 7.112, 1903). In cases of hearsay in which reports do not suffice to give us testimony-based knowledge, we may be permitted to reject such testimony. In those cases, we may have "some positive reason for disbelieving it" (CP 7.112, 1903). For instance, Peirce rejects the testimony about paranormal experiences of some people on the grounds that they are members of a group known to be "superstitious, ignorant, and credulous" (W 6:78, 1887).

Peirce maintains that crude inductions are truth-approximating because if the conclusion is false, experience will eventually compel us to reject the premise upon which it is based. For instance, natural philosophers, having never observed a stone fall from the sky, concluded that no stones fall from the sky. But eventually experience proved the contrary, as asteroid strikes are examples of stones falling from the sky. Peirce writes, "[t]he justification of this [i.e., crude induction] is that it goes by such light as we have, and that truth is bound eventually to come to light; and therefore if this mode of reasoning temporarily leads us away from the truth, yet steadily pursued, it will lead to the truth at last" (CP 7.111, 1903). Peirce thinks that such arguments are very weak, and it is in our interest to use different sorts of inductive inferences when we can. Nevertheless, the utility of crude induction is that sometimes the only information we have is the absence of any evidence to the contrary: "It goes upon the roughest kind of information, upon merely negative information; but that is the only information we can have concerning the great majority of subjects" (CP 7.111, 1903).

One will rightly wonder, at this juncture, why we should think the counterexamples to such cases will ever come within the ken of our experience. We are, after all, humans who live on some far-flung planet on the edge of a galaxy. Our records go back only some 10,000 years, to be exceedingly generous. We will someday go extinct. Why think that in such a short time, we should ever experience very rare events? Peirce's answer is that we must regard the community of inquirers—and with it the mass of experience—as extending beyond our own race, our own geological epoch, and across all time. The community of inquirers, "must not be limited, but must extend to all races of beings with whom we can come into immediate or mediate intellectual relation. It must reach, however vaguely, *beyond this geological epoch, beyond all bounds*" (ILS 116, 1878, emphasis added). With such an expansive community of inquirers, if there is counterevidence to be had, it will eventually come within the ken of our experience.

Note that for Peirce the community of inquirers is indefinitely and infinitely wide. It does not merely consist of us now. It includes those inquirers among all of rational life, even alien rational life (should any exist) and rational life which might evolve after humans have gone extinct. Peirce maintains that for inductive inferences—all species of inductive inference—to be valid, we must hope that the community of inquirers will exist forever, have faith in future inquirers to carry inquiry further, and love the community such that we neither restrict our inquiries to local and trivial concerns nor allow our inquiries to be biased by self-interest. Only if there is some such community can inductive inference be valid. Yet Peirce does not claim to have any proof or argument that there is or will be some such community. Rather, he argues that there can be no argument against it and logic requires that we hope there will be such a community. As I shall have much more to say about validity and the community of inquirers in this chapter and the next, I shall set the matter aside for the moment so as to explain the other species of induction.

Hypothetic Induction

I turn now to hypothetic induction. In Chapter 2, I distinguished between hypothetic induction as a mode of inquiry and hypothetic induction as a species of inference. The latter is a part of the former. When we make hypothetic inductions as a species of inference, we eliminate, corroborate, or confirm hypotheses based on testing predictions of those hypotheses. Hypothetic induction as a mode of inquiry includes the preparatory steps of forming those hypotheses and ranking them (properly abduction) and tracing out the conceivable practical consequences of those hypotheses so we can put them to the test (properly a part of deduction, as briefly discussed in the previous chapter under the heading of logical analysis).

In his more mature work, Peirce calls hypothetic induction qualitative induction. He does so to distinguish this species of induction from quantitative induction. Whereas for quantitative induction the inference "depends on measurements, or on statistics, or on countings" (CP 6.473, 1908), for qualitative induction the inference is not based "upon a collection of numerable instances of equal evidential values, but upon a stream of experience in which the relative evidential values of different parts of it have to be estimated according to our sense of the impressions they make upon us" (CP 2.759, 1905). The last clause is important. The premises of qualitative inductions can still have numbers assigned to them, only these numbers will not be measurements of the thing studied, such as the number of white beans in a bag. Rather, the numbers assigned will be rough estimations of the evidential support each theory has relative to how we feel about it. This suggests, and I shall argue for it more thoroughly momentarily, that hypothetic induction may include something like updating degrees of belief in the ways contemporary subjective Bayesians would recommend. First, though, let us get a better sense of how hypothetic induction as a mode of inference works in the more standard cases.

Hypothetic Induction in Standard Cases: Suppose we observe some phenomena and frame various hypotheses which might explain them. We will want to put those hypotheses to the test. To do so, we will consider what we should expect to occur were the hypothesis true. This may require little more than continued observation. If, say, I hypothesize that there is a bird resting in a nearby bush, I may need only to wait and to watch. If I see a bird fly away from the bush, my observation will confirm the hypothesis. But other times, we will have to design experiments in order to test hypotheses. Such experiments may require travel, as Arthur Eddington traveled to Principe to observe a solar eclipse and test Einstein's theory of relativity. Experiments may also require the design and construction of equipment, as with the construction of the Laser Interferometer Gravitational-Wave Observatory (LIGO) to observe gravitational waves and corroborate the theory of relativity.

In both the straightforward observational case and the complex experimental cases, we have some conditional statement the antecedent of which is the hypothesis. The consequent of that conditional will be another conditional. The antecedent of that other conditional will describe the conditions of observation or experimentation. The consequent of that conditional will describe the expected observation under those conditions. This second conditional may be termed the prediction. In the bird case just mentioned, the main conditional is that *if there is a bird resting in that bush, then if I continue to watch the bush attentively, I should see a bird fly away from it.* The prediction is the consequent of that main conditional: *if I continue to watch the bush attentively, I should see a bird fly away from it.* The antecedent of the prediction is the experimental conditions: *I continue to watch the bush attentively.* The consequent of the antecedent is the expected observation: *I see the bird fly away.*

The experimental conditions described in other cases may be much more involved. In the case of Eddington's observation of the eclipse, for example, it would be along the following lines: *if Einstein's theory of relativity is true, then if I observe a solar eclipse and take images of the eclipse using a telescope with lenses which have no aberrations, set on a steady surface with a firm footing and on a mount with no slippage in the hinges, on photographic plates that are not flawed, on a noncloudy day, and so on, then I should observe that the degree of deflection of light around the sun should be approximately 1.75 arcseconds.* And of course, Einstein's theory of relativity had a competitor, Newton's theory of universal gravitation. It can be tested in the same way, and it has a corresponding conditional: *if Newton's theory of universal gravitation is true, then if I observe a solar eclipse and take images of the eclipse using a telescope with lenses which have no aberrations, set on a steady surface with a firm footing and on a mount with no slippage in the hinges, on photographic plates that are not flawed, on a noncloudy day, and so on, then I should observe that the degree of deflection of light around the sun should be approximately .875 arcseconds.*

It is significant that these conditionals involve also a description of the experimental conditions. One may doubt that the experimental conditions have been satisfied.[3] Perhaps I had been inattentive, in spite of my best efforts, and failed to see the bird fly away. Perhaps there were flaws in the photographic plates, or perhaps atmospheric conditions also affect the angular deflection of light. In such cases, we might also inquire into whether the experimental conditions were in fact satisfied. We might subject the telescope to further scrutiny, checking the lenses. We might investigate the process by which the photographic plates have been made and then the images developed. Very frequently, there will be room for doubt that these experimental conditions have all been satisfied. But typically, there will be just as much room to dispel those doubts by probing whether the experimental conditions were in fact satisfied. As I shall have more to say about this in my discussion of inductive strength, I shall set the matter aside for now.

Once the conditional has been made sufficiently clear and the tests performed, we will find that the observations we make are either consistent with the prediction or not. If the results are not consistent with the prediction and granting the experimental conditions are satisfied, then we may eliminate the hypothesis. Whenever I refer to the elimination of a hypothesis, I always mean 'elimination' in an attenuated sense that includes elimination entirely from further consideration, elimination from consideration in its present form, or elimination from consideration for the moment. Note that we may defer the entire elimination of a hypothesis until we have probed whether the observational conditions were in fact satisfied.

Alternatively, the observations may be consistent with what was predicted. In such a case, the hypothesis is corroborated or confirmed, again granting the experimental conditions are satisfied. By 'corroboration' here I do not mean Popperian corroboration but the supplying of corroborating evidence in the legal sense, that is, evidence that backs up the hypothesis. If we think of a hypothesis as a sort of story we tell about the way things are, then a hypothesis is corroborated when we gain evidence to back up the story. By 'confirmation' I mean that the evidence we have suffices to remove all genuine doubt we may have regarding the hypothesis. For example, Edmond Halley's contemporaries thought that comets may travel in straight lines or in sling-shot orbits around the sun. Halley proposed that they also travel in elliptical orbits. He surmised that comet sightings recorded by Paolo Toscanelli in 1456, Peter Apian in 1531, and Johannes Kepler in 1607, as well as the comet Halley himself saw in 1682, were in fact sightings of the very same comet. If some comets in fact travel in elliptical orbits, Halley maintained, then there should be another comet sighting in about 1758. When a comet was observed in 1758, this corroborated—if it did not confirm—Halley's hypothesis.

Why have I claimed this observation "corroborated—if it did not confirm" Halley's hypothesis? Peirce states that a proof is "[a]n argument which suffices to remove all real doubt from a mind that apprehends it" (CP 2.782, 1901). Such a proof may be a mathematical proof, a probable deduction, or an "inductive, i.e., experimental, proof" (CP 2.782, 1901). We could take a weak view of confirmation on which a hypothesis is confirmed just in case what is predicted is observed or just in case the prior probability increases. In such a case, we may say that the hypothesis has passed a test. But it is important to note both (1) that a hypothesis may pass one test though there are many others and (2) that a hypothesis may pass a test while there are other hypotheses still on the table and yet to be eliminated. If we take a stronger view of confirmation on which a hypothesis is confirmed only when it has been proved in the Peircean sense such that any real doubt is removed, then we ought to claim that passing just one test is usually only corroboration of the hypothesis, in the sense noted earlier, not confirmation of it. Yet if that one test is severe and that prediction is unique to the one hypothesis, then even a single test may suffice to confirm a hypothesis.

Note that in appealing to corroboration, I am not endorsing the Popperian view that hypotheses cannot be confirmed by hypothetic induction. I am rather claiming that such a confirmation occurs only when all real doubt has been removed by the mind that apprehends the proof. Furthermore, note that it is only real doubts that need to be removed, not fictitious doubts. Evil demons, spontaneous and inexplicable hallucinations, alternative hypotheses which have yet to be conceived but may at some time be conceived and then compete with the one remaining hypothesis among those currently conceived, or other such scenarios do not suffice to instill real doubts. For any given hypothesis that has been confirmed, it may be the case that it is wrong. But mere maybes do not constitute real doubts any more than the fact that I may quantum tunnel through the car speeding toward me instills a real doubt as to whether I ought to jump out of the way.

To return to the example of Halley's comet, plainly, Johann Palitzsch's observation of the comet confirms Halley's hypothesis in the weak sense that the hypothesis passed the test proposed. But this is only one test. The question is whether passing this test also suffices to remove all real doubt in the minds of those who apprehend the argument. If it does, then the hypothesis is confirmed. If it does not, then the hypothesis is only corroborated in the sense that we have gained evidence which backs up the hypothesis. Now in fact, the test developed was quite demanding, or severe. Those who understood the argument for Halley's comet traveling in an elliptical orbit surely found it decisive. And insofar as they did, the argument confirmed the hypothesis.

Yet if we consider other cases, there are times when it is plain a hypothesis is at best corroborated and not confirmed by experimental evidence. Galileo's defense of heliocentrism could hardly be said to remove doubts about the hypothesis, even among those who understood his arguments. There were too many problems still to be ironed out, among them the failure to observe stellar parallax as the earth orbited the sun and the fact that the stars themselves would have to be enormously larger than our own sun. Though we now can remove those doubts (we can observe parallax with modern instruments and some stars are in fact so large), Galileo's contemporaries could not. Consequently, heliocentrism was at best corroborated and not confirmed in his lifetime.

It is important, though, to distinguish between hypothetic inductions that suffice to remove real doubts in the minds of those who understand the inference and those inferences which have not been understood. At the time Eddington confirmed Einstein's theory of relativity, Einstein's theory was not yet well-understood. Many astronomers were not in a position to understand the experimental proof Eddington provided. But Eddington's experimental argument for Einstein's theory was no less a proof because people failed to understand it. Eddington, after all, understood the proof. Accordingly, theorists were not justified in pooh-poohing his observations any more than they were Davy's study of St. Januarius's blood. That said, it may also be doubted that Eddington's experiment was in fact a proof, for even among those who understood the inference made, some theorists still had real doubts about the truth of the theory of relativity. In particular, astronomers had doubts about the accuracy of the observations and wanted additional corroborating evidence. More light deflection measurements were demanded in order to confirm Einstein's theory. Since these doubts were not removed, Eddington's experiment amounted only to a corroboration of Einstein's theory and not a confirmation of it.

Peirce's Bayesian Sympathies: We are now pushing up against a claim that I made earlier, namely, that qualitative induction may involve updating our credences in a manner similar to what contemporary subjective Bayesians recommend. I have just been describing hypothetic inductions as confirming hypotheses by removing any real doubts about them. The reference to real doubts, however, suggests that we may regard persons as having some degree of belief in a hypothesis. The hypothesis may be regarded as (strongly) confirmed when that degree of belief is sufficiently high,[4] and as corroborated (or weakly confirmed) when the prior probability is increased. Such an account seems to place Peirce in bed with subjective Bayesians, where probabilities measure degrees of belief. This is strange company, though, for Peirce explicitly disavows conceptualism, the nineteenth-century forerunner of subjective Bayesianism. The conceptualists hold that probability is "the degree of belief which ought to attach to a proposition" (ILS 133, 1878). But Peirce insists this is a mistake and probability is rather to be regarded "as a matter a *fact*, i.e., as the proportion of times in which an occurrence of one kind is accompanied by an occurrence of another kind" (ILS 133, 1878). This is Peirce's frequentism, and indeed, Peirce is sometimes cast as the arch-frequentist.

But that Peirce is opposed to Bayesianism in toto and subjective Bayesianism in general misapprehends Peirce's mature position. First, it is important to note that having doubts about Bayesianism is not to reject Bayes's Rule. Like anyone who accepts the axioms of the probability calculus (whether frequentists or Bayesians), Peirce has no option but to endorse Bayes's Rule; it is a consequence of his other commitments. He is committed to the axioms of the probability calculus, and those axioms imply Bayes's Rule.

Second and more importantly, while it is true that Peirce forcefully defends frequentism in his early works, especially the *Illustrations of the Logic of Science* series from 1877 to 1878, it is also true that Peirce continued to think deeply about probability over the course of his life. One way in which Peirce's mature theory of probability changes from his early work is that he more clearly distinguishes between probabilities or frequencies as they have been ascertained and probabilities or frequencies as they are in fact. Ascertained frequencies are, as the phrase suggests, those which have been established by observation, sampling, measurement, or the like. But it does not follow that the frequency we have ascertained is the frequency as things are in reality. This is obvious in cases in which our sampling has been slight, but it may also be true when our sampling has been large. Even in these latter cases, it is always possible that the real frequency—that is, the real uniformity or probability in nature—is different from what we have ascertained from sampling. For that reason, we state our conclusions in terms of confidence intervals.

Real frequencies may be of two varieties. They may be real propensities in things, such as the frequency with which some die would show a six. In the case of a die's propensities, the frequency in fact consists in "a certain habit or disposition of behaviour, in its present state of wear" (ILS 276, 1910) such that were the die rolled an infinite series of times it would show six some definite proportion of times. But this supposes that "the result of no throw [has] the slightest influence upon the result of any other throw" (ILS 125, 1910), which is plainly false insofar as the very act of rolling the die will cause the die, however so slightly, to change shape. Repeated rolling will cause the sides to wear even more. It follows that our attempts to ascertain what the frequency with which some die would in fact show a six when rolled can at best give us only an approximation of the die's propensity to show a six, since the very act of rolling the die may affect the wear of the die and so the propensity.

Second, real frequencies may be regularities in nature independent of the propensities of things. For example, there is some frequency with which I sneeze when Jupiter is in the house of Virgo. There is some frequency that my family members are killed by a comet on condition that I exist. But neither of these two examples about sneezing and death by comet are real habits or tendencies in things. Other cases are unclear. There is some real frequency with which it rains when the clouds are dark and heavy and the atmospheric pressure has dropped. If we could not identify some real frequency with which it rains on condition of certain atmospheric conditions, there would be no such thing as modern meteorology. But it is unclear whether these should be treated as propensities. I will not pretend here to decide that question. The key point for the moment is that no matter how it is decided, there can be real frequencies which involve no such causal relations and ought not to be regarded as propensities, such as in the Jupiter and comet cases. To be certain, the scientist will want to separate out mere coincidences and correlations from laws of nature. As Peirce states in a later work, "[t]he problem of how an accidental regularity can be distinguished from an essential one is precisely the problem of inductive logic" (CP 3.605, 1903). Nonetheless, there is some real frequency to accidental regularities, as Peirce calls them.

In addition to refining his account of probability, Peirce also endorses a position adjacent to contemporary accounts of subjective Bayesianism provided (and these are two important provisos!) we (a) rightly apprehend what we are doing and (b) restrict its application to appropriate cases. He is most explicit about endorsing such a view in later writings:

> [W]hat I attack is the method of deciding questions of fact by weighing, that is by algebraically adding, the feelings of approval produced in the mind, by the different testimonies and other arguments pertinent to the case. I acknowledge that this method is supported, under abstract conditions, by the doctrine of chances, and that there are cases in which it is useful. But I maintain that those conditions are not often even roughly fulfilled in questions of ancient history; so that in those investigations it commonly has no value worth consideration. (HP 2:711, 1901)

What I wish to stress in this quotation are Peirce's claims (1) that the method is supported under abstract conditions by the doctrine of chances, (2) that there are cases in which it is useful, but (3) that there are other cases in which is inappropriately applied.

Let me begin by focusing on what the method, or procedure, is and how it is supported under abstract conditions. As Peirce indicates, the procedure in question is a method for deciding questions of fact. One way to decide questions of fact is to probe hypotheses in the way mentioned earlier. We formulate hypotheses. We make predictions on the basis of those hypotheses. We then test whether those predictions are borne out. Another way of deciding questions of fact is by quantitative induction. We may wish to know the proportion of white beans to beans in a bag. To do so, we take some samples and ascertain approximately what the proportion is.

Yet a third way of deciding questions of fact—the one at stake in Peirce's comments—is by "weighing . . . the feelings of approval produced in the mind." This procedure may be described qualitatively or quantitatively. I begin with the qualitative account. We may think of belief as a switch which turns on when the evidence favors it and off when the evidence disfavors it.[5] Depending on how strong or weak the evidence is, the intensity of the feeling of belief (our degree of conviction) may change. In his earlier writings, Peirce appeals to two laws concerning belief. These two laws pertain to belief and to the intensity of the feeling of belief. The first law is that "[w]hen the chance becomes less, then a contrary belief should[6] spring up" (ILS 135, 1878). Peirce conceives of chance as the proportion of favorable to unfavorable cases; chance differs from probability in that probability is the proportion of favorable cases to all cases. If our evidence indicates there is some greater chance of hypothesis H being true than $not\text{-}H$, then our belief that H turns on. If our evidence indicates there is some greater chance of $not\text{-}H$ being true than H, our belief that H turns off, which is to say that our belief that $not\text{-}H$ turns on. If our evidence indicates the chance is even, then we suspend belief.

The second law is "Fechner's psychophysical law . . . that the intensity of any sensation is proportional to the logarithm of the external force which produces it" (ILS 136, 1878). Peirce remarks, "[w]e generally know when we wish to ask a question and when we wish to pronounce a judgment, for there is a dissimilarity between the sensation of doubting and that of believing" (ILS 53, 1877). We hold our beliefs with more or less conviction. This degree of conviction is caused by an external force, which is the course of investigation or stream of experience as it bears on our feelings of conviction.

On the Peircean view, we do not have degrees of belief (we either believe H or we believe $not\text{-}H$), but our feelings of conviction do come in different strengths. This feeling of conviction is determined by our impressions of the evidence: "the relative evidential values of different parts of [the stream of experience] have to be estimated according to our sense of the impressions they make upon us" (CP 2.759, 1905). The greater the impression of the evidential value, the greater our conviction, that is, the intensity of the sensation of belief.[7]

Now suppose we have some stream of experience and evidence in that stream pertains to H. In the course of experience, our feeling of conviction as to H will vary. If we come across a lot of evidence suggestive of *not-H*, our feeling of conviction that H is true will decrease. If in our estimation the evidence drops below an even chance, then we will believe *not-H*. Alternatively, we may come across a lot of evidence suggestive of H. If we do, then our feeling of conviction that H is true will increase logarithmically. If our estimation of the evidence is at an even chance, we suspend belief.

The foregoing describes Peirce's psychophysics of belief. Of course, there may be pathological cases such that people believe in the teeth of the evidence, but Peirce takes the foregoing to be generally descriptive of the actual belief-formation practices of people. It is an account of our doxastic machinery, so to speak. Suppose we believe H with a degree of conviction C. We get in some evidence concerning H. That evidence makes some impression on us, altering our feeling of conviction as to the truth of H. If that feeling of conviction increases, then we will continue to believe H with even greater conviction. If that feeling of conviction decreases, we may continue to believe H, though our conviction that H is true may drop. If our feeling of conviction declines precipitously such that we think there is a less-than-even chance that H, then we will believe *not-H*.

Now it is not hard to see that the qualitative description of our doxastic machinery just given can be made more definite by assigning actual numbers to our degree of conviction and updating our beliefs accordingly. Suppose I start by believing that H with a conviction of .6 on a scale from zero to one.[8] This will be my prior probability that H. Suppose also that I get some new evidence E and my feeling of conviction that $P(E|H) = .2$, whereas my feeling of conviction that $P(E) = .9$. It follows that $P(H|E) = .133$. Now given the new evidence, my feeling of conviction that H has dropped below the threshold of chance. The contrary belief that *not-H*, as Peirce puts it, springs up.

The preceding paragraph is a straightforward application of Bayes's Rule. Bayes's Rule states that $P(H|E) = [P(E|H) \times P(H)]/P(E)$. The rule is provable from the axioms of the probability calculus. Because the rule is a deductive consequence of those axioms, Peirce concedes that the procedure of deciding questions of fact is supported under abstract conditions by the doctrine of chances, as quoted earlier. If we abstract away from how we are getting evidence and simply imagine people being perfect Bayesians—updating their beliefs as the evidence comes in—the mathematics supports such a procedure. There is nothing wrong with following it under those abstract conditions, and Peirce concedes as much.[9]

Peirce's Attacks on the Bayesian Procedure: But abstract conditions are not the actual conditions of inquiry, and here we get to the heart of Peirce's attack on the procedure. In fact, we find in Peirce's writings five criticisms of the Bayesian procedure.

Peirce's first criticism is that the procedure of balancing reasons by algebraically adding feelings of approval and disapproval may be used only when the arguments (i.e., the evidence) are independent. It is, however, very difficult to ascertain when the arguments are independent. Peirce notes that in card games, it is usually thought that cards are thoroughly shuffled and each deal is independent of the previous. But he remarks "it is often an exceedingly nice question whether two events may be considered as independent with sufficient accuracy or not" (ILS 121, 1878). When Peirce asked a friend to keep track of deals in the card game whist, he found there was imperfect shuffling. Very often arguments are mutually supportive or there are subtle connections among the phenomena under consideration. Consequently, there will be cases when balancing reasons is incorrectly employed, if we wrongly think our arguments are independent when they are not.

A second criticism Peirce raises concerns how we set our initial degrees of conviction when we have no evidence on which to go. Some Bayesians will appeal to the principle of indifference. A standard version of the principle of indifference states that if there are n physically possible outcomes of some event E, the probability of any given outcome should be set at $1/n$ in the absence of any reason to think otherwise. A flipped coin will land heads or tails. Following the principle of indifference, the probability of a coin landing heads should be set at 1/2. The probability of drawing the two of hearts from a standard deck of cards should be set at 1/52. Peirce objects to the principle of indifference. He proffers three arguments against it.

First, in "The Probability of Induction," Peirce presents the following objection to the principle of indifference. Suppose we have a panel of possible colors. We circle a part of it. We ask: Will the first aliens we meet have a hair color in the circle or not? By the principle of indifference, we should assign a probability of 50% to them having a hair color within the circle and 50% to them having a hair color without the circle. If we circle another area of the panel, we should assign a 50% probability to them having a hair color in that second region and 50% to them having a hair color outside of it. If we then circle a third region that encloses just those first two regions, then the probability of aliens having that color of hair should be the sum of those probabilities, or 100%. I quote from Peirce:

> In the conceptualistic view of probability, complete ignorance, where the judgment ought not to swerve either toward or away from the hypothesis, is represented by the probability ½.
>
> . . . Let us enclose such an area with a closed line, and ask what is the chance on conceptualistic principles that the color of the hair of the inhabitants of Saturn falls within that area? . . . [O]n conceptualistic principles . . . [t]he answer can . . . only be one-half, since the judgment should neither favor nor oppose the hypothesis. What is true of this area is true of any other one; and it will equally be true of a third area which embraces the other two. But the probability for each of the smaller areas being one-half, that for the larger should be at least unity, which is absurd. (ILS 137–138, 1878)

This is not a good argument against the principle of indifference because the reference class shifts. For each circle, including the third one, the principle of indifference states we should assign a 50% probability to the alien's hair color being within the circle and a 50% probability to it being without the circle. Peirce, however, calculates the probability for the third circle as the sum of the probabilities for the first two circles it encloses.

A second and better objection is to ask why we should assign a 50% probability to the claim that the alien's hair color will be within the circle in the first place. By the principle of indifference as stated by Peirce, it would follow we should assign a probability of 50% whether the circle is very small or very large. If a circle has the radius of one nanometer, we should assign the claim that the first aliens we encounter will have a hair color within the circle a probability of 50%. Similarly, if it has a radius of one centimeter, we should a sign it a probability of 50%.

Peirce argues that the principle of indifference requires us to assert that there is an even chance of a totally unknown event. The problem is that "[t]here is an indefinite variety of ways of enumerating the different possibilities, which, on the application of this principle, would give different results" (ILS 139, 1878). The principle of indifference requires us to start by assuming that nature is arranged in some equiprobable way so that the principle of indifference can be applied. But Peirce denies that we have any right to talk about "[t]he relative probability of this or that arrangement of Nature" as if "universes were as plenty as blackberries, if we could put a quantity of them in a bag, shake them well up, draw out a sample, and examine them to see what proportion of them had one arrangement and what proportion another" (ILS 141, 1878). There is no privileged way of divvying up possibilities so as to apply the principle of indifference. For example, is nature arranged so that the outcomes of rolling a die are equiprobable across all the faces of the die (a 1/6 probability of rolling a three by the principle of indifference)? Or are the possible outcomes whether I roll a three or not-three (a 1/2 probability of rolling a three by the principle of indifference)? Similarly, suppose I have a ten-by-ten grid of possible hair colors of aliens. I may divide the grid in half along the center. In that case, the area of the resulting rectangle will be fifty square units. But I may also take a compass, place its point in the center of the grid, and draw a circle with a radius of 2.5 units. In that case, the area of my circle will be over 30 square units smaller than the area of the grid divided in half. However, by the principle of indifference, I ought to assign a 50% probability to it being in the rectangle and a 50% probability to it being in the circle.

Peirce drives this point home most forcefully in a lecture from 1866. Suppose we are drawing balls from an urn. We might assign the probability of the first ball being blue as .5. Peirce then notes that if there are three possible outcomes, we should rather assign it a probability of .33. But then the possibilities can be listed as an "infinite number of kinds," so that the "the probability of drawing a ball which is blue is equal by that principle, to the number of shades of blue, divided by the number of all shades. And as both these numbers are infinite, the probability is infinity divided by infinity which is we know not what" (W 1:402, 1866). Colors are on a continuous spectrum. Suppose we demarcate some part of that spectrum as blue. That part may itself be infinitely subdivided so that there are an infinite number of colors within that range. But there is also an infinite number of colors on the spectrum altogether. Accordingly, the proportion of colors in the blue range to colors altogether is infinity over infinity, which is undefined.

Peirce raises a third argument against the principle of indifference. Suppose we do not know how religiously devout persons of European ancestry behave in Tahiti. We may assign a probability of 50% to the person acting responsibly and 50% to not. Peirce asks, "how could an insurance company fare who should try to do business on such a basis? A basis for business has got to be knowledge and not ignorance" (NEM 3:173, 1911). Were such persons when in Tahiti to become quite morally lax to their own detriment, the insurance company may go broke. "A probability, if it is correct," Peirce remarks, "is a basis for business. But there can be no such basis except experience and the idea of deducing any quarter of fact from anything but knowledge is absurd" (NEM 3:173, 1911).

These criticisms of the principle of indifference raised, however, Peirce does not regard the objection as decisive against subjective Bayesianism. First, one may be a Bayesian without endorsing the principle of indifference. Second, even if our prior probabilities are set in peculiar ways, assiduously applying the procedure of Bayesian updating will correct our degrees of conviction in the long run, provided we are following appropriate procedures of inquiry more generally. (I shall have more to say about that last clause momentarily.) Whether we embrace the principle of indifference or not, the mathematics bears out the fact that if we update our degrees of conviction appropriately, our beliefs will eventually settle on how things are.

A third objection Peirce raises to the procedure of Bayesian updating is that the subjective Bayesian often misuses her procedure. When she does, her argument is invalid because she has not followed the procedure she intends to employ. In particular, the Bayesian will sometimes attach probabilities to the hypotheses, when in fact probabilities should be attached to our degrees of conviction in the hypothesis. Suppose H is the hypothesis that an asteroid strike caused the extinction of the dinosaurs. We ought not to claim that there is an 80% probability that an asteroid strike caused the extinction of the dinosaurs. The hypothesis is either true or false, and there is no probability about it. We ought rather to conclude that our degree of conviction that an asteroid strike caused the extinction of the dinosaurs is .8. Peirce expresses this point when he discusses laws of nature: "It is nonsense to talk of the probability of a law, as if we could pick universes out of a grab-bag and find in what proportion of them the law held good. Therefore, such an induction is not valid; for it does not do what it professes to do, namely, to make its conclusion probable" (CP 2.780, 1901). There is not a 90% probability that all stars of a certain specified mass will be supernovas. Either they will or they will not. If we are following the procedure of Bayesian updating, what we are justified in claiming is that for any given star, if it is of such-and-such a mass, our degree of conviction that it will be a supernova is 90%.

Two points follow from this consideration. The first is that the quantitative presentation of the procedure of Bayesian updating given earlier is based on the fiction that hypotheses have degrees of probability. For our purposes, we may treat the hypothesis as having some definite probability. But in cases of Bayesian updating, we err if we draw the conclusion that *there is a such-and-such probability that H*, when those probabilities represent our degrees of conviction. To be sure, the fiction may be convenient. But a convenient fiction is a fiction nonetheless.

The second point is that our degrees of conviction do not mean much unless the way in which we have been convinced is respectable. This brings us to Peirce's fourth and fifth critiques of the procedure of Bayesian updating. The fourth critique is that there will be inquiries for which the procedure of updating is inapt. Peirce thinks this is especially so in cases of historical inquiry. He quips:

> if by "probability" be meant the degree to which a hypothesis in regard to what happened in ancient Greece recommends itself to a professor in a German university town, then there is no mathematical theory of probabilities which will withstand the artillery of modern mathematical criticism. A probability, in that sense, is nothing but the degree to which a hypothesis accords with one's preconceived notions; and its value depends entirely upon how those notions have been formed, and upon how much objectivity they can lay a solid claim to. (HP 2:715, 1901)

Our degrees of conviction will be shaped by what we regard as appropriate or likely in our circumstances. In cases of ancient history, these prejudices will cause us to incorrectly weight the evidence. We might dismiss an ancient report merely because we regard it as incredible by our contemporary standards.[10]

Indeed, the historian can preserve her preconceived notions by simply keeping all counterevidence out of view. She may dismiss counterevidence as forgeries, discredit the testimony as lies, or treat claims as the confabulations of nonscientific minds. This brings us to Peirce's fifth critique of the Bayesian procedure, the one that is the most damning. The procedure of Bayesian updating, taken by itself, is consistent with forming our beliefs according to the method of tenacity. If we can manage to keep all counterevidence out of view, then we will never be compelled to change our beliefs.

Peirce grants that "belief ought to be proportional to the weight of evidence" (ILS 135, 1878). What Peirce objects to is letting our beliefs be dictated by the subjective Bayesian's procedure alone. In "The Fixation of Belief," Peirce distinguishes among four methods of fixing belief. They are the method of tenacity, the method of authority, the a priori method, and the scientific method.[11] According to the method of tenacity, one clings to one's belief come what may by keeping all counterevidence at bay. The problem for the subjective Bayesian is that by controlling one's environment, one may avoid any evidence that would require updating one's belief in ways contrary to it. Peirce recounts a story of a man telling him not to read certain literature about free trade because it may change Peirce's mind. Refusing to read that literature would enable the subjective Bayesian to retain full conviction in her beliefs. If Peirce only listens to those persons who agree with him, his opinion will not change. Of this method, Peirce remarks:

> A man may go through life, systematically keeping out of view all that might cause a change in his opinions, and if he only succeeds—basing his method, as he does, on two fundamental psychological laws—I do not see what can be said against his doing so. It would be an egotistical impertinence to object that his procedure is irrational, for that only amounts to saying that his method of settling belief is not ours. He does not propose to himself to be rational, and, indeed, will often talk with scorn of man's weak and illusive reason. (ILS 59, 1877)

In this passage, Peirce does not explicitly tell us what these two psychological laws are. I suggest that they are the two psychophysical laws mentioned earlier. By blocking any counterevidence, the person who adheres to the method of tenacity can ensure that her belief will not give way to the contrary belief. If she is especially good at keeping counterevidence bay, her degree of conviction in the initial belief will remain as high as ever.

Notice that Peirce does not claim that the method of tenacity is irrational. To the contrary, it is fully consistent with the procedure of Bayesian updating. The method of tenacity implemented so as to keep all counterevidence out of view can keep beliefs settled. The only drawback is that it will not settle on how things actually are. As a consequence, the belief state is not expected to remain settled should one let down one's guard against counterevidence.

As Jeff Kasser remarks, we can "measure the extent to which our cognitive state settles what we are to expect (this is reflected in measures like degree of belief). But we can also measure how settled the expectative state itself is" (2019, 15). The conceptualist who keeps out of sight all contrary evidence will have very high degrees of conviction. But the expectative state itself will not be settled well because it does not rest on a strong footing.

Plainly, any procedure of settling questions of fact that can allow for the method of tenacity is contrary to sound scientific practice. For Peirce, that the method of tenacity and subjective Bayesianism are consistent is the problem with subjective Bayesianism. For the subjective Bayesian, "the conjoint probability of all the arguments in our possession, with reference to any fact, must be intimately connected with the just degree of our belief in that fact" (ILS 136, 1878). Peirce denies that it is so intimately connected. The Bayesian is at risk of embracing weak and illusive practices of reasoning. To make her inferences strong, she needs resources beyond those which Bayesian updating alone offers. The practice of Bayesian updating, while supported by the mathematics under abstract conditions, must be supplemented with and monitored by quantitative inductions which aim to ascertain the real frequencies or uniformities in nature. Only then can she be assured that her inferences will represent things as they actually are. In her discussion of subjective Bayesianism, Mayo hits the nail on the head. She insists that the Bayesian's practice "does not fit the actual procedure of inquiry" wherein we perform severe tests to probe a hypothesis for error (Mayo 1996, 99). Those who think we can perform such error-probing tests do not think reason is weak and illusive. Rather, they maintain we can make our inferences strong even in the short run. Nothing in the Bayesian procedure alone gives us resources to do that. I shall have more to say about inductive strength later in this chapter, but what I have said here suffices to get to the point: In cases in which it is apt to be used, subjective Bayesianism must be supplemented with and monitored by good practices of inquiry. Otherwise, it is consistent with tenacious believing.

Qualitative Induction: Thus far, I have explained Peirce's standard account of the procedure of hypothetic induction. I have, moreover, shown that Peirce supports subjective Bayesianism under abstract conditions. Yet he also notes that the abstract conditions are not our actual conditions of inquiry. Consequently, it would seem that subjective Bayesianism is not an appropriate procedure for our actual practices. And yet, when he characterizes qualitative induction, Peirce claims that it proceeds on the basis of "a stream of experience in which the relative evidential values of different parts of it have to be estimated according to our sense of the impressions they make upon us" (CP 2.759, 1905), which strongly suggests in favor of some version of subjective Bayesianism articulated earlier. Let us see if we can make peace among these commitments.

The key is to focus in on the idea of a stream of experience. In Peirce's mature works, he defines experience as "that state of cognition which the course of life, by some part there of, has forced upon the recognition of the experient, or person who undergoes the experience, under conditions due usually, in part, at least, to his own action" (LI 344, 1906). When we make hypothetic inductions in Peirce's ordinary way, we form predictions, which involve a description of the experimental conditions of some test as well as an expected observation. The result of performing that experiment is an experience.

That experience, in turn, will alter our degrees of conviction in the hypothesis. We will either (a) eliminate the hypothesis in the attenuated sense of rejecting it, tabling it, or refining it; (b) regard it as corroborated in the sense noted earlier; or (c) regard it as confirmed. In accordance with Peirce's psychophysics of belief, our belief will track the degree of our feeling of conviction, as determined by our sense of the nature and quality of the evidence. These degrees of feeling may be fictitiously treated as probabilities, but in fact they are the qualitative force the evidence makes on us. We feel convinced of some hypothesis H, but we do not ordinarily assign numbers to our feelings of conviction even if the feelings are in principle measurable. As Peirce states, in actual practice, they are but estimations according to the sense of the impressions the parts of experience (i.e., what is evident to us in the stream of experience) make on us.

Notice, though, that the procedure just described does not consist merely of setting a prior and then updating. As Peirce's fifth critique of Bayesian updating demands, our degrees of conviction will be sensitive to sound practices of inquiry. Otherwise, our procedure is consistent with adhering to the method of tenacity. What we must do is supplement the bare procedure of updating with the standard procedure of hypothetic induction. We must probe the hypothesis by making predictions from it and putting those predictions to the test.

This requirement also holds in cases of historical inquiry, when the method of Bayesian updating comes up short. Peirce maintains that even the historian must make predictions from her hypotheses. Peirce employs this method in three ways. The first way is by defending the ancient account of how we came to have the works of Aristotle, *pace* its rejection by German critics (CP 7.233–255, 1901). The second way is by giving a chronology of the life of Plato (HP 2:763–791, 1901). The third way he employs the method is by giving an account of the life of Pythagoras (HP 2:791–800, 1901). Because of the scant evidence in this third case, Peirce declares that there can be "no possibility of testing any hypothesis. In such a case, we must renounce at the outset any attempt to reach anything approaching to certainty" (HP 2:791–792, 1901). Nonetheless, he maintains that it is possible to "to embody or rather, to ensoul, all the pertinent facts, that is to say, the facts *that those writers make those statements*, in such a hypothesis as best unifies them, and will serve as a source of experiential predictions, whenever, in the future, it may be in our power to verify or refute any predictions on the subject" (HP 2:791–792, 1901). That is, Peirce aims to perform one of the preparatory steps to making a hypothetic induction. That preparatory step is to frame a hypothesis about the life of Pythagoras from which experimental predictions may be drawn. Peirce himself will not test his hypotheses, but he hopes future inquirers will.

We are now in a position to understand the nature of qualitative induction as a species of inference. When we make qualitative inductions, we will have some hypothesis or hypotheses which we have framed and from which we have deduced predictions. We will likely have some antecedent feeling of conviction that some single hypothesis is true, whereas the others are false. As we put those hypotheses to the test, our feelings of conviction will be affected by the results of those tests. Our conviction that some hypothesis H is true will rise or fall depending on whether H passes the tests set for it or not. These degrees of conviction are feelings, or qualities, produced by the relative force of the evidence upon us. For the purposes of making things more definite, we may treat the degrees of these feelings of convictions as probabilities. But these probabilities are only convenient fictions. At best, they are estimations as to which hypothesis we think is true at present. Convenient, fictitious, and estimative though the numbers may be, our degrees of conviction are not arbitrary. Much to the contrary, our degrees of conviction are sensitive to the results of the experiments we have performed, at least in nonpathological cases. The evidence makes some impression upon us, and our beliefs track that evidence. Such Bayesian updating is perfectly acceptable and our beliefs will eventually incline toward the truth, so long as the procedure is supplemented with and monitored by the ordinary practices of sound inquiry by which we test what our hypotheses predict.[12,13]

This account explains, moreover, why qualitative induction has the truth-producing virtue of being truth-approximating in the long run. Provided the true hypothesis is among those we have framed to put to the test, the procedure of eliminating hypotheses will eventually lead us to reject the false one and believe the true one.[14] Such a procedure may not be accomplished in our lifetimes. It may take generations to perform some tests, and some data may only be gathered by future inquirers. But that is no objection to the procedure, so long as the community of inquirers keeps at the task. Of course, we will have needed to frame the hypothesis in the first place, and we will need to have correctly deduced the predictions of such hypotheses. These tasks, however, are part of hypothetic induction as a mode of inquiry and not of hypothetic induction as a species of inference. I have already discussed deduction in Chapter 4. I will have more to say about abduction in Chapter 7.

Before proceeding to an examination of quantitative induction, one more word on qualitative induction is in order. In his earlier works, Peirce thinks of hypothetic induction and statistical induction as essentially the same: "All probable inference, whether induction or hypothesis, is inference from the parts to the whole. It is essentially the same, therefore, as statistical inference" (W 2:268, 1869). Peirce is apparently thinking of the predictions of a hypothesis as a population that is, in some sense, sampled by testing. So far as it goes, there is no error in such a view. But it only goes so far. We will be misled by such a claim if we think testing any given prediction is not importantly different from testing any other. But that is a mistake for two reasons. First, some predictions from a hypothesis may be true of many other hypotheses. Consequently, some single test might corroborate several hypotheses or lead us to eliminate several hypotheses. Peirce holds that if we can find some testable prediction from a set of hypotheses that is shared by exactly half of them, we will proceed most efficiently to test that prediction. Suppose, for example, we have six hypotheses. If hypotheses one through three entail prediction P and four through six entail that not-P, then testing for P will be an efficient way to narrow our field of hypotheses. Once we narrow down our hypotheses to three, we will want to ascertain which consequences are unique to each of those three so that we can put them to the test individually. Peirce writes, "suppose that there are thirty-two different possible ways of explaining a set of phenomena. . . . The most economical procedure, when it is practicable, will be to find some observable fact which, under conditions easily brought about, would result from sixteen of the hypotheses and not from any of the other sixteen" (CP 6.529, 1901). Also, we "should split up a hypothesis into its items as much as possible, so as to test each one singly" (HP 2:761. 1901). Second, the probative value of tests will vary greatly depending on whether the test is severe. I shall have more to say about severity in the discussion of inductive strength later in this chapter.

Quantitative Induction

I turn now to quantitative induction. In the early works, by an 'induction' Peirce means a statistical inference that proceeds by sampling a population: "[i]nduction . . . assumes that that is true of a whole collection which is true of a number of instances taken from it at random. This might be called statistical argument" (W 2:217, 1868). As Ian Hacking makes clear, population sampling is the context in which statistical inference arose and "[s]ampling . . . was Peirce's model for induction" (1990, 210). Peirce's example is this: Suppose we want to know the frequency with which certain letters appear in English script. In order to calculate this, we will take "half a dozen other English writings" and count the occurrences of each letter. As we do so, "the relative number of e's approaches 11 ¼ *per cent* of the whole, that of the t's 8 ½ *per cent*, that of the a's 8 *per cent*" and so on (W 2:216, 1868). This is quantitative induction that proceeds by sampling a set of books for the frequency of letters.

Whenever one might want to know the relative frequency of some event given some other events, we will make a quantitative induction. For example, one might want to know the relative frequency with which it rains when the clouds are dark and heavy and the atmospheric pressure drops. Or one might want to know by what percentage some candidate is favored to win the next election. Or one might want to know the relative frequency with which some die shows a six. For statistical induction, the project of inquiry will consist in sampling so as to ascertain these ratios. A completely thorough sampling will consist of a sampling of every instance. That is typically impossible; a die might be tossed an infinite number of times. In fact, Peirce denies that a complete enumeration of instances constitutes an induction at all (see W 1:263, 1865; CP 2.757, 1908). It is rather a deduction, since the procedure by which the inference is made (sorting a population) is truth-preserving. Since a complete enumeration is usually impossible, we must resort to sampling a population. When we do so, we make a quantitative inductive inference. Even in his late work, Peirce tends to present quantitative induction as consisting of sampling. In 1908, he writes that such inferences begin by

> collecting, on scientific principles, a "fair sample" of the S's, taking due account, in doing so, of the intention of using its proportion of members that possess the predesignate character of being P. This sample will contain none of those S's on which the retroduction was founded. The induction then presumes that the value of the proportion, among the S's of the sample, of those that are P, probably approximates, within a certain limit of approximation, to the value of the real probability in question. I propose to term such reasoning *Quantitative Induction*. (CP 2.758, 1905)

Note that the conclusion of a quantitative induction from sampling is not that, for example, *50% of the beans in this bag are white*. Rather, the conclusion is that the ratio ascertained probably approximates, within a certain range, the value of the real probability. This aligns with and anticipates the use of confidence intervals, such that the right way to state the conclusion of an inference based on sampling is that some unknown quantity (the real probability) lies in an interval the estimation of which is determined according to a method that is right within some confidence interval. What induction does is "commence a proceeding which must in the long run approximate to the truth" (CP 2.780, 1901). As sampling increases, the interval shrinks, and over time we better approximate the real probability.

Yet other passages suggest a broader conception of quantitative induction, as when he writes that quantitative induction "depends on measurements, or on statistics, or on countings" (CP 6.473, 1908). Suppose I wish to know the most likely path a hurricane will take. The calculation of that path will be made by considering a number of variables, including the current atmospheric conditions, the geography of the area, the temperature of the water, and so on. It will also draw on past data about hurricanes. The calculation of the most likely path that a hurricane will take plainly depends on measurements, but it would be peculiar at least to regard such calculations as paradigmatically just a case of sampling a population. To be sure, I will draw on the past data, and those past data are a sampling (in some sense) of data from all past hurricanes. Yet the calculation of the most likely path depends on much more than just that past data. Even if quantitative induction is paradigmatically inference made by population sampling, it ought not to be regarded merely as inference made by population sampling. Rather, quantitative induction includes the full arsenal of statistical inferencing and modeling.

Consider another example. Suppose I wish to know the average speed at which the moon appears to move across the sky. I take, say, two measurements per night over thirty nights by measuring the amount of time it takes the leading edge of the moon to move from occluding one star to occluding another, where I already know the apparent distance between those two stars in the night sky. Such an inference is a quantitative induction. It is, however, peculiar to regard it as an inference based on sampling, except in an attenuated sense of sampling some select nights rather than others or sampling some set of observations I could have made but did not. In either case, it is clear that the sampling (in whatever sense of that word might be appropriate) in the moon example is different from sampling beans from a bag of beans. There is a definite quantity of beans in the bag, but there is a not a definite quantity of measurements of the moon's apparent speed that could be taken. Also, the calculation in the moon case gives me an average speed, not a proportion of a part to a whole as in the bean case.

It remains to be asked why such procedures are, in general, truth-approximating in the long run. I stress, again, we can only claim that the procedures in general are truth-approximating in the long run. In specific contexts of inquiry, various statistical techniques may be inappropriate to use. Plainly, for quantitative inductions, any given sample or measurement may not be representative of the whole or of the real probability. But an indefinitely continued, infinitely wide, self-sacrificing community of inquirers which continues to take measurements would be able to balance out the errors of any given sample. Here is Peirce making the point in "On the Theory of Errors of Observation":

> the general nature of induction [i.e., statistical induction] is everywhere the same, and is completely typified in the following example. From a bag of mixed black and white beans I take out a handful, and count the number of black and the number of white beans, and I assume that the black and white are nearly in the same ratio throughout the bag. If I am in error in this conclusion, it is an error which a repetition of the same process must tend to rectify. It is, therefore, a valid inference. (W 3:116, 1873)

Moreover, he remarks, "[i]t must never be forgotten that h [the precision of an observation] is a statistical quantity [calculated from a series of observations]; not one which belongs to a single observation, but one which belongs to an infinite series of observations" (W 3:132, 1873). Whereas we will calculate probabilities and make measurements based on observations in the short run, those observations are members of a set of an infinite series of observations that could be made by the community of inquirers.

This point may be made from a different direction. In "A Theory of Probable Inference," Peirce notes that Whately, Mill, and their ilk are committed to the thesis that nature is uniform and treat that as the major premise of all inductions. As we shall see in the next chapter, Peirce denies inductive inference proceeds on such an assumption. In contrast, his theory only requires one to admit the "negative fact" that "supernal powers withhold their hands and let me alone, and that no mysterious uniformity or adaptation interferes with the action of chance" (W 4:445, 1883). However, he proceeds to note that even if there were such supernal influences, inductive inquiries would still lead us to approximations of the truth so long as "their influence were not too overwhelming" (W 4:445, 1883). It is the indefinite continuance of measurements and sampling that ensures their influence is not too overwhelming, for "the universe of marks is only limited by the limitation of human interests and powers of observation. Except for that limitation, every lot of objects in the universe would have (as I have elsewhere shown [viz., in "The Order of Nature," as shall be explained in the next chapter]) some character in common and peculiar to it" (W 4:446, 1883).

The idea just expressed in the previous paragraphs—that with increased sampling, errors will balance out—is known in statistics as the law of large numbers. The law of large numbers states that the sample mean tends toward the population mean. For example, if blackjack favors the house at .5%, the house might lose over the short run. But over the long run, the house will win approximately 50.5% of the time. The divergences from this value will balance out over the long run. Peirce scholars have proposed three different accounts of the role of the law of large numbers in Peirce's theory of induction. James Liszka (1996, 75) claims the law of large numbers is a corollary of Peirce's account of induction. This cannot be right, however. The law is a theorem of probability theory, which is deductive in Peirce's sense. Others appeal to the law to justify statistical deduction. But in that case, the law is not applicable to statistical induction as such, only to inferring probability-that statements from probability-of statements.

Aaron Wilson (2016, 273) claims the law of large numbers underwrites the validity of induction. This, however, is too strong of a claim. To understand why, we must look at Peirce's account of induction both from a formal point of view and from a practical point of view. From the formal point of view, statistical induction would be useless if the law of large numbers were false. On a contemporary and formal approach to inference, the law of large numbers is a necessary condition for the validity of inductive inference. But from a practical point of view, when we consider statistical induction as a technique used in inquiry, the law of large numbers is insignificant unless the community of inquirers is indefinitely enduring and embraces the entire community of inquirers. It does not matter that quantitative induction is truth-approximating in the long run unless there is a long run in which the truth can be approximated. What is requisite is that we have a sufficient amount of data on the basis of which to draw an inference. Although inquiry may settle on the correct frequency at any given time, we cannot be assured at that time that it is the correct frequency. We can only be assured that in the long run we will approximate it.

The validity of induction does not merely rest on the law of large numbers. Rather, it rests on the community of inquirers bringing it about that we have a sufficiently large number of samples so that the errors can balance out in the course of inquiry. This is why, as we shall see more clearly later, Peirce claims that the validity of induction depends on the parts being parts of the whole but further depends on the community of inquirers. The law of large numbers corroborates our inductive practices. If it were false, then we should have no hope that our inductive practices as pursued by the unlimited community of inquirers will ever reach a close approximation of the real uniformities of nature. That is, if it turned out the law were false, then our hope would be delusive. But we still need a community of inquirers to approximate the real frequencies in the long run of inquiry.

An indefinitely enduring and infinitely large loving community of inquirers will be able to approximate the real probabilities, the uniformities of nature. We individually, or even currently as a community, may well be wrong. But the community of inquirers will be able to correct our current errors. Peirce remarks, "[i]nduction . . . would be valid if it should at first lead us into a hundred errors for one truth for even then truth would prevail in the long run" (W 1:433, 1866). This is the reason that induction is self-corrective. As Isaac Levi (1980, 138) rightly emphasizes, Peirce's claim that induction is self-corrective is not the claim that we will reach some end of inquiry wherein we do ascertain what the uniformities of nature, the real probabilities, are. Rather, it is the claim that we will either be correct or additional inquiry, additional sampling, will get us closer to alighting on the truth and rooting out error (although at any given time we may be led further from the truth). Moreover, it is why Peirce likens inductive inquiries to insurance companies. Insurance companies "do not know what will happen to this or that policy-holder; they only know that they are secure in the long run" (W 2:269, 1869). Similarly, we do not know what will happen with respect to our current results. Future inquires may give us the same or similar statistic, or they may lead us to wildly different statistics. Our only security lies in the community of inquirers carrying inquiry forward into the future.

Mayo distinguishes between two different conceptions of the long run of inquiry. One conception consists in "asymptotic self-correction" according to which we imagine "the sample size increases toward infinity" and as it does so "one gets better and better estimates of some feature of the population" (2005, 311). The second conception consists in the error probabilities of a test, wherein "one has a sample of size n, say 10, and imagines hypothetical replications of the experiment—each with samples of 10" (2005, 311). She maintains that Peirce embraces the second position. But there is no need to choose between them. The long run of inquiry may include ever-increasing sample sizes and repeated testing and sampling. In fact, it should include both. We want to approach as complete an enumeration as possible. We also want to repeat our testing and sampling. The community of inquirers can do both.

When we reason from the sample to the population as a whole, if we take more samples, the errors will balance out. Nicholas Rescher remarks, "[b]y constantly readjusting the estimate of the population-frequency in light of the actual sample-frequency, the scientist is bound eventually to get things right" (1978, 3). Rescher ought rather to have written "approximately right" instead of "right." Also, he should have written "the community of inquirers," not "the scientist." As Cornelius Delaney points out, "[t]o resist this line of thought is to entertain a conception of the population of the whole which will never manifest itself in the samples or the parts" (1993, 28). Continued sampling over the long run will give us an approximation of the population mean.

Nonetheless, a word of caution is in order here. Peirce also holds that the final opinion—that is, the one that will hold in the long run—has already been reached on diverse topics. He remarks that even "throwing off as probably erroneous a thousandth or even a hundredth of all the beliefs established beyond present doubt, there must remain a vast multitude in which the final opinion has been reached. Every directory, guide-book, dictionary, history, and work of science is crammed with such facts" (W 5:226, 1885). Any opinion may be overturned in the future, but there are no grounds for thinking every opinion will be.

In the preceding comments, I have been hinting that these considerations concerning the community of inquirers pertain not merely to quantitative induction but to qualitative induction as well. If we think of knowledge as something possessed not merely by individuals but by a community of inquirers, then even for qualitative induction the hypotheses that that community accepts will ultimately approximate the truth as false hypotheses are eliminated and true or approximately true hypotheses are corroborated and confirmed. Larry Laudan (1973), however, has argued that Peirce fails to show qualitative inductions are self-corrective. Three points are in order. First, Laudan is incorrect insofar as qualitative inductions will oftentimes draw on quantitative induction in order to test hypotheses. If quantitative induction is self-corrective, it follows that qualitative inductions will be self-corrective so far as they draw on quantitative inductions. As Rescher states it, "the performance of scientific induction as a whole, qualitative induction specifically included, can be monitored by statistical techniques, which in their turn are self-corrective in the narrower sense" (1978, 7).[15]

Second, Peirce regards qualitative induction to be a sort of sampling. As already explained, Peirce holds that hypotheses are (a) distinguished from one another by their sets of predictions and (b) provable from their sets of predictions such that any false hypothesis will also have a false prediction. If Peirce is correct, then so long as the population of predictions continues to be tested, the true hypotheses will be sorted out from the false ones over the long run. As Peirce remarks in a later work, the mind has the power to originate true ideas, but these true ideas are "almost drowned in a flood of false notions" (EP 2:154, 1903). In order to be rid of the false notions, "that which experience does is gradually, and by a sort of fractionation, to precipitate and filter off the false ideas, eliminating them and letting the truth pour on in its mighty current" (EP 2:154, 1903). The continued testing of the predictions of a hypothesis filters out false hypotheses.

Third, there is one respect in which qualitative inductions may not be self-corrective. Namely, we may never hit on the true hypothesis (or a hypothesis that is approximately true). In that case, we will never be correct. This problem is not, however, a problem concerning qualitative induction. Rather, it is a problem concerning abduction, the mode of inference which generates and ranks hypotheses. I shall have more to say about abduction in Chapter 7.

Yet so long as we do have the (approximately) true hypothesis among those we have conceived, qualitative induction is self-corrective. It will enable us to sort those theories which are (approximately) true from those which are not by testing their predictions.

Furthermore, the claim induction is self-corrective is different from the claim that the scientific enterprise as a whole is self-corrective. Qualitative induction is self-corrective because we will find out which hypotheses are in error by putting them to the test. Quantitative induction is self-corrective because increased sampling will decrease our margin of error. As induction is part of the scientific enterprise, scientific inquiry may be regarded as self-corrective in a secondary and subsidiary way. It is self-corrective insofar as inductive inquiries are a part of it. But in the case of scientific inquiry more broadly, we must root out deductive fallacies and errors. We also have to continue to develop our curiosity, to seek out new or improved theories. Peirce marvels at the

> power of genuine scientific reasoning to correct itself—that is, to correct its own previous conclusions by the admission of additional evidence, to correct its very premises, as it constantly does in the most exact sciences, and even to correct its own fallacies, of which there are many historical examples—and owing to the fact that this scientific procedure is nothing but the self-development of man's original impulse of curiosity or interest. (CN 2:150, 1897)

Paul Forster rightly insists that it "is wrong to assume that Peirce needed, or intended, to show that every inference is warranted in the same way; specifically, that it is self-corrective" (1989, 423).

The Validity of Induction and the Community of Inquirers

We have seen in the preceding exposition of the species of inductive inference that each must make an appeal to a community of inquirers. That community is to correct our errors over the long run. Although we have no grounds for claiming that there will be some such community of inquirers, Peirce maintains that neither is there any reason for denying that there can be. Rather, we must hope that the community of inquirers will extend indefinitely into the future, we must have faith in them to continue our inquiries, and we must love them such that we put aside our local and trivial concerns as well as our biases. That community which we hope for, have faith in, and love is what underwrites the truth-approximating virtue of induction. In the long run, the community of inquirers will approximate the truth.

Nonetheless, one may wonder why the community of inquirers should play such a central role in Peirce's account of induction's validity across all of the species of inductive inference. Peirce appeals to the community of inquirers because our observations, testing, and sampling are quite limited. Are we in fact able to learn anything from induction? Or are our inductions in fact worthless, vitiated by our limited sampling and testing? Can induction enable us to unlock the secrets of nature? Peirce's answer is in the affirmative, and that answer is at the heart of his account of induction's validity more generally.

Summarily stated, his argument is this: Let us suppose that we could not learn anything by induction. If so, "the order of things (as they appear in experience), would then undergo a revolution" (W 2:269, 1869). The samples would vary wildly, never leading to any fixed ratio. We could confirm some hypothesis H by testing its prediction P one day but on the next day testing P might lead us to eliminate the hypothesis. We could replicate an experiment sometimes but not others. In such a universe, Peirce claims, "the order of the universe should depend on how much men should know of it" (W 2:269, 1869). However, over the long run we would be able to discover that the order of nature depends on how much we know of it, as I shall explain momentarily. Therefore, he concludes, "in such a universe there would be nothing which would not sooner or later be known; and it would have an order capable of discovery by a sufficiently long course of reasoning" (W 2:269, 1869).

Peirce's argument is briefly and tightly stated; it needs to be unpacked. Induction is a strategy for unlocking nature's secrets. What Peirce is claiming is that for inductive inferences to fail—for us to be incapable of learning anything from crude, qualitative, or quantitative induction—we may suppose that nature is a competitor trying to thwart our inferences. In fact, Peirce proposes we think of induction as engaging in a game with nature. He suggests the game of twenty questions. "When the questions put to nature will only be answered by yes or no," Peirce remarks, "he will advance with the greatest rapidity (as in the game of twenty questions) by asking questions an affirmative answer to which is equally probable with a negative one" (W 3:5–6, 1872).[16] When asking what it would take for inductive inference to fail, we are to suppose that nature and inquirers are locked in a game of pure competition. What would it take for the strategies of inductive inquiry, in a game of pure competition, to fail? To shed some light on that question, it will help to think of ourselves as locked in a competition with nature such that we are trying to unlock its secrets.

Playing Competitively: Let us start by considering a game of pure competition, such as matching pennies. In matching pennies, two players—Rose and Colum—have an enormously large canister of pennies. At the same time, each player takes a penny out of the canister and shows the other player a side of the penny. If Rose and Colum show the same side of the penny (HH or TT), then Rose gets Colum's penny. If they show different sides of the penny (HT or TH), Colum gets Rose's penny. The payoff matrix is represented in Table 5.1. If the game is played just once, then Rose and Colum each has a 50% chance of gaining a penny.

Now suppose that the game is iterated. Rose and Colum repeatedly play the game according to the rules specified. Moreover, suppose that Rose is keeping track of Colum's plays. And let us suppose that Rose and Colum are both ideally rational players. If Rose finds Colum tends to play, for example, heads 60% of the time, she has a strategy to win. All she must do is show heads all of the time. Colum will lose 60% of the interactions. Furthermore, if Colum follows some pattern in playing the game—for example, H,T,H,H,T,T,H,T,H,H,T,T—all Rose must do is mimic the pattern by playing H,T,H,H,T,T,H,T,H,H,T,T. In sum, if Colum shows a preference for one side of the coin to another, Rose has a strategy to win. If Colum shows a pattern, then Rose has a strategy to win.

What would it take for Rose to have no winning strategy? For Colum to foil Rose's efforts, he will have to show heads 50% of the time and randomly. If he fails to show heads exactly 50% of the time, Rose will be able to win by exploiting Colum's preference. Even if Colum shows heads just 51% of the time, Rose will win over the long run. The preference can be quite low and still be in Rose's favor. Casinos work on the same principle. The house's edge

Table 5.1 Matching Pennies

	Heads	Tails
Heads	1, −1	−1, 1
Tails	−1, 1	1, −1

in blackjack can be as low as .5%. Also, if Colum fails to show heads randomly, Rose will be able to win. If Rose can guess with better than even odds what Colum will play next, then she will be able to win. What is requisite for Rose to have no winning strategy is that on any given play, the probability of Colum showing heads must be equal to the probability of him showing tails. Importantly, though, Rose will be able to discover that she has no such winning strategy.

We are now in a position to draw the analogy to induction. Suppose now that Colum is the personification of nature (Nature) and that Rose is a human inquirer (Inquirer). To have a strategy to win is to be able to learn from induction. Let us suppose that Nature is trying to thwart Inquirer from learning anything. Now posit that Nature shows a preference for heads, playing it 60% of the time. Provided Inquirer is keeping track of Nature's plays, she can discover this preference and learn from induction that Nature plays heads 60% of the time. Inquirer now has a strategy for winning. She has learned from induction. Alternatively, posit that Nature operates according to a pattern. Provided that Inquirer is keeping track of Nature's plays, she can learn from induction that Nature follows a pattern. Inquirer now has a strategy for winning. She has learned from induction. What we must posit for Nature to have a winning strategy is that, as Peirce states, the order of things (as they appear in experience) would undergo a revolution. That is, we must posit that Nature shows heads exactly 50% of the time and randomly. But Inquirer can now discover that Nature follows this rule. She will not have a strategy to win, but she will have learned from induction that there is a 50% chance Nature will next play heads. And that is still to have learned something by induction.

A more logical example may be helpful. Let us imagine not pennies but premises. Our premise (the penny) will be *the clouds are dark and heavy and the atmospheric pressure has dropped*. On each side of our premise is written a conclusion, *it rains* or *it does not rain*. What will it take for Nature to beat Inquirer at this game, such that Inquirer cannot discover how Nature operates? If Nature shows *it rains* 90% of the time, then Inquirer will be able to conclude that 90% of the time when the clouds are dark and heavy and the atmospheric pressure has dropped, it rains. Inquirer will have learned something by induction. If Nature alternates between *it rains* and *it does not rain* according to some pattern, then Inquirer will discover this pattern. She will have learned something by induction. If, on the other hand, Nature shows *it rains* 50% of the time and randomly, then Inquirer will have discovered at least that. She will be able to conclude that the connection between *the clouds are dark and heavy and the atmospheric pressure has dropped* and *it rains* is random and unpredictable.

It may be tempting to remark here that when the clouds are dark and heavy and the atmospheric pressure drops, any number of events may occur. It may rain. A cow may moo. A star may explode. A Greenlander may sneeze. This is obviously true, but it does not matter. Suppose we let one side of the coin be *a cow moos* and the other be *a cow does not moo*. Either that event will occur when the clouds are dark and heavy and the atmospheric pressure drops or it will not. If it does, Inquirer will be able to learn that (say) 60% of the time when the clouds are dark and heavy and the atmospheric pressure drops, a cow moos. The point of this objection is not whether we can learn from induction but whether we are finding mere correlations or real laws of nature. As noted earlier, however, Peirce concedes that there may be regularities in nature that are not laws. The question of whether the sort of connection between premises and conclusion is merely coincidental has not to do with validity but with inductive strength as well as abductive validity. Questions of inductive validity must be kept separate from questions of inductive strength and of abductive validity. I shall have more to say about inductive strength later in this chapter. In Chapter 7, I shall take up the matter of abductive validity. For the moment, it will suffice to note that the atmospheric conditions do not suggest to us an explanation for when or whether a cow will moo.

Three Problems: Nevertheless, there is an important disanalogy between Rose and Colum's game of matching pennies and Inquirer and Nature's game of premise and conclusion. The disanalogy is that any given inquirer lives for but a finite time, observes but a little part of nature, and has her own peculiarities and prejudices. It is one thing to play matching pennies against another person. It would be quite different to play matching pennies against a god. It is one thing to know about what will occur in one's little corner of the world over one's lifespan. It is quite different to know about the whole of nature. We inquirers live for a finite time at the outskirts of a far-flung galaxy. We have biases and peculiarities. How could we possibly play a game like matching pennies against Nature?

In order to avoid this problem, Peirce offloads the responsibility of induction. It does not rest on the shoulders of any single inquirer or even a local group of inquirers but on an infinitely expansive and enduring community of inquirers. Peirce states, "the validity of induction depends simply upon the fact that the parts make up and constitute the whole. This in its turn depends simply upon there being such a state of things, that any general terms are possible. But it has been shown [in "Some Consequences of Four Incapacities"] that being at all is being in general" (W 2:268, 1869). Peirce's initial point is the one already canvassed. The conclusion of an induction follows from the premises just in case the parts stated in the premises really are parts of the whole stated in the conclusion. Nonetheless, Peirce is clearly claiming in this passage that the validity of induction does not merely rest on the fact that the parts make up the whole. While that is true, the observation of a mere part of the whole does not make an inference valid. The validity of induction further "depends" on the fact that being is general.

Crucially, Peirce's reference to "Some Consequences" is to the passage wherein he introduces the very idea of a community of inquirers. Peirce introduces the idea of a community of inquirers in distinguishing between the real and the unreal. He holds the real "is that which, sooner or later, information and reasoning would finally result in, and which is therefore independent of the vagaries of me and you" (W 2:239, 1868). The problem with figuring out what is real and unreal, however, is the problem just indicated. Our own lifespans—the lifespan of Rose, for instance—are finite and localized. Our viewpoints may be clouded with personal concerns and idiosyncrasies. Accordingly, we must not regard the real generalities (the laws and uniformities) into which scientists inquire as what any one person concludes or believes. As Peirce complains of Descartes, "to make single individuals absolute judges of truth is most pernicious" (W 2:212, 1868). Rather, the distinction between the real and the unreal "essentially involves the notion of a COMMUNITY, without definite limits, and capable of an indefinite increase of knowledge" (W 2:239, 1868).

As an individual inquirer, Rose's inductive inferences confront three problems. The first is the problem of the short run. Consider Rose qua Inquirer playing against Nature. Imagine that Rose lives for only seventy years, whereas Nature endures for 14 billion years. Now it is conceivable that Nature is following the strategy of playing heads exactly 50% of the time and randomly but that Nature has a long streak of showing heads. Considering Nature endures for 14 billion years, it is even conceivable that that a long streak of playing heads lasts for seventy years, the entire timespan of Rose's life. Seventy straight years of showing nothing but heads would be just a blip over 14 billion years for Nature. It is perfectly conceivable that the regularity Rose observes in nature for her lifespan is just a blip.

The second problem is the problem of localization. Imagine now that Nature is playing matching pennies against another million players, none of whom know each other. Suppose, moreover, that Nature is following the strategy of showing heads exactly 50% of the time and randomly across all players. It may turn out that so far as Nature plays against Rose, Nature appears to favor heads. If averaged out over all players, however, Nature may not show any such preference. Rose would be led to a false belief about Nature's strategy because Rose does not have knowledge of how Nature plays against the other players. The problem is that any given human inquirer can observe only a small portion of the universe. There may be parts of the universe that are unobservable by us. There are parts of the universe not always observed by us. As a consequence, it may be that the uniformity we observe is only apparent. Possibly, across all the universe, what we expect to occur occurs only 50% of the time and randomly.

One scenario that deserves particular mention is that it is possible a frequency oscillates. For example, it may appear that the frequency with which heads shows on a flipped coin oscillates between ¼ during a certain span of time and ¾ during some other (see Lenz 1964, 157–158). Not much stock should be put in this worry, however. The reason is that probabilities-as-ascertained belong to inferences. Suppose I find that 90% of the time, when the clouds are dark and heavy and the atmospheric pressure drops, it rains. *It rains on the condition that the clouds are dark and heavy and the atmospheric pressure drops* together with *the clouds are dark and heavy and the atmospheric pressure has dropped* concluding to *it will rain* is one inference, which will lead me to the truth 90% of the time that I make it. But now suppose that I spend half of my time on the windward side of a mountain and the rest of my time on the leeward side. I may find that the frequency oscillates such that on the windward side the inference leads me to a true conclusion 99% of the time, whereas on the leeward side it leads me to a true conclusion 81% of the time. These, however, are in fact different inferences. The first is the inference that *when on the windward side of the mountain, the clouds are dark and heavy, and the atmospheric pressure drops, it rains.* The second is the inference that *when on the leeward side of the mountain, the clouds are dark and heavy, and the atmospheric pressure drops, it rains.* The probabilities of these inferences must be separately calculated. If the probability appears to oscillate, one will seek some additional characteristic or circumstance that explains the oscillation. This will lead to different inferences with (perhaps) different probabilities.

The third problem is the problem of self-interest. Suppose that much depends for humanity on Rose figuring out the strategy that Nature employs.

Rose will win accolades and become enormously wealthy if she ascertains Nature's strategy. If so, Rose may allow various biases to influence her interpretation of the data. Whether intentionally or not, she may wrongly record Nature's plays. She may purport to find a pattern in Nature's plays, perhaps by suppressing some data or by treating some plays as insignificant outliers. Rose wins her accolades, but the theory she promotes is false. Moreover, the consequences may be dire for Rose if she fails to find a pattern. She may labor in obscurity for her whole life in a field abandoned by future researchers. If an inquirer would benefit from promoting her pet theory, from limiting her inquiries to whatever will bring her fame or wealth, from toeing the line of other influential and wealthy people, from misreporting, fudging, or suppressing evidence—in short, if an inquirer were biased by self-interest or by her local and trivial concerns—the result would be that opportunities for discovering the truth would be lost and false theories embraced. The consequence is that self-interest may prime us to "find" the uniformity we purport to observe when there is in fact none there. Self-interest may compel us to embrace a theory even when the evidence weighs against it.

The Community of Inquirers: There are regularities, or uniformities, in nature, as hardly anyone doubts (though as we shall see in the next chapter, Peirce has an argument for the claim). Granted there are such uniformities, they are discoverable. They are discoverable because, in a game of pure competition, there is always a strategy to win, or to learn that there is no strategy to win. The problem we now confront is whether these regularities are discoverable by us, finite inquirers that we are. The answer is that they may not be discoverable by any single person. We live for a limited period of time. We see but a corner of the knowable universe. We are subject to biases. These three problems—the problem of the short run, the problem of localization, and the problem of self-interest—show that while we may well settle on true conclusions in the short run of inductive inquiry, any assurance that our inductions are valid must rest on the shoulders of inquirers in the long run. The inductions of a fallible human with a short lifespan (cosmically speaking) who lives on a planet orbiting a star at the edge of the Milky Way galaxy have limited value.

But Peirce asks us to think of induction more broadly than limited to a single inquirer. In fact, he asks us to think of induction much, much more broadly. While it is true that Rose herself will engage in inductive reasonings, the inferences drawn could also be made by others. This is the significance of Peirce's claim that being at all is being in general: The uniformities or regularities of nature are discoverable by any inquirer, not just some particular inquirer on some particular occasion. To draw the analogy to matching pennies, it is true that Rose herself will play but for a finite period of time. Yet we might also suppose that, in spite of Rose's short life span, another player takes her place. That player receives Rose's list of Nature's plays and continues playing. And we can suppose that Rose's successor has a successor and so on. In this case, while it is true that Rose will not live 14 billion years, her successors might. Rose would then be part of a community of inquirers whose researches continue indefinitely into the future. She may put her faith in those future inquirers to continue her work.

A similar point may be made about localization. While it is true that Rose observes but a little bit of nature, Rose can hope to be part of a community of inquirers that extends beyond her little corner of the universe. Wild though the thought may be (I shall provide some quotations showing Peirce in fact has such wild thoughts in mind), suppose humanity should go extinct but orcas evolve rational capacities of the sort necessary for scientific inquiry. Suppose, moreover, that the rational orcas should engage in archaeological endeavors and find the record of Rose's observations, which contribute to their own inductive inquiries. Though at that future time the inductive inquiries should be carried out by rational orcas, Rose is a part of the community of inquirers. By the same token, suppose we come into mediate or immediate contact with rational aliens from other parts of the universe. If we communicate our scientific results to them and they take up the same or similar lines of inquiry, we together form a community of inquirers. Humans may go extinct. It does not follow that our inquiries die with us.

The third problem—that of self-interest—arises because Rose may have ulterior motives that lead her (consciously or not) to wrongly evaluate the data or embrace false hypotheses. Also, Rose may commit herself to a line of research that will have no positive outcome, as serious researchers in phrenology or astrology once did. Accordingly, Rose must put aside her self-interested concerns. She must take the risk of spending her life in pursuit of answering a question with no tangible results. Rose might spend her life studying Nature's plays. She may take the risk of making inductive inferences on the basis of her observations only to be wrong. Induction requires accepting this sort of risk. One must love the truth and one's fellow inquirers enough to sacrifice one's efforts, one's time, one's reputation, in pursuit of answering the questions one has.

The upshot of these considerations is this: To address the worry that Rose's observations and inductive inquiries are temporally limited, localized, and self-interested, what is requisite is that inquiry continue indefinitely into the future. Inquiry must embrace the whole community of inquirers, whether the humans of today, the rational orcas of the future, or whatever alien species there may be. Moreover, inquirers must strive to put aside their own interests. Instead, they must pursue the truth for its own sake. In "Some Consequences," Peirce describes the community of inquirers as "without definite limits, and capable of an indefinite increase of knowledge" (W 2:239, 1868). In "Grounds of Validity of the Laws of Logic" he claims that the community "may be wider than man" (W 2:271, 1869). The community of inquirers embraces all rational life, is unlimited, indefinite in scope, and continues forever. As Cheryl Misak rightly notes, Peirce "did not hold that truth depends on some finite collection of finite beings happening to agree that p is the case at some point in time" (2016, 26). Rather, our inductive inquiries approximate the truth in the long run, where "our" is an indefinitely continued, infinitely wide, self-sacrificing community of inquirers. Peirce puts the matter beautifully when he writes of the scientific person devoted to discovering the truth that "science to him must be worshipped in order not to fall down before the feet of some idol of human workmanship. Remember that the human race is but an ephemeral thing. In a little while it will be altogether done with and cast aside. Even now it is merely dominant on one small planet of one insignificant star, while all that our sight embraces on a starry night is to the universe far less than a single cell of brains is to the whole man" (PSR 54, 1905).

Bayesianism, Redux: We find the argument just developed in the preceding sections primarily in Peirce's "Grounds of Validity of the Laws of Logic" published in 1869. As late as 1910, Peirce continues to endorse the account from his earlier essay, writing that there were two key points well-made in this early essay. The first is that "no man can be logical whose supreme desire is the well-being of himself or of any other existing person or collection of persons" and the second is that "probability never properly refers immediately to a single event, but exclusively to the happening of a given kind of event on any occasion of a given kind" (ILS 122). In "Grounds," we have seen that Peirce's defense of induction's validity can be stated in a game-theoretic context. But Peirce also discusses the community of inquirers in his 1878 essay "The Doctrine of Chances," in which he continues to endorse these two key points. But in "The

Doctrine of Chances," Peirce comes at the matter through a consideration of Bayesianism. I have already argued that Peirce's theory of qualitative induction is not opposed to subjective Bayesianism, when it is supplemented with and monitored by sound practices of inquiry. Here I turn to arguing that even the Bayesian (of whatever sort) must appeal to a community of inquirers. The aim in the present context is to show that both the Bayesian and the Peircean are in the same boat. Both must posit an unlimited community of inquirers, one that embraces all of rational life and continues indefinitely into the future.

Broadly characterized, Bayesianism is the position that "every solid inference may be represented by legitimate arithmetical operations upon the numbers given in the premises" (ILS 112, 1878). The main difference between the account of probability Peirce endorses and Bayesianism is as follows. Peirce holds that probabilities are primarily and non-derivatively the probabilities of arguments. He contends that "to speak of the probability of the event B, without naming the condition, really has no meaning at all" (ILS 114, 1878). The Bayesian denies this. She assigns probabilities to other claims without there being a prior inductive inference. Peirce can only assign probabilities to probability-that statements based on probability-of statements through the use of statistical deduction.

Bayesians initially assign probabilities in ways not derived from prior arguments in order to prime the engines of updating. Bayes's Rule gives us a way to calculate a conditional probability if we have certain other information. In order to calculate that probability, we will need some prior probability with which to begin our calculations. In the most straightforward cases, this is typically the probability of the hypothesis (note that this is not the same as the probability of the conclusion given the premises), which is ordinarily called the prior probability. The issue is how we get that prior probability. For Peirce, we will only be able to get it from a prior statistical induction or (in the case of a qualitative induction) a course of experience which makes some evidential impression upon us. That is because all probabilities derive ultimately from such conditional probabilities. We derive categorical probabilities by statistical deduction from conditional probabilities already ascertained. In contrast, for some Bayesians, we will follow the principle of indifference, Peirce's critique of which has already been explained. Other Bayesians will claim that the prior will simply be the degree of our antecedent feeling of conviction that the hypothesis is true, however that may have come about. And yet for still other Bayesians, we can essentially pull the prior probability out of thin air. We can simply make it up, and the mathematics will do the rest.

An illustration of how updating works may be helpful. Suppose I see a bright light in the sky at night. I wonder if it appears every night. I form my hypothesis: The bright light appears every night. I arbitrarily decide to set the probability of this hypothesis at 50%, that is, P(C) = .50. This implies both that P(¬C) is .50 and that P(A|C) is 100%; that is, there is a probability of 1 that if the light appears every night, the light will appear tonight. Of course, even if my hypothesis is false, the light might appear tonight. I set the probability of P(A|¬C) at .50. Now all we must do is repeatedly apply Bayes's Rule to these numbers and update. Suppose I observe the light the next night. P(C) will be updated to P(C) = .667. If I observe it again the night after that, P(C) increases to .80. If I see it a third day, P(C) ups to .89, the fourth to .94, the fifth to .97, and so on, approaching 1. A Bayesian proof of my hypothesis, as Peirce claims, proceeds by mere arithmetic.

I have already argued that Peirce has some sympathies with the Bayesians who will assign a prior based on an antecedent feeling of conviction arising from a course of experience, provided that we supplement and monitor our use of Bayes's Rule with sound procedures of inquiry. In the present context, I shall argue that Peirce and the Bayesian (of whatever stripe) are in the same boat. Both must appeal to the community of inquirers to ensure that the probabilities they establish through inquiry approximate the real uniformities in nature. Peirce has no objection to Bayes's Rule as such; it follows from his own axioms of the probability calculus. Neither does he have any objection to Bayesian updating, provided the procedure is supplemented with and monitored by sound practices of inductive inquiry and appropriate to the domain of inquiry. However, whether one is a Bayesian or a Peircean, one has to admit that the problems of the short run, of localization, and of self-interest affect all inductive inference and can only be addressed by positing an unlimited, indefinitely extended community of inquirers embracing all possible rational life. Peirce's way into defending this thesis is with the Single Case Problem.

The Single Case Problem: I have already presented Peirce's defense of the claim that probabilities attach specifically to inferences: There is no sense in saying that there is a 90% probability of rain unless we specify under what conditions it will rain, for what the probability refers to is the frequency with which we will be right when we infer that it will rain from some set of premises. But this account faces a very significant problem, the Single Case

Problem. If probability attaches to arguments, and granted that the premises and conclusion of an inference must be true or false, then probability can have no meaning in reference to single cases. In a given case, it either will rain or it will not.

Peirce's example of a single case is dramatic. He asks us to imagine that a man must choose one card from one of two decks. The first deck contains twenty-five black cards and one red one. The second deck contains twenty-five red cards and one black one. Now if he draws a red card, he shall be transported to eternal felicity. If he draws a black card, he shall be transported to eternal woe. From which deck of cards should he draw? The answer is obvious. The man ought to draw from the second deck, where he has a much better chance of drawing a red card. Suppose he does so but draws the one black card. Peirce asks, "what consolation would he have?" (ILS 114, 1878). If we say that he made the most reasonable choice since he increased his odds of drawing a red card, "that would only show that his reason was absolutely worthless" (ILS 114, 1878). After all, the choice condemned him to eternal woe. The problem of the single case comes in here: "He could not say that if he had drawn from the other pack, he might have drawn the wrong one, because an hypothetical proposition such as, 'if A, then B,' means nothing with reference to a single case" (ILS 114–115, 1878). This, Peirce argues, is because truth requires some real fact to which the true proposition corresponds. But since the man did not draw from the other pack, there is no real fact corresponding to what would have happened had he drawn from the other deck. The Bayesian, however, may not seem to face the same problem. She can assign probabilities directly to the conclusion. From the start, what the Bayesian would ordinarily do is assign a probability of 1/26th to the claim that the man will draw a black card from the second deck and a 25/26th probability to the claim that the man would have drawn a black card from the first deck. These will serve as prior probabilities. The selection of the priors will be determined, in this case, by the principle of indifference: Of a set of possible outcomes such that we have no reason for thinking any one outcome is more likely than any other, we should assign each outcome an equal probability. The Bayesian would seem to be in a better position than the Peircean to account for such a probability assignment.

Peirce's First-Pass Reply: Peirce first concedes that in reference to single cases probability has no meaning: "there can be no sense in reasoning in an isolated case, at all" (ILS 115, 1878). But the single cases we are talking about here are instances or tokens of inferences. Each token is an individual inference, but probability is assigned to the type of inference. Take again the example of the inference that it will rain when the clouds are dark and heavy and the atmospheric pressure drops. On any given occasion, either it will rain or it will not rain. On the occasion when one makes the inference from the fact that the clouds are dark and heavy and the atmospheric has dropped, it either rains or does not. That tokened argument has no probability. Probability can only be assigned to the type of such inferences. Namely, those who make such inferences will be right 90% of the time. There is no probability as to whether this particular inference on this occasion will be right or wrong, for it either rains or does not. Probability attaches to the type of inference, how frequently it holds good that when the clouds are dark and heavy and the atmospheric pressure drops, it rains.

Admittedly, for the particular man who draws the black card, there is "no parallel [case] as far as this man is concerned, [for] there would be no real fact whose existence could give any truth to the statement that, if he had drawn from the other pack, he might have drawn a black card" (ILS 115, 1878). There is no parallel case to the rain example because the man will not have the opportunity to draw a card again. But what is at issue with respect to probability is not this individual man's concerns. Independently of that man's concerns, there is still some probability as to the frequency with which those who draw a card the from the second pack will go to eternal felicity. That frequency is independent of what has happened to this particular man. Over the long run, the probability of persons drawing from the second pack going to eternal felicity is about 96%. Those drawing from the first pack go to eternal woe with the same frequency. As Edward Madden rightly remarks, "while his own choice may end in disaster by getting the only black card in the deck, nevertheless he must take heart in the fact that in the long run, following the same procedure, a large number of his fellow men will be saved" (1964, 128). That is all the inductive inference permits. It does not permit us to make any claim about the single case except by way of an additional statistical deduction. Once we have determined the probability over the course of sampling, we can use statistical deduction to conclude that there is a 96% probability that that man will enjoy eternal felicity. The statistical induction, however, must be made prior to this statistical deduction. This is Peirce's answer to the problem of the single case. He accepts the consequence. From a single case, we can calculate no frequency. However, from several cases, we can calculate a probability. Moreover, by statistical deduction, we may make inferences regarding individuals. Madden, whom I just quoted, finds this result to be cold comfort to the man who draws the black card. He misses the point (which Levi 1995 does not) that we can still assign probabilities to the single case by statistical deductions from the ratio of those who do go to heaven.

The Bayesian Riposte: Peirce is not content with this point. He carries the matter further: "These considerations appear, at first sight, to dispose of the difficulty mentioned. Yet the case of the other side is not yet exhausted" (ILS 115). The Bayesian is the other side. She now generalizes from the problem of the single case to the Peircean position altogether. The problem is that whereas a type of inference may have a definite probability, even so the instances of the inference will be finite. "Now the number of risks, the number of probable inferences, which a man draws in his whole life, is a finite one," Peirce notes, "and he cannot be absolutely *certain* that the mean result will accord with the probabilities at all. Taking all his risks collectively, then, it cannot be certain that they will not fail, and his case does not differ, except in degree, from the last one supposed" (ILS 115, 1878). If (a) probabilities are assigned to the type of the inference based on the individual inferences and (b) the individual inferences are only made a limited number of times, how can we be certain that this ratio will accord with the actual frequency with which the two events are associated in reality?

Let us admit that probabilities are assigned to types of inference. Consider again the inference that when the clouds are dark and heavy and the atmospheric pressure drops, it rains. The individual inference will be made a finite number of times. The probability of the type of the inference will be the number of times the argument leads us to the truth to the number of times it is made. If 9/10ths of the time when the premises are true it rains, then the probability of the inference is 90%. However, it is possible that the probability calculated from the finite cases does not match the actual probability of the events such as they are in nature. Perhaps future inquiry would find that the inference holds well only 88% of the time. To take another example, suppose I have a coin that really is fair though I do not know it is fair. I flip it one hundred times. It lands tails 60% of the time. I conclude that the coin is not fair, even though it is. The evidence favors the position that 60% of the time when the coin is flipped it lands tails. In fact, though, the coin is fair and the probability is 50%. What guarantees that the sample is reflective of the facts? The Bayesian is pushing the Single Case Problem further by claiming that the problem is but a special instance of a more general problem. The more general problem is that our experiences are finite. We only see some little corner of the universe for a brief period of time. There is no assurance that the number of cases on which the inference is drawn has settled on the actual frequency, that the sample mean really does approximate the population mean.

Peirce's Second-Pass Reply: Peirce's reply is to turn the tables on the Bayesian. He argues that both he and the Bayesian are in the same boat. We have already seen Peirce argue in "Grounds" that the solution to the problems of the short run, localization, and self-interest with respect to induction is the community of inquirers. He now argues that the Bayesian must make recourse to the same solution. Both must hold that logic inexorably requires that our interests shall not be limited. They must not stop at our own fate but embrace the whole community of inquirers. This community is not limited but extends to all races of beings with whom we may come into immediate or mediate intellectual relation. As Peirce states, the community of inquirers

> must not be limited, but must extend to all races of beings with whom we can come into immediate *or mediate* intellectual relation. It must reach, however vaguely, *beyond this geological epoch, beyond all bounds.* He who would not sacrifice his own soul to save the whole world, is, as it seems to me, illogical in all his inferences, collectively. Logic is rooted in the social principle. (ILS 116, 1878, emphases added)

Why must Bayesians appeal to a community of inquirers? Bayesians are sometimes criticized on the grounds that their prior probabilities are arbitrarily selected. Though there is sometimes truth to this objection, that fact is irrelevant to Peirce's reply. His reply is that the very process of Bayesian updating also requires reference to a community of inquirers. The arbitrariness is not the problem. The problem is that no matter where we set our priors, we will need additional inductive inferences to get our priors to line up with the real uniformities of, the real regularities in, nature.

On the one hand, the elegance of Bayesian updating is that given enough testing and sampling, we will eventually arrive at the right probability. On the other hand, this is also why the Bayesian faces the same problem as the Peircean: We need to get enough samples, enough data. What makes Bayesian updating elegant is precisely why the Bayesian and the Peircean are in the same boat. Either (a) the probabilities have been arbitrarily set or (b) they have been improved or supported by updating. If they have been arbitrarily set, then we have no grounds to think they are accurate. If they have been improved or supported by updating supplemented with and monitored by sound practices of inquiry, the result in the long run will be a probability assignment that hews more closely to the uniformities of nature (though at any given time, the results may not reflect the real regularities in nature).

These considerations can be made more evident if we again take the case of observing the bright light at night. Let us doubt that it appears every night. Now I will set the hypothesis that it appears every night to one over a nonillion to the nonillionth to the nonillionth. With such a low prior, it will take a very long time to reach the actual frequency with which the light appears. Updating does not guarantee that in our finite lives we will ever ascertain the real uniformities of nature. Bayesianism also requires a long run of updating. This is why the Bayesian and the Peircean are in the same boat. For the Peircean, the number of times the inference is made will be finite. Based on our tokened or individual inferences, we may never ascertain the frequencies that in fact obtain. However, the community of inquirers will be able to approximate the actual frequencies in the long run. Similarly for the Bayesian, the prior may not represent the actual uniformity of nature. However, updating will approximate it in the long run. The only reason even the Bayesian can have to believe that updating will lead to a posterior probability that represents the uniformities of nature is if she embraces an unlimited community of inquirers. Otherwise, she has grounds to worry that the probabilities on which her inquiries settle will not represent the real uniformities of nature because her observations are limited to the short amount of time we have existed in this universe and the small corner of the universe in which we live and because we are subject to self-interested biases.

As Deborah Mayo rightly stresses, the Bayesian's insistence that convergence will be reached "is irrelevant to the day-to-day problem of evaluating the evidential bearing of data in science" (1996, 84). Over the long run, those who start with different priors will eventually reach the same or similar results. But in the short run, they will not be concerned with the long-term validity of their inferential practices. Rather, in the short run, their interest will be in the trustworthiness of the results they have. That is, they will want to know whether their inferences are strong. Mayo is correct that practicing scientists will not care about their degrees of belief in cases of disagreement in the short run. They will instead seek out severe tests of their hypotheses and apply error-statistical methods for "unearthing a given error" (1996, 7). I shall broach the topic of inductive strength in the next section.

One might object that the example I have given is quite outrageous. Who would set the probability of the light appearing nightly so low? This objection does not help matters in the least. First, while Bayesians might appeal to the principle of indifference, it is quite difficult to formulate an adequate statement of the principle of indifference. Second, even a more modest prior may require far too long of a period of updating to ascertain the real uniformity of nature. Third, even so, the sample will only give us a snapshot of the actual probability of the inference. Perhaps if a coin is tossed 30 times, it lands heads 15 times. But if it were tossed 1 trillion times, it may land heads only 499 billion times. Or perhaps it lands heads only those 15 times out of a trillion. Fourth, suppose the probability of an event is in fact one over a nonillion to the nonillionth to the nonillionth. If we start by assigning a probability of one-half according to the principle of indifference, our hopes of attaining the truth must again rest in the future community of inquirers.

However initial probabilities are assigned, they will need to be updated in any ordinary course of scientific inquiry. Our observations are limited in scope. There is no contradiction in supposing that throughout a person's entire life, she updates her beliefs and yet is led to a probability that does not reflect the actual uniformities of nature. Her experience is limited to her short life in a little corner of a far-flung galaxy. A diversity of predictions of hypotheses needs to be tested, and those tests need to be replicated. She may test some in her finite life, but not all. She will have the time to replicate only some of the studies. By all means, Bayesians should update their beliefs. The Peircean will revise her ratios in light of further inquiries, too. The Peircean may even use Bayes's Rule to do so, as I have argued one may sometimes do with respect to qualitative induction. Nevertheless, both the Bayesian and the Peircean will need a community of inquirers to address the limited scope of our inductions. Peirce remarks, "death makes the number of our risks, of our inferences, finite, and so makes their mean result uncertain. The very idea of probability and of reasoning rests on the assumption that this number is indefinitely great" (ILS 116, 1878).

Summary: In sum, Peirce's defense of induction's validity in general turns on the fact that an infinitely wide, enduring, self-sacrificing community of inquirers will approximate the truth over the long run. In cases of crude induction, the mass of our social experience will ultimately establish or fail to establish the claims in question. To use Peirce's example, if some instance of genuine clairvoyance is established, the community of inquirers will overturn the inference from *No instance of a genuine power of clairvoyance has ever been established* to *So I presume there is no such thing*. In the case of qualitative induction, so long as the hypothesis to be probed is among those to be tested, the community of inquirers will eventually be able to home in on the true hypothesis and eliminate those which are false. In the case of quantitative induction, the law of large numbers together with the continued observations and inquiries of the community of inquirers ensures that our inductions will approximate the truth in the long run.

The Strength of Induction

In the preceding section, I examined Peirce's account of induction's validity. According to Peirce, inferences are valid just in case the procedures used in making the inference have some truth-producing virtue and the person making the inference in fact adheres to the procedure. Inductive inferences, when valid, have the virtue of being truth-approximating in the long run. Crude induction, which consults the mass of past experience to determine whether there are any counterexamples to some claim, approximates the truth because the mass of experience is not just one's own experiences but the entire community of inquirers' experiences. For qualitative induction, that community probes hypotheses over the long run and, by a slow process of fractionation, sloughs off those theories which are false and corroborates or confirms those which are true. Quantitative inductions are valid because the community of inquirers would be able to make better and better approximations of the real values, as determined by using sampling, measuring, counting, and other statistical techniques.

In addition to making valid inductive inferences, we will want to make strong inductive inferences. There is a familiar quip that in the long run we are all dead. To be sure, all of us alive today will be dead in the long run. But for the Peircean, we do hope that the community of inquirers continues forever, and induction is valid not because of us but because of them. Nonetheless, such long-run faith in the community of inquirers is little comfort to us in the short run. We will often rely on induction to make decisions, and those decisions may have life-or-death consequences. An engineer who builds a suspension bridge will want to know the tensile strength of the cables she plans to use. The strength of those materials will have been determined by induction. The company that makes the cables will want to be certain of their quality. To determine whether the cables have been well-made, the company will use induction. But these are short-run, not long-run, concerns. Millions of dollars, and the lives of persons who cross the bridge daily, are at stake.

Similarly, we will often want to gain information about some current state of affairs. In such cases, while the fact that our procedure is valid is important, the fact that we will not be making long-run inquiries is just as noteworthy. Suppose a political action group wishes to know the preferences of voters at some time. To be sure, the more they poll and the more they sample, the closer their results will be to the facts in the long run. But the political action committee cannot keep polling indefinitely, and they have limited resources to conduct the poll at all. The political action committee will be concerned not merely with the long-run validity of their procedure but the short-run strength of the results.

The strength of a statistical induction based on sampling will depend on the manner in which the sample is obtained. If sampling from a bag of beans, a good sampling procedure will mix the beans thoroughly and draw samples from different places in the bag. One might think that we could simply specify some such rules for obtaining samples. But sampling beans from a bag is a highly idealized procedure. Strategies for making inductive inferences strong will be partly determined by various factors. Among these factors are the question at hand, the object of our study, our tolerance for error, and the apparatuses we have available for performing tests.

These points just raised also pose a problem for subjective Bayesianism. As I have argued, Peirce holds that so long as the Bayesian's procedures are supplemented with and monitored by good practices of inquiry, there is no harm in following the procedure. But we will need to supplement the Bayesian procedure not merely with techniques for making valid inferences but techniques for making strong inferences. As Peirce remarks, conceptualism "is only an absurd attempt to reduce synthetic to analytic reason, and . . . no definite solution [of out of all possible states of things how many will accord with the results from updating] is possible" (ILS 142, 1878). Once again, to think that induction can proceed merely by "legitimate arithmetical operations upon the numbers given in the premises" (ILS 112, 1878) is a mistake.

The impossibility of specifying rules for making inductions strong in every area of inquiry does not preclude making general recommendations about how to ensure our inductions are strong. Throughout his writings, Peirce makes many such recommendations. In this section, I survey some of those general recommendations. I take up qualitative induction first and then turn to quantitative induction. Before proceeding, two points are in order. The first point is that I shall only gather together some recommendations which Peirce makes and briefly comment on them. A more robust investigation of Peirce's scientific work deserves another book. Moreover, Deborah Mayo (1993, 1996, 2005, and 2018) has already done much of the heavy lifting to develop a Peircean account of testing and quantitative induction. The second point is that many scientific inquiries will use quantitative inductions in support of or in tandem with qualitative inductions. Consequently, recommendations that apply to making qualitative inductions strong may also apply to quantitative inductions, and vice versa.

Qualitative Induction

Qualitative induction is inference from consequence and consequent to antecedent with the end of establishing whether the antecedent is true. For qualitative inductions in a scientific context, the antecedent is typically an explanatory hypothesis. Peirce holds that hypotheses are (a) distinguished from one another by their predictions and (b) demonstrable as true or false from their predictions. Let hypothesis H_1 have a set of consequences $P_I = \{P_1, P_2, \ldots P_n\}$ and H_2 have a set of consequences $P_{II} = \{P_1, P_2, \ldots P_n\}$. If $P_I = P_{II}$, then H_1 and H_2 are equivalent hypotheses. Moreover, if H_1 is false, there will be some member of P_I that, if put to the test, will be found to be not true. Although this position may raise the hackles of contemporary theorists from a philosophical point of view, it is much less clear that the practicing scientist who aims to confirm some theory over others would disagree with Peirce so far as it concerns her specific researches. As explained momentarily, she will have some finite set of hypotheses which have suggested themselves. By a hypothesis suggesting itself, I mean that something in us prompts us to take the hypothesis into consideration. As I shall argue more fully in Chapter 7, our minds have a bent to understanding nature, instilled in us through natural selection. We have some instinctive prompting, honed by practice and training, for guessing at some answers rather than others, and those we guess at are those which suggest themselves. The person making a hypothetic induction will then begin the labor of sorting through those hypotheses so as to put them to the test. If she has no way to experimentally differentiate her hypotheses, then she has no way to eliminate, corroborate, or confirm one hypothesis with respect to some others.

Gather as Many Hypotheses as Suggest Themselves: First, for whatever issue is under consideration, we ought to gather together as many hypotheses as suggest themselves to us for explaining the phenomena. It is a separate question as to how hypotheses suggest themselves or what properties make hypotheses suggestive (e.g., simplicity, elegance, non-fine-tuning, etc.). What is involved in hypotheses suggesting themselves concerns abduction. For the moment, it is sufficient to note that some hypotheses do not suggest themselves. The sneeze of a Greenlander does not suggest itself as a hypothesis to explain the bending of light around the sun. Other hypotheses do suggest themselves. That I have developed an allergy to tree pollen might explain why I have started having the sniffles in the spring when I never previously did.

We should take care to bring into consideration as many hypotheses as do suggest themselves. Peirce remarks on the work of Louis Pasteur in this respect:

> Somebody at the Academy of Medicine was one day urging the importance of a certain hypothesis. "Oh," exclaimed Pasteur, "as for hypotheses, we fetch them into our laboratories by the armful." Mind, he did not absurdly pretend to keep them out, as some do; nor want them only in homæopathic quantities, like many more. No, he would order large supplies of them; he would consume them in armfuls. And what would he do with his armful of hypotheses, when it was fetched? He would begin by sorting them over, with unwearied industry. (CN 3:65, 1902)

The reason we need to collect as many hypotheses as suggest themselves is that sometimes hypotheses that seem unlikely turn out to be true. We cannot antecedently determine which hypothesis will be the true one. That is the point of experimentation.

Make Ideas Clear and Hypotheses Explicit and Definite: Second, for any given hypothesis, we need to make our conception of it as clear, explicit, and definite as possible. We make our ideas clear. We make our hypotheses explicit and definite. "The hypothesis," Peirce states, "should be distinctly put as a question, before making the observations which are to test its truth. In other words, we must try to see what the result of predictions from the hypothesis will be" (ILS 175, 1878).

As I explained in Chapter 4, Peirce acknowledges at least three ways in which our ideas may be made clear. Our ideas have clarity so far as we know how to use them. They have distinctness so far as we can provide a definition of them. They are pragmatically developed so far as we can identify certain marks or indicators of them (as smoke is an indicator of fire) or specify procedures for isolating the object of our inquiry (as the procedure Peirce gives for isolating lithium at EP 2:286, 1903). They are pragmatically adequate when they have been sufficiently pragmatically developed for our purposes. If, for example, an object is green, we might wish to make our conception of what green is clearer. We will get an initial grasp on the idea by considering how people use the word 'green.' For example, they will claim that grass is green, as are emeralds and praying mantises. Further, we will want to define what green is. On this topic, philosophers and color scientists may disagree. Some will identify green with microphysical properties of objects. Others will identify green with relational properties. Some deny that green has any real definition, that there is anything that it is to be green. Moreover, we will want to pragmatically develop our idea of green. For that, we ascertain marks of something being green. For example, something is green only if we could match its color by mixing blue and yellow. Or something is green only if it absorbs and reemits such-and-such wavelengths of light, as determined by a spectrometer.

We should also make our hypotheses as explicit and definite as possible. Peirce notes that many of our most primitive beliefs "are very vague indeed (such as, that fire burns) without being perfectly so" (CP 5.498, 1905). That fire burns is consistent with many different hypotheses. It is consistent with both the theory of phlogiston (that fire burns by releasing a rarefied substance called phlogiston) and our current theory (that fire burns because of oxidation, oxygen atoms binding—in the case of a wood fire—with carbon and with hydrogen). No one doubts that fire burns. But if we want to put an explanation for why fire burns to the test, we will have to be more explicit and definite about what we are testing.

If our hypotheses are not explicit and definite, we may (a) mistakenly take some prediction to follow from a hypothesis when it does not, or (b) fail to recognize two hypotheses we take to be different are really identical, or (c) fail to recognize that what we take to be some single hypothesis is ambiguous between two different hypotheses. If we go about testing a hypothesis by its consequences, we need to ensure that the predictions we test for are in fact predictions of the hypothesis. Moreover, we need to be able to distinguish between our hypotheses. It is an important fact, for example, that on Einstein's and Newton's theories of gravity light is predicted to be deflected around the sun at different angles. Recognizing this fact is one of the ways in which we are able to put the theories to the test.

Peirce recounts one amusing episode from the life of Louis Pasteur illustrating the importance of making our hypotheses explicit and definite (see CN 3:65–66, 1902). Pasteur remarked at a meeting that fowls do not take anthrax. Another professor, Gabriel Colin, objected that a hen may easily be given anthrax. Pasteur challenged him to do so. When Colin failed, he admitted that he had erred and adverted that a hen cannot be given anthrax. Pasteur replied that in fact a hen can be given anthrax. He proceeded to do so by first cooling the body temperature of a hen and then exposing it to anthrax. Colin had confused the thesis *a hen does not take anthrax* with *a hen cannot be given anthrax*. As the highest temperature at which anthrax can live is a few degrees below the normal body temperature of a hen, it cannot take anthrax. However, if a hen is first given an ice bath to lower its body temperature a few degrees, the hen can be given anthrax.

Do Not Disregard Unlikely Hypotheses That Have Suggested Themselves: Third, we ought not to reject any hypothesis merely because it seems unlikely to us. Although likelihood may affect the order in which we set about testing hypotheses, it is not alone grounds for rejecting a hypothesis. In his late work, Peirce distinguishes among plausibility, verisimilitude, and probability (see ILS 123–124, 1910). In its narrowest sense, probability pertains specifically to quantitative inductions. It is the ratio of successful arguments to total arguments, the ascertained probability. Nevertheless, Peirce does sometimes use 'probability' to mean a higher degree of verisimilitude. Plausibility and verisimilitude are contrasted in that when a hypothesis is plausible, (a) the hypothesis has not been tested but (b) some surprising phenomena have been observed which the hypothesis would explain if it were true. Prior to the testing of Einstein's theory of relativity, his theory was plausible because there were some surprising phenomena it explained, such as the precession of Mercury's perihelion. In contrast, a verisimilar hypothesis is one (a) we do not yet have sufficient support to assert is true to the facts but (b) which the evidence we do have supports and (c) from what we can tell, the remainder of the conceivable possible evidence, if put to the test, would conclusively prove its truth. For example, prior to the observation of the Higgs Boson, we could not claim that the hypothesis there is some such elementary particle is true. Nevertheless, some of the predictions were confirmed. Furthermore, it seemed that the remainder of the evidence was likely to bear the hypothesis out. Accordingly, the hypothesis was verisimilar.

Importantly, both plausibility and verisimilitude come in degrees. A hypothesis will strike us as more or less plausible; sometimes, it will be so plausible that we irresistibly believe it. If a hypothesis is highly plausible, Peirce claims, it will "justify us in seriously inclining toward belief in it, as long as the phenomena be inexplicable otherwise" (ILS 123, 1910). I shall have more to say about plausibility in Chapter 7. If a hypothesis is to rise to the level of verisimilitude, we must put it to the test. Verisimilitude comes in degrees depending on the number and nature of predictions of the hypothesis that have been put to the test. "Strictly speaking," Peirce remarks, "matters of fact never can be demonstrably proved, since it will always remain conceivable that there should be some mistake about it" (ILS 124, 1910). That is, strictly speaking, we may never claim that that a hypothesis is beyond doubt and so demonstrated. We may only claim that its degree of verisimilitude is exceedingly high or probable, albeit not in the statistical sense. As Peirce states in a lecture from 1898, scientists do claim that there are established truths, only what they mean is that there are propositions "to which no competent man today demurs" (RLT 112).

Peirce sometimes uses 'likelihood' synonymously with 'verisimilitude,' but in claiming that we should not reject unlikely hypotheses, I mean that we should not reject hypotheses merely because they have a low degree of plausibility or a low degree of verisimilitude. Even if a hypothesis is unlikely, that alone is not grounds for rejecting it out of hand. Provided it has suggested itself to us as an explanation of some surprising phenomena, it is plausible even if not highly plausible. Similarly, provided it has passed some tests even if not many severe tests, it may be verisimilar even if not highly verisimilar. In a letter to William James, Peirce claims that "the cuneiform inscriptions could never have been deciphered if very unlikely hypotheses had not been tried" (CWJ 8:244, 1898). Even unlikely hypotheses may turn out to be true or useful to try because doing so may bring us closer to the truth. He claims that the scientific inquirer "will not even refuse to entertain a grossly improbable hypothesis, so long as it possesses the one merit of being the theory which is at the moment most conveniently and economically compared with observation. A successful investigator occupies the larger part of his time with theories of all degrees of improbability; for he expects to reject the larger proportion of those which he cordially receives on trial" (HP 1:478, 1896).[17]

Furthermore, Peirce would advise us against rejecting hypotheses just because they are unlovely. In the literature on theoretical virtues, there are many proposals for sorting through hypotheses on the grounds of their loveliness, simplicity, elegance, whether they require fine-tuning (i.e., precise adjustment of a model's parameters to fit observations), and so on.

Peirce would not regard any of these as grounds for rejecting a hypothesis outright. At best, they are grounds for ranking our hypotheses in the order in which we might set about testing them. Peirce notes, "every advance of science that further opens the truth to our view discloses a world of unexpected complications" (EP 2:444, 1908). There are no a priori assurances that the world is lovely, simple, elegant, or the like. We might prefer to spend our time testing those hypotheses that are lovely, and so on. But those preferences are not grounds for rejecting alternative hypotheses out of hand.[18]

These considerations are fully consistent with adopting less extraordinary hypotheses in the interim and preferring to put those to the test. As Peirce claims, we are not to explain facts "*by a hypothesis more extraordinary than those facts themselves; and of various hypotheses the least extraordinary must be adopted,*" yet he hastens to add that this "maxim is rather too vague to make a part of a logical theory, but it is sufficiently precise to be readily applied in practice" (W 1:452, 1866). These considerations tie into Peirce's claim that the validity of induction reposes on the community of inquirers. Suppose we prefer testing lovely hypotheses first. In the long run, we may find that all of our lovely theories are false. Inquiry need not and ought not to cease. We should then take up the unlovely hypotheses and put those to the test. If, however, we had already discarded those unlovely hypotheses, we would have nothing to put to the test. Tabling a hypothesis for later investigation is different from discarding a hypothesis altogether. We may discard hypotheses if we find that the experimental evidence conclusively does not bear them out. But when a hypothesis is merely unlikely, that is no grounds for discarding it. At best, it is grounds for tabling the hypothesis and putting our faith in the work of future inquirers.

Search for Unique Predictions: Fourth, once we have gathered together those hypotheses that suggest themselves and made them as clear, explicit, and definite as possible, we will also want to search for those predictions from hypotheses that are unique to them individually or to sets of them: "The testing of the hypothesis proceeds by deducing from it experimental consequences almost incredible, and finding that they really happen" (CP 7.83, 1902). Furthermore, "the first thing that will be done, as soon as a hypothesis has been adopted [for probation], will be [to] trace out its necessary and probable experiential consequences" (HP 2:733, 1901). Also, Peirce notes, "Mr. Mill's four methods" are "extremely valuable and should not be neglected" (ILS 175, 1878). In searching out the cause of an event or events, we will seek to identify commonalities and differences among them. The more occurrences we have to compare, the better.

The search for unique predictions may involve identifying those predictions that are shared between some set of hypotheses and not others as well as those that are individually unique to the hypothesis. And, of course, we will want to trace out as many predictions of some given hypothesis as we can. We "should split up a hypothesis into its items as much as possible, so as to test each one singly" (HP 2:761, 1901). Suppose, for example, we have six hypotheses. If hypotheses one through three entail that *P* and four through six entail that *not: P*, then testing for *P* will be an efficient way to narrow our field of hypotheses (see CP 6.529, 1901). Once we narrow down our hypotheses to three, we will want to ascertain which predictions are unique to each of those three so that we can put them to the test individually.

The sorts of tests that we will be especially keen to perform are those which Mayo calls severe tests. Mayo holds that some evidence is good grounds for a hypothesis to the extent that the hypothesis has passed a severe test with respect to that evidence (see 1996, 177). A hypothesis passes a severe test "just to the extent that it is very improbable for such a passing result to occur, were [the hypothesis] false" (1996, 178, italics removed). When a hypothesis H passes a severe test, it will have passed a test that "is highly capable of probing the ways in which *H* can err" (1996, 9). Two inconsistent hypotheses concerning the same subject matter may pass some single test, even if very many other hypotheses do not. In that case, the test would corroborate both hypotheses. What we must further seek out are the different ways in which those two remaining hypotheses can be probed for error. When we are probing for error, severity and uniqueness go hand in hand. We need unique tests to discriminate hypotheses. We need severe tests to confirm them, to regard them as proven in the sense that all real doubt has been removed.

Ensure One's Instruments Are in Good Order: One will often use instruments in order to put hypotheses to the test. Instruments are not just microscopes, telescopes, and the like but such things as maps and tools for measurement. Peirce holds that scientific progress is made by the development of apparatuses, such that "a given piece of apparatus quickly teaches us all it can teach, so that to make new progress new apparatus is required" (HP 1:46, 1893). Peirce takes 'apparatus' broadly to include any tool or equipment that might be used to make something apparent or open to direct observation (see HP 1:59, 1893, and HP 1:135, 1894). Modern science, he maintains, has made its advances because inquirers have spent their lives "in their laboratories and in the field ... not gazing on nature with a vacant eye, that is in passive perception unassisted by thought—but have been *observing*—that is perceiving by the aid of analysis,—and testing suggestions of theories" (W 2:315, 1869).

Instruments can introduce their own errors. Aberrations in lenses or mirrors, for example, can introduce errors into one's data. Peirce had first-hand familiarity with such problems. To give a few examples, he found that slight flexures in the arms of pendula vitiated the data gathered from grav-itation experiments. He performed experiments to ascertain the degree of flexure so as to correct the errors (W 3:217–229, 1877; W 4:515–528, 1883; W 5:262–278, 1885). He also designed a new sort of pendulum that minimized flexure. Once when performing pendulum experiments, he discovered that the screws of the base had been tightened by hand only, causing some move-ment when the pendulum swung (W 5:10, 1884). When observing a solar eclipse in 1869, he found that adjusting a binding-screw on the telescope caused the telescope to shift. He had to correct for the problem (see W 2:290, 1869). He discovered that ghost lines that appear in spectroscopic images were from diamonds creating a bur on the glass. He developed a way to treat the glass so as to eliminate the ghost lines (W 4:4–5, 1879).

One may discover that one's instruments are not in good order when dif-ferent methods of observation or the use of different instruments delivers divergent results. Aberrations in mirrors will not likely be uniform across telescopes. Therefore, observing the same galaxy using two different mirrors with different aberrations will be likely to introduce differences in observa-tion that can be used to correct one another. Similarly, for the ghost lines in spectroscopic images, the lines did not appear in the spectra from a prism. This fact suggested the lines were an artifact of the diffraction plates.

'Instruments' includes all sorts of aids to observation and calculation. In this broader sense, maps and tables are instruments. Accordingly, we sometimes find Peirce complaining about the carelessness some writers take with respect to their work. He writes of a star atlas that the author has made many errors which "are a source of great inconvenience" (W 3:1, 1872). In his capacity at the Coast Survey's Office of Weights and Measures, Peirce was particularly concerned with how accurate were measures of the pound and yard. For instance, weights made of some materials were apt to be worn down from use, and variations in temperature would affect the length of bars used in measurements (see his testimony before Congress at W 5:149–161, 1885).

Use Diverse Methods: A sixth recommendation is to use several different methods to investigate some question. These methods may be used to fortify each other, so that any error introduced by the use of one method may be corrected or discovered by the use of another. Peirce, for example, uses a different method from John Herschel to infer the ellipticity of the earth. "The principles of my procedure being different from his," Peirce states, "I have thought it worth while to complete my work and see how far Major Herschel's and my results would agree" (W 4:529n1, 1881). As Peirce states, "it [is] extremely advantageous in all ampliative reasoning to fortify one method of investigation by another" (W 4:429, 1883). This might involve the use of diverse instruments to investigate some phenomenon, such as pendula with different designs. With respect to quantitative induction, there are different methods for sampling a population. We want our sampling to be random, but how do we ensure it is? We may use different methods for selecting members of a population at random. For a survey, we may use phone records to generate a list of names. Alternatively, we may use census records to generate a list of names. Further, we might use tax records to generate a list of names. Moreover, the manner in which we decide on a subset of those names to survey may vary.

Carefully Evaluate One's Data: Distraction may introduce errors into observation. Carelessness may cause one to miswrite some number (say 35 instead of 53). Persons may be dishonest or have reasons to hide important information. Environmental factors may affect one's instruments. Even if one's instruments are in good order, data may need to be discarded because wind, changes in temperature, or unstable foundations for instruments make the data useless. Not all data are equal. Sometimes one is well-advised not to take some of the data into consideration.

One case highlighting the importance of data evaluation is Peirce's controversy with the authors of a book which purported to prove spontaneous telepathy, or the existence of ghosts. Their argument is one that involves both qualitative induction (confirming the existence of ghosts) and quantitative induction (they sample a population). Peirce argues that much of the data is vitiated. Many reports of people who claim to have seen ghosts should not be trusted (see W 6:74–81 and 101–141, 1887). Sometimes the data are outside the range of time that is supposed to be under consideration. Other times, the person admits to being in poor health or intoxicated. Peirce also rejects some testimony on the grounds the person is affiliated with a group of people known to be "superstitious, ignorant, and credulous" (W 6:78, 1887). He also argues that the interviewers sometimes led interviewees to the desired answer by asking suggestive or leading questions (see W 6:133, 1887). Another problem with the data was the sampling procedure.

A second interesting case is the numerous corrections Peirce makes to his data gathered from pendulum experiments. His report on gravity, written in 1889, makes corrections to data based on such factors as changes in temperature, flexure, atmospheric pressure, differences in timepieces, and the like. In another report, he discards some data because he finds the sandstone piers on which the pendulum had been set bent under the oscillating pendulum (see W 6:216–217, 1888). These diverse factors require making corrections to some data and discarding other data. Data ought not to be taken at face value. Sometimes a datum should be discarded. At other times, a datum should be weighted less than other data. On other occasions, corrections must be made to the data based on other known factors affecting the observations.

As a third case, Peirce frequently mentions census data. He notes that according to the data, the number of persons who are twenty-one quite outstrips those who are twenty, even though "in all other cases, the ages expressed in round numbers are in great excess" (CP 5.576, 1898). The reason is that it is sometimes to one's advantage to be thought older than one is. Those who are twenty lie and claim to be twenty-one. On other occasions, it is desirable to be thought younger than one is. People find it to their advantage to round down their ages. A careful evaluation of the data reveals these trends. The person making inferences on the basis of the data will need to make corrections to that data.

As a fourth case, the observer is herself subject to error. She may be subject to illusions under certain conditions of observation. She may be careless or insufficiently cautious. Her judgments and the ways she goes about gathering data may be marred by racial or gender prejudices. The observer must be on guard against introducing errors into the data as a consequence of her own perceptual and cognitive biases. "We all know, only too well, how terribly insistent perception may be," Peirce observes, "and yet, for all that, in its most insistent degrees, it may be utterly false" (CP 7.647, 1903). Insistent though an illusion may be, "when one knows the right trick it will be curious to see how easily it is downed" (CP 7.647, 1903). Peirce maintains that an educational course can give us control over some of these misleading appearances. Such educational courses also enable us to make more precise observations, as early astronomers were trained in judging the magnitude of stars (see CP 7.258, c. 1900).[19] Of Lavoisier, Peirce claims that he "never made an original experiment in his life, but he saw which the decisive experiments were, he repeated them with such high precision and such circumspect precautions that confidence was commanded, he coördinated them with a clear, scientific logic of which no contemporary was the master" (CN 3:89, 1902).

Elimination of Hypotheses: As we set about testing hypotheses, we will begin by eliminating those that are inconsistent with other known facts. By 'eliminate' I mean we reject, refine, or table a hypothesis. For the remaining hypotheses, we will devise unique and severe tests to start eliminating them "by a skilful alternations of excogitation and experimentation" (CN 3:65, 1902). Moreover, if "a hypothesis would necessitate an experimental result that can be cheaply refuted if it is not true, or would be greatly at variance with preconceived ideas, that hypothesis has a strong claim to early examination" (CP 7.83, 1902). Once we have narrowed down our hypotheses to a favored hypothesis, that hypothesis must then be "treated as an enemy ... [by] trying to find something which the hypothesis would require to be true, but which experiment should refute" (CN 3:65, 1902). We should regard no hypothesis as precious and established. Rather, we should throw our efforts into endeavoring to refute it. This is part and parcel to Peirce's "first rule of logic" that "in order to learn you must desire to learn and in so desiring not be satisfied with what you already incline to think" (EP 2:48, 1898). Perhaps all of the hypotheses that have suggested themselves are false. In that case, we may need to begin our inquiries anew.

Quantitative Inductions

In his *Illustrations of the Logic of Science* series from the late 1870s, Peirce recommends that for quantitative inductions the "respect in regard to which the resemblances are noted must be taken at random" and second that the "failures as well as the successes of the predictions must be honestly noted" (ILS 175, 1878). But in addition to these brief comments, we find in Peirce's writings discussions of diverse statistical techniques for improving the strength of inductions.

Ensure One's Techniques are Applicable: Although there are many statistical techniques for improving the strength of our inductions, we must make certain that the technique is appropriate to use in our specific context of inquiry. Peirce was familiar with the limitations of some methods for improving the strength of inductions. His 1873 essay "On the Theory of Errors of Observation" has the aim of "showing what the limitations to the applicability of the method of least squares are, and what course is to be pursued when that method fails" (W 3:114). The method of least squares is appropriate to use when one might err on both sides when making an observation, say clicking one's stop watch a little too soon or a little too late. But when observing the appearance of a star that has been occluded by the sun or moon, one will not click one's stopwatch too soon since one will not click the stopwatch until one sees the star. Consequently, the method of least squares is not appropriate to use in that context of inquiry. Sometimes heuristics for improving the strength of our inductions have limited applicability.[20] As another example, historians must be on guard against allowing their preconceived notions to affect their interpretation of the historical data. As explained earlier, one of the limitations of subjective Bayesianism is that it cannot be applied to ancient history. Also, historians will not often find it useful to apply the method of least squares.

Correlations Are Not Always Noteworthy: Second, we should be cautious of concluding that just because we discover some correlations that those correlations are theoretically noteworthy. Peirce claims that there is "no greater nor more frequent mistake in practical logic than to suppose that things which resemble one another strongly in some respects are any the more likely for that to be alike in others" (ILS 174, 1878). His example is "comparative mythologists" who seek to find parallelisms between "solar phenomena and the careers of the heroes of all sorts of traditional stories" (ILS 174, 1878). Peirce mentions that another scholar—the French physicist Jean-Baptiste Pérès (see ILS 182n13)—has lampooned such arguments. Pérès satirically uses them to show that Napoleon was a merely mythical, not historical, figure. That celestial events are correlated with historical events does not imply that the celestial events caused them.

Predesignation Prevents Data Dredging: Third, to make a strong statistical induction, we should predesignate those properties for which we are going to take a sample. If we simply gather data willy-nilly, we will be able to sort through the data in order to find something of interest. This is known as data dredging or p-hacking. Peirce gives many such examples of it. He illustrates how it might be done with respect to the ages of poets at the time of their deaths. He takes the first five poets listed in a biography of famous persons and finds three different relationships that hold for all of their ages (see ILS 156, 1878). Another example is Bode's Law about the distances of planets from one another. Bode identified a numerical relation among the distances of the then-known planets from the sun. The law predicted the orbit of Uranus, but it was disproved by the discovery of Neptune.

Peirce also criticizes Cesare Lombroso, the Italian criminologist, on this score. Lombroso argues that genius is a sort of insanity. Peirce accuses Lombroso of first identifying some mark of insanity and then "search[ing] high and low for instances which may look as if some men of genius have had that symptom" (W 8:279, 1892). Peirce writes, "suppose we were to draw our inferences without the predesignation of the character P; then we might in every case find some recondite character in which those instances would all agree" (W 4:434, 1883). Mayo (1993 and 1996, Ch. 9) has explored these concerns surrounding predesignation from a Peircean point of view. She notes that Peirce does make an exception to predesignation for "honest hunters" who set about the task of "finding ways of showing that the overall error probability is fairly low" (1996, 313). What Peirce requires in such cases is that a large number of samples be found to have some notable characteristic in common.

No Optional Stopping: In addition to predesignating the characteristics for which we are going to sample, we should decide in advance how many samples we will take. The reason is that we could conceivably keep sampling until we get the proportion we want. Peirce expresses both the requirement of predesignation and the prohibition on optional stopping when he writes, "in sampling any class, say the M's, we first decide what the character P is for which we propose to sample that class, and also how many instances we propose to draw" (W 4:434, 1883).[21]

Random Sampling: Fifth, we must endeavor so far as we can to ensure that our samples are randomly taken. As Peirce states, "the sample should be drawn at random and independently from the whole lot sampled" (W 4:427, 1883). If we can mix that from which we will take our sample, we should mix it thoroughly. We ought also to draw our samples from diverse places, if we can. Peirce pioneered the use of randomization in testing, as Ian Hacking (1988) has explained. In experiments he designed with his student Joseph Jastrow in order to ascertain whether persons could discern small differences in the sensations of different weights, they used the colors of well-shuffled playing cards to decide whether they would first increase and then decrease the weight or first decrease and then increase the weight (see W 5:130, 1884).

Mixing a sample is not always possible, and even random sampling is not always possible. Peirce remarks, "[r]andom sampling and predesignation of the character sampled for should always be striven after in inductive reasoning, but when they cannot be attained, so long as it is conducted honestly, the inference retains some value" (W 8:116, 1891). But honesty is a tricky matter. "The drawing of objects at random is an act in which honesty is called for," Peirce notes, "and it is often hard enough to be sure that we have dealt honestly with ourselves in the matter, and still more hard to be satisfied of the honesty of another" (W 4:428, 1883).

Sample as Much as Possible: Sixth, Peirce maintains that we should sample as much as we reasonably can. We should not only draw large samples, but we should also repeatedly draw the quantity of samples determined by our stopping rule. That is, if we can draw a sample of one hundred beans from a bag rather than just ten beans, we should do so. If we can repeatedly draw samples of one hundred beans, we should do that as well. The more we sample, the more confident can we be that the probability ascertained hews close to the actual uniformities of nature, though at any given time our ascertained probabilities may be in error. "Do not take as a sample a picked specimen nor a small one" (W 1:448, 1866), Peirce exhorts. He proceeds to remark, "[i]t is by violating this maxim that figures are made to lie. Medical statistics in particular are usually contemptibly small, as well as open to the suspicion of being picked" (W 1:449, 1866).

Record Both Failures and Successes: Peirce maintains that we should record our failures as well as our successes: "The failures as well as the successes of the predictions must be honestly noted. The whole proceeding must be fair and unbiased" (ILS 175, 1878). A test might fail to give us the results we want on the first, second, third, or however many times. Nevertheless, if persisted in, one may well get the result one hopes for on some occasion. For example, one may have a bag of one hundred pennies and one hundred dimes. Ignorant of this fact, one may hope to show that the proportion of pennies to dimes is six to four. Sampling twenty coins, one will eventually get a sample of twelve pennies to eight dimes, the very proportion one hopes for. It is noteworthy, though, if to get that proportion one had to redo the sampling procedure fifty times. (Of course, one might even get that proportion on the first try. That possibility underscores the need for sampling as much as we reasonably can.)

Establish Efficient Sampling Procedures: We will also want our sampling techniques to be designed in such a way that they can more efficiently and easily increase the scope of our sampling. This can be most readily highlighted by considering the paradox of the ravens. *All ravens are black* is equivalent to *all non-black things are non-ravens.* However, we can much more readily increase the scope of our sampling relative to the population by first finding ravens and checking them for blackness than by examining all non-black things and checking them for non-ravenness. The number of non-black things in existence is many times larger than the number of ravens. Accordingly, the better procedure is to go about sampling the population of ravens than to go about sampling the population of non-black things. Whichever way we go about sampling, the community of inquirers will eventually find a non-black raven (viz., one that is albino or leucitic). But we will likely correct our errors more quickly by predesignating that we will check ravens for blackness than by predesignating that we will check non-black things for non-ravenness. I shall have more to say about the paradox of the ravens in Chapter 6.

Uniformities, Laws of Nature, and Guiding Principles

Peirce's appeals to inductive strength and the community of inquirers address another problem. Recall that Peirce holds there are many uniformities in nature, but a uniformity is not always a law. Correlations are not causation. Regularities may be essential or accidental. Many of these uniformities are mere coincidences, such as that none of my family members have been killed by a comet since I was born. There is no causal or explanatory power in the correlation of my birth with family members not being killed by a comet (I presume). My birth has not caused comets to avoid striking my family members. How, then, is it that we sift among the worthless uniformities to find the ones that represent laws of nature? Peirce wonders, "why men are not fated always to light upon those inductions which are highly deceptive" (W 2:269, 1869).

Uniformities such as *my family members have not been hit by a comet since I have existed* will approach a probability of one. But we do not just seek uniformities; we seek explanations for these uniformities. Let us suppose that no one has ever been struck by a comet (some archaeological evidence suggests this is false). Do we explain this by the birth of a child? Or do we explain this by the fact that comet strikes are incredibly rare? What is at issue is which gives us the stronger explanation for the phenomenon under consideration. Does birthing a child somehow prevent death by comet? Or is death by comet highly unlikely in the first place? Since we do not find childless persons repeatedly dying of comet strikes (if they did, this would be a very good argument for procreating!), we conclude it is the latter.

Furthermore, Peirce does not deny that there are guiding principles which appertain to specific lines of inquiry, such as that what is true of the magnetic properties of one sample of copper is true of any other. But he denies that the validity of induction depends on the generic principle that nature is uniform. A man might infer from a series of unlucky Fridays that Fridays are unlucky days. This inference uses the guiding principle that what is true of these Fridays is true of all Fridays. This, according to Peirce,[22] is a valid inference. The problem with the inference is not with its validity but with its strength. The grounds for the guiding principle are exceedingly weak insofar as the sample is limited to the man's own Fridays. Moreover, putting the hypothesis to the test would surely undermine it. There will be found other explanations for the man's ill luck than the fact that it is Friday (perhaps, for example, that he whittles his Fridays away in casinos). For these reasons, the induction is weak albeit valid.

In contrast, Europeans who have only ever seen white swans and only ever heard reports of white swans have a valid and strong inductive inference in favor of the conclusion that all swans are white. This inference employs the guiding principle that what is true of these swans observed with respect to their color is true of all swans. It just so happens that the inference fails. Inductive inferences only approximate the truth. The community of inquirers, with improved sampling, found the proportion of white swans to all swans is less than one. Guiding principles will be used in our inquiries. But these pared-down, specific guiding principles can be put to the test in a course of inquiry. If there are grounds for rejecting or modifying the guiding principle, we hope for and have faith in the loved, unlimited community of inquirers to find those grounds.

Plainly, these considerations put us in view of various objections to the validity of induction. Among such objections are whether there are any laws of nature at all, as well as the circularity objection raised by David Hume. I shall take up those objections in the next chapter.

Conclusion

Peirce recognizes three different species of inductive inference. These species are crude induction, qualitative induction, and quantitative induction.

He maintains that each of these species of inductive inference has the truth-producing virtue of being truth-approximating in the long run. Our inductive inferences are valid because our errors will be corrected by the community of inquirers in the long run. We hope that the community of inquirers will endure into the long run, we put our faith in them to continue such inquiries, and we love them in ways that support such inquiries. Although the validity of induction makes reference to the long run of inquiry, we can also make our inductions strong in the short run. Peirce makes various recommendations concerning how we can strengthen our inductive inferences, but he acknowledges that the specific techniques for strengthening inductions will be dependent on the specific subject of investigation.

6

Induction

Objections and Replies

In the previous chapter, I examined Peirce's account of induction's validity as
well as how we can make our inductions strong. Peirce holds that induction is
valid because it has the truth-producing virtue of being truth-approximating
in the long run. Accordingly, objections to the validity of induction will be
objections to the claim that it is in fact truth-approximating in the long run.
Moreover, as we saw in the previous chapter, Peirce's defense of induction's
validity appeals to an infinitely enduring and loving community of inquirers
in which we put our faith. If there can be no such community, then Peirce's
defense of the validity of induction is in doubt. In this chapter, I consider various objections to the validity of induction and Peirce's (or Peircean) replies
to them. There are twelve such objections. As we shall see, some of these
objections pertain to all the species of induction, whereas others pertain only
to some single species of induction. Among the objections, some were perfectly familiar to Peirce, and he directly addresses them. Other objections,
however, only came to the forefront of philosophical concerns after Peirce's
death, such as Goodman's new riddle of induction. In these latter cases, I shall
endeavor to extrapolate from Peirce's writings and contemporary writers a
reply to such problems.

The Chance-World Objection

The Objection

One objection to the validity of induction is that we may live in a chance
world. Perhaps our universe is a universe of pure chance. Peirce defines
chance as the ratio of favorable cases to unfavorable cases; it is what we today
call the odds. Whereas ascertained probabilities are the ratio of the number
of times the premises of an inference are true and the conclusion is true to the

Peirce on Inference. Richard Kenneth Atkins, Oxford University Press. © Oxford University Press 2023.
DOI: 10.1093/oso/9780197689066.003.0007

number of times the premises are true, chance is the ratio of the number of times the premises of an inference are true and the conclusion is true to the number of times the premises are true and the conclusion is false. If the probability of an inference is 10%, then the chance a person who makes it will be right is one to nine. If the probability of an inference is 90%, then the chance a person who makes it will be right is nine to one.

If this world were a world of pure chance, one might worry that induction could not possibly be valid. John Stuart Mill claims that "any one accustomed to abstraction and analysis . . . will . . . find no difficulty in conceiving that . . . events may succeed one another at random, without any fixed law; nor can anything in our experience, or in our mental nature, constitute a sufficient, or indeed any, reason for believing that this is nowhere the case" (1869, 338). Yet if we live in a chance world, then there must not be any uniformities in nature. And if there are no uniformities in nature, then there are no laws to discover. It follows that using induction to unlock nature's secrets is bound to fail: There may be no secrets to unlock. Induction is supposed to be truth-approximating in the long run. But if there are no truths to approximate, then a fortiori neither can induction approximate the truth in the long run.

Peirce's Reply

Peirce argues that the hypothesis of a universe of pure chance is absurd. There cannot fail to be some uniformities in nature. In fact, we can tease out three arguments for this claim, one from "Grounds of Validity of the Laws of Logic" and two from "The Order of Nature." In the latter, he provides two separate but mutually supporting arguments. I here explain these arguments.

His Reply in "Grounds": Peirce insists that while nature is not uniform, "the special laws and regularities are innumerable" (W 2:264, 1869). This is a very strong claim, applicable perhaps to our universe, but Peirce's argument only shows that there are at least some uniformities. To explain why, he illustrates the point with a checkerboard. We are to imagine an enormously large checkerboard with the squares painted an enormously diverse number of colors. Now suppose that "myriads of dice were to be thrown" on the checkerboard (W 2:264, 1869). Although this is an entirely chance-like event, Peirce argues there is bound to be some uniformity in the distribution of dice. He gives two specific possibilities. Either (a) "upon some color, or shade of color, out of so many, some one of the six numbers should not be uppermost on any die"

(W 2:264, 1869); for example, the number three may show up on no shade of blue. Or (b) "on every color every number is turned up" (W 2:264, 1869). Clearly at least one of (a) or (b) must occur, though not both will. Regardless of which one does occur, however, there will be some uniformity in nature. Both would be uniformities: either all numbers turn up on every color or, for example, no number three turns up on blue. So there cannot fail to be at least one uniformity in nature. As noted earlier, a uniformity in nature may be merely coincidental, merely fortuitous. A uniformity is not the same as a real, operative law of nature. Mere chance could give rise to uniformities (see also W 6:206–207, 1887–1888).

Next, Peirce turns to a consideration of different kinds of universal propositions. He does so because uniformities in nature are stated as A- or E- propositions, such as that all numbers turn up on every color of the checkerboard or, for example, no number three turns up on blue. E-propositions can always be restated as A- propositions; *No S is P* is equivalent to *All S are not P* (which is not to be confused with *Not: all S are P*). Peirce's position is that laws of nature are to be stated as universal claims. However, nature may only conform to that law some proportion of time, for which reason we may still make probabilistic claims about whether some event will occur.

A- propositions are universal propositions of the form *All S are P*. But there are different kinds of universal propositions. We have standard universal propositions, such as that all mammals are animals. There are also plural universal propositions, such as that all wrens and robins are winged, feathered, and beaked. And there are particular universal propositions, such as that all things that are George Washington are things that wear ivory teeth (i.e., George Washington wears ivory teeth).

There can be no true standard universal propositions in a universe of pure chance, for those would be uniformities. Neither, Peirce argues, can there be true plural universal propositions. If there were true plural universal propositions, then arguments from analogy would be successful. For example, I could make this inference:

(1) All wrens and robins are winged, feathered, and beaked.
(2) Jays are winged and feathered.
(3) Therefore, jays are beaked.

One might object that this inference already sneaks laws of nature into both premises. But it can be constructed from individuals, too:

(1) This wren and this robin are winged, feathered, and beaked.
(2) This jay is winged and feathered.
(3) Therefore, this jay is beaked.

Yet we are supposed to be conceiving of a world where induction fails. Recall that this is the challenge Mill sets: to conceive of a world of pure chance wherein induction fails. In this world of true plural universal propositions, induction will not fail because such inferences from analogy would be better than guesses. "[I]f there were a plural universal proposition," Peirce remarks, "inferences by analogy from one particular to another would hold good invariably in reference to that subject. So that these arguments might be no better than guesses in reference to other parts of the universe, but they would invariably hold good in a finite proportion of it, and so would on the whole be somewhat better than guesses" (W 2:266, 1869).

Neither, Peirce argues, could there be true particular universal propositions in a world of pure chance. Let us suppose there is some true particular universal proposition, such as that George Washington wears ivory teeth. It follows that there is nothing which is George Washington and does not wear ivory teeth. But that is equivalent to: All non-ivory-teeth wearing things are non-George-Washingtons. And that is a true universal proposition, a uniformity of nature discoverable by induction. In order to avoid this, what we must imagine is that both combinations of properties appear. In that case, there will be both an ivory-teeth-wearing George Washington and a non-ivory-teeth-wearing George Washington. However, George Washington cannot at the same time and in the same respect be both ivory-teeth wearing and not-ivory-teeth wearing, on pain of contradiction. Accordingly, Peirce claims that "[t]here could, also, be no individuals in that universe" (W 2:266, 1869). Every object must be supposed to be a set of general properties.

Granted that objects can only be sets of general properties, we will also have to suppose that general properties appear in combination with one another with equal frequency. If some combination did not appear and do so with equal frequency, that would be a uniformity knowable by induction. For example, if nothing is A, B, and C, it is a uniformity that everything is non-A-B-C. And there's the rub: In a chance world such as Mill asks us to imagine, every combination of properties will have to occur and occur with equal frequency. But if every combination of properties occurs and occurs with equal frequency, the universe will be utterly systematic. Consequently, "we can suppose [there is a chance world in which induction fails] in general terms, but we cannot specify how it should be other than self-contradictory" (W 2:267, 1869). In other words, a person may aver she is conceiving of a world in which there are no uniformities. But if she is pressed on how she is conceiving that world to be, we will soon find that there are some uniformities in that world.

Peirce's First Argument in "The Order of Nature": In "The Order of Nature," Peirce provides an example akin to the checkerboard case. He draws on the writings of John Tillotson, who was the Archbishop of Canterbury in the early 1690s. Tillotson asks us to imagine a bag full of letters poured out on the ground. Clearly, the letters will not fall into the form of a poem; they will be jumbled and disordered. Peirce, however, claims that "some elements are orderly and some are disorderly," for "[t]he laws of space are supposed ... to be rigidly preserved, and there is also a certain amount of regularity in the formation of the letters" (ILS 153, 1878). Were the letters thoroughly disordered, we should find none of them lying on the same geometric plane or none of the letters appearing the same.

Peirce contends that even a thoroughly disordered and nonuniform arrangement would be systematically ordered. We already have a hint of such a claim from the checkerboard and dice example. It will be a uniformity if on some geometric plane no letter lies, and it will be a uniformity if on all planes some letter lies. Peirce asks us to imagine what an arrangement of total disorder would look like. He first notes that all laws are stated in the form of A- or E- propositions: "Any uniformity, or law of Nature, may be stated in the form, 'Every A is B'; as every ray of light is a non-curved line, every body is accelerated towards the earth's centre, etc. This is the same as to say, 'There does not exist any A which is not B'" (ILS 152, 1878).

When Mill asks us to conceive of a chance world, he is asking us to conceive of a world in which, given any event E_1, some subsequent event E_2 is as apt to occur as not. But this is just to say that the chance of the event is one to one. If the chance of E_2 on condition of E_1 is one to one, then the probability of the argument that E_2 on condition of E_1 is 50%. But then, while we could make no sure predictions about whether E_2 will occur on a given occasion, (a) we will have discovered that the probability is 50% by reckoning together all of our observations, (b) it will be a uniformity of nature that we can make no such predictions better than chance, and (c) ex hypothesi, it is a uniformity of nature that the chance is one to one.

Let us try to imagine a world of pure chance. Peirce regards objects as sets of properties. Let us suppose that there are just three independent properties, A, B, and C. I use 'property' very broadly to mean whatever may be predicated of some thing. I stress these are independent so that there is no contradiction in supposing an object to have any one property without supposing it to have another. For example, both *being cold* and *being hard* may be predicated of ice. An object will consist of a set formed of these properties or their lack, including the null object which has no properties and therefore represents nonexistence.[1] The formula for the total number of possible sets formed from n properties is 2^n, or 8 when $n = 3$. We then will have eight possible objects. (There will be seven possible existent objects, since the null object will be nothing. The numbers could be calculated without the null object, but there will be no need to fuss with it.) Those objects are as follows:

- Object I, which is not-A, not-B, not-C (the null object)
- Object II, which is A, not-B, not-C
- Object III, which is not-A, B, and not-C
- Object IV, which is not-A, not-B, and C
- Object V, which is A, B, and not-C
- Object VI, which is A, not-B, and C
- Object VII, which is not-A, B, and C
- Object VIII, which is A, B, and C

These objects can, in turn, be formed into their own sets or groups of possible objects. If it is helpful, a group of objects might be regarded as an event conceived of as an arrangement of objects.[2] In addition to groups of objects, we can have groups of groups of objects. We might call this a world or nature. In short, properties are collected into objects, which are collected into groups (or events), which are collected into a world (or nature).

Scientific inquirers are mainly interested in the uniformities to be found in this world, not any possible world. As such, they are interested in groups of objects (events). The more objects there are in a group, the more likely it will be that there are some uniform relations, and there must always be some uniformity. However, with the increase of objects, the nonuniformities will increase more quickly than will the uniformities. Note that a world can only have one of each object. If there were two Objects V in some world, then those two Objects V would have the same properties and so be identical (assuming the principle of the identity of indiscernibles). Limiting our considerations to a universe of eight objects, the total number of possible groups (events) will be 2^8, or 256. Any group will have some uniformities among its objects, but it will also have some nonuniformities among them.

For example, say Group 7 consists only of Objects I and VIII. The uniformity will be that the objects share nothing in common. Three nonuniformities will be that Object I differs from Object VIII with respect to A, with respect to B, and with respect to C. However, suppose that Group 39 consists of Objects II and V. In that case, there will be a uniformity in that Objects II and V will be alike with respect to having A and not-C but unlike with respect to having B.

However, the more objects we introduce into a group, while there may be increased uniformities, the nonuniformities will increase more quickly. Suppose we have Group 212 consisting of Objects II, III, and IV. These uniformities will obtain:

- Objects II and III will be alike in being C-less.
- Objects III and IV will be alike in being A-less.
- Objects II and IV will be alike in being B-less.

But the nonuniformities will be greater:

- Objects II and III will be unalike with respect to A.
- Objects II and III will be unalike with respect to B.
- Objects II and IV will be unalike with respect to A.
- Objects II and IV will be unalike with respect to C.
- Objects III and IV will be unalike with respect to B.
- Objects III and IV will be unalike with respect to C.

The more objects we introduce into a group, approaching infinity, the greater will be the uniformities in that group. However, the nonuniformities will outstrip the uniformities. I shall return to this point momentarily and support it with more mathematical rigor.

For the moment, the key point is this. Mill asks us to start by imagining a thoroughly disordered arrangement of objects. In order to imagine a thoroughly disordered arrangement of objects, we will first need to tally up the possible properties in that world. Let n be the total number of properties in the world. I have just supposed $n = 3$, where the three properties are A, B, and C. Any object may either have that property or not have that property, giving us a total of eight possible objects, or 2^3. More generally, the total possible objects for n properties would be 2^n.

Peirce argues that a chance world would have to have all eight of these objects. For our imagined arrangement to be thoroughly disordered, all eight objects would have to exist in it.[3] The reason every object must exist is that if there were no objects which have both A and C, for instance, then that will be a uniformity of nature. It will be the uniformity that "nothing of the sort A is of the sort C, or everything of the sort A is of the sort non-C" (ILS 152, 1878). Each possible combination of properties will have to be instantiated in the chance world. If, say, Object VIII were missing from this arrangement, then it would be a uniformity that nothing is A, B, and C. If every object exists, however, it is patent that our supposedly chance world is actually quite ordered. Peirce writes, "certainly nothing could be imagined more systematic" (ILS 153, 1878). The arrangement of objects is systematic because every possible object is instantiated and every property is instantiated an equal number of times. These are uniformities. A chance world, then, is not an arrangement in which there are no uniformities.

One might wonder: What if the uniformities of nature are themselves changing, such that the probability is 60% at one time, 40% at another time, 10% at a third time, and so on. First, at any given time, it would follow that we are not in a chance world. Second, over all times, if probable inferences were to fail, nature would have to be disordered in such a way that these probabilities ultimately even out to 50%.

Peirce's Second Argument in "The Order of Nature": Peirce carries his first argument from "The Order of Nature" further by offering us a purely mathematical argument for the conclusion that a thoroughly chance world of groups of objects would be impossible. His argument as written is obscure to the extreme. I have found only two other scholars who have tried to explain how Peirce arrives at his claim.[4] I am not confident, however, that either quite plumbs the depths of Peirce's argument. Here is my understanding of it, on which the argument proceeds in four steps.

The first step of Peirce's argument is to recognize that there are properties. As noted, I am using 'property' broadly to include anything that might be predicated of some subject, such as that ice *is cold* and that ice *is hard*.[5] We will presume these properties to be independent. Recall also that Peirce holds properties are collected into objects, objects are collected into groups of objects (events), and events are collected into a world. Although from any set of properties there are many possible worlds, induction ascertains the uniformities of only some world, the actual world. As Peirce states, "if universes were as plenty as blackberries, if we could put a quantity of them in a bag, shake them well up, draw out a sample, and examine them to see what proportion of them had one arrangement and what proportion another," we would have a right to talk of the probability of arrangements of nature as a whole (ILS 141, 1878). But he adds, "even in that case, a higher universe would contain us, in regard to whose arrangements the conception of probability could have no applicability" (ILS 141, 1878).

Let there be n properties. Objects will be sets of properties. Accordingly, the total number of possible objects will be 2^n. The total number of possible groups of objects (events) will be 2 raised by 2^n. Suppose there are three properties. In that case, there are eight possible objects and 256 possible groups of objects. These are the numbers presented earlier. Peirce's example is slightly different. He imagines there are five properties. In that case, there are 32 possible objects and just shy of 4.3 billion groups of objects (exactly: 4,294,967,296 groups of objects).

The second step is the most important of Peirce's argument. He introduces it by noting that "we can never get to the bottom of this question [of the order of nature] until we take account of a highly-important logical principle" (ILS 153, 1878). Peirce argues that for any three objects, there can always be found some property between two of those three objects which they share in common but the third lacks. Peirce states the theorem as "any plurality or lot of objects whatever have some character in common (no matter how insignificant) which is peculiar to them and not shared by anything else" (ILS 153, 1878). Note that this theorem requires a minimum of three objects, the two objects that form the plurality and the other object from which they differ. That object may be the null object. This property they have in common will not always be a simple property, but it may be a compound property. For example, suppose that there are three objects: Objects II, III, and IV, as listed earlier. Object II is but A, III is but B, and IV is but C. IV will differ from II and III in being A-B-less. But if IV is A-B-less, then II and III together are un-A-B-less. Peirce concludes, "[i]t is obvious that what has thus been shown true of two things is, *mutatis mutandis*, true of any number of things. Q.E.D." (ILS 154, 1878). It follows that there must be some uniformity in nature, for every collection of objects must have some property in common, even if that property is insignificant or trivial.

For any set of properties distributed among three possible objects, there is only a certain number of unique arrangements of those properties among those three objects. This is found by calculating the permutations of those properties among the objects. The formula for calculating a permutation with replacement is O^n, where O is the number of objects and n is the number of properties those objects might have. So the total number of unique ways of arranging three properties into three objects is 3^3, or 27. The total number of unique ways of arranging five properties into three objects is 3^5, or 243. If there are eight objects and four properties, the total number of unique ways of arranging them is 4,096.

For step three of the argument, we now neatly line our formulae up:

(1) Let n = the number of properties
(2) The total number of possible objects = 2^n
(3) The total number of possible groups of objects = 2 raised by 2^n
(4) The total number of unique arrangements of n properties among three objects = 3^n

Recall that in a totally disordered arrangement of objects, every possible object must be instantiated. Similarly, for a totally disordered arrangement of groups of objects, every possible group must be instantiated. This will be a thoroughly chance world. It may be helpful to bear in mind the increasing order of properties constituting objects, objects constituting groups of objects (or events), and groups of objects constituting worlds. In imagining a world of pure chance, we must imagine a world in which every object and every group of objects appears, for otherwise there would be some uniformity, as noted earlier.

In keeping with our earlier example, let us suppose there are three properties. Then using our formulae, we get the following:

- Total number of properties = 3
- Total number of objects = 8
- Total number of groups of objects = 256
- Total number of unique arrangements of three properties among three objects = 27

Clearly, the total number of unique arrangements of three properties among three objects (27) is less than the total number of possible groups of objects (256). Yet in a chance world, all 256 groups of objects must be instantiated. Therefore, among those 256 groups, there must be some repetition of these unique arrangements of properties among three objects. But if there is some repetition of these unique arrangements among those 256 groups, there are some uniformities in nature.

Peirce, I noted earlier, runs the argument by imagining five properties. That gives us the following numbers:

- Total number of properties = 5
- Total number of objects = 32
- Total number of groups of objects ≈ 4.3 billion
- Total number of unique arrangements of five properties among three objects = 243

Patently, 243 is much less than 4.3 billion. So there must be some repetition of those unique arrangements of five properties among three objects. Peirce argues, "in a world of 32 things, instead of there being only 3^5 or 243 characters [i.e., unique arrangements of five properties among three objects], as we have seen that the notion of a chance-world requires, there would, in fact, be no less than 2^{32}, or 4,294,967,296 characters [i.e., groups of objects], which would not be all independent, but would have all possible relations with one another" (ILS 154, 1878). The sentence is convoluted and ambiguous. I take it I have shown (a) how Peirce arrives at these numbers and (b) why the emphasis should be laid on the concluding clauses that in the purely chance world of roughly 4.3 billion groups of objects there would be "all possible relations with one another." A thoroughly chance world would require no less than roughly 4.3 billion groups of objects (events), but there are only 243 possible unique arrangements of five properties among three objects. Some of those unique arrangements will have to repeat.

We can run the formulae with other numbers. Let us suppose there are four properties. Moreover, suppose that while there could be sixteen objects, there are only ten. According to our suppositions, this will not be a chance world since not every object is instantiated. The total number of groups of objects will be 2^{10}, or 1,042. The total number of unique arrangements will be 81. Even though this is not a world of pure chance, there will still be a repetition of unique arrangements of the properties among the objects. Alternatively, we could conceive of a world of eight properties and, say, three objects. In that case, the total number of possible groups of objects will be eight, yet the total number of unique arrangements of eight properties among three objects will be 6,561. The possible groups of objects will be less than the total number of unique arrangements of the properties among three objects. However, in this case, we do not have a pure chance arrangement of objects. Consequently, some A- or E- proposition will be true, such as that nothing is A.

What about a universe of only one object? In that case, the universe will be perfectly uniform: Everything will be identical to that one object. What about a universe of only two objects? It will be a uniformity that all objects A are non-B. Similarly, all objects B are non-A.

Finally, one might raise the objection that these uniformities are useless, trivial, or uninteresting. What should it matter if some things share the property of un-A-B-lessness? This brings us to the fourth and final step of Peirce's argument. Having argued that nature has some uniformities, Peirce proceeds to "attempting to imagine a world... in which none of the uniformities should have reference to characters interesting or important to us" (ILS 155, 1878). Peirce makes several points. First, if none of the properties were important to us, there would be "nothing to ask" about. There would be nothing to ask about because there would no uniformities of interest to us so far as we experience the world. Second, no action would have any important consequences for us. In that case, we would be "perfectly free from all responsibility" (ILS 155, 1878). If actions had foreseeable consequences that implied personal or collective responsibilities, these would be important consequences. Third, we would have no motivation, will, thought, or memory, for these would require "a law of our organization" (ILS 155, 1878). Fourth, any objects of sense would be "absolutely transitory" since we would not remember them (ILS 155, 1878). Peirce describes such a world as being like the world experienced by "the mind of a polyp" (ILS 155, 1878) He claims the "interest which the uniformities of Nature have for an animal measures his place in the scale of intelligence" (ILS 155, 1878). Of course, we do have such an interest in those uniformities. Moreover, our intelligence is much greater than that of a polyp.

These considerations point to another way of responding to anyone who doubts that there are uniformities in nature. No one seriously doubts that there are uniformities in nature. Anyone who did would be at pains to explain why she makes plans for the next day, why she reaches for a coffee cup without doubting whether she will be able to pick it up, and why she thinks her memory of having once been burned by a fire is accurate. In "Some Consequences of Four Incapacities," Peirce exhorts us not to "pretend to doubt in philosophy what we do not doubt in our hearts" (W 2:212, 1868). In "The Fixation of Belief" he similarly claims that inquiry requires "real and living doubt, and without this all discussion is idle" (ILS 56, 1877). There is no room for a real and living doubt as to whether there are uniformities in nature of interest to us.

Hume's Circularity Objection

I have just been presenting Peirce's arguments for the conclusion that there are uniformities in nature. Note that this conclusion is different from the claim that nature is uniform. There may be uniformities in nature though nature as a whole is not uniform. Yet some theorists have claimed that every inductive inference turns on the assumption that nature is uniform. Perhaps no argument has exercised the intellects of philosophers thinking about induction more than Hume's objection to the validity of induction on the grounds that any attempt to provide a rational justification for the use of inductive inference is bound to be circular. Peirce addresses this objection directly, and my earlier comments hint at how he replies to the objection. In this section, I shall briefly state Hume's argument. Next, I will examine proposed solutions to the problem, solutions which Peirce rejects. Lastly, I will explain Peirce's preferred solution.

The Objection

As I noted in the previous chapter, induction faces a problem of scope. Our observations are limited to a little corner of a far-flung galaxy for a short period of time. Peirce's appeal to the community of inquirers is intended to address that problem. Yet if we presume to make claims about nature as a whole, we need some way to bridge the gap between the limited scope of our observations and the universe as a whole. In order to do so, it has been thought, we will need to appeal to some principle, such as that nature is uniform. Let us call this the principle of uniformity. Yet once we make an appeal to the principle of uniformity, we will find that inductive inferences are circular. Circular arguments are not logically acceptable arguments, and so inductive inferences must not be valid either.[6]

Induction will require some principle that links our limited testing and sampling to the whole. For example, we will need some way to connect the particular claim that this sample of copper has such-and-such magnetic properties to the universal claim that every sample of copper has such-and-such magnetic properties. But we will not want our guiding principle to be restricted to claims about copper alone. We will need a principle that also allows us to link our observations about gravity in the here and now to gravity in general. Similarly, we will need a principle that allows us to link our observations about the structure of DNA in the here and now to the structure of DNA generally. The principle of uniformity will enable us to make that connection.

The principle of uniformity could be stated in more than one way, but it is a generic claim about nature or the universe as a whole. It may be that nature is uniform, that the future will continue to resemble the past, or that what is true of some part of nature is true of the whole. What is important here is the universality of the claim as applying to the whole of nature both in the here and now and in the past and future.

Overemphasizing the temporal formulation—that the future will continue to resemble the past—can be misleading as not every inductive inference involves such temporal indexes. If a geologist takes a core from an ice sheet and regards it as a representative sample of the surrounding acre, induction on the core does not turn on the assumption that the future will continue to resemble the past. To be sure, the inference implies that if the geologist were to take more cores, those new samples would be uniform with the previous sample. Nonetheless, the inference only turns on the claim that the core is representative of the surrounding area. Peirce realizes this, for which reason he phrases the argument as questioning how by examining a part of a class we can know what is true of the whole of the class, and by study of the past we can know the future. The point is also made by Mill: "It is not from the past to the future, *as* past and future, that we infer, but from the known to the unknown; from facts observed to facts unobserved; from what we have perceived, or been directly conscious of, to what has not come within our experience" (1869, 184). Following Peirce, I shall treat the principle of uniformity as the claim that nature is uniform.

Problems arise when we ask why we should accept the principle of uniformity. Any defense of the principle of uniformity will be either a deductive or an inductive inference. However, the inference patently cannot be deductive. It cannot be deductive because it is conceivable things are otherwise than our limited testing and sampling indicate. Therefore, the principle of uniformity must be supported by an inductive inference. But induction is the very mode of inference that needs to be validated by the principle of uniformity. Consequently, any inductive inference will be circular.

Here is Hume's statement of the circularity objection, quoted at length:

All reasonings may be divided into two kinds, namely, demonstrative reasoning, or that concerning relations of ideas, and moral reasoning, or that
concerning matter of fact and existence. That there are no demonstrative
arguments in the case seems evident; since it implies no contradiction that
the course of nature may change, and that an object, seemingly like those
which we have experienced, may be attended with different or contrary
effects. . . .

If we be, therefore, engaged by arguments to put trust in past experience,
and make it the standard of our future judgement, these arguments must be
probable only, or such as regard matter of fact and real existence according
to the division above mentioned. But that there is no argument of this kind,
must appear, if our explication of that species of reasoning be admitted as
solid and satisfactory. We have said that all arguments concerning existence
are founded on the relation of cause and effect; that our knowledge of that
relation is derived entirely from experience; and that all our experimental
conclusions proceed upon the supposition that the future will be conform-
able to the past. To endeavour, therefore, the proof of this last supposition
by probable arguments, or arguments regarding existence, must be evidently going in a circle, and taking that for granted, which is the very point
in question. (2007, 36–37)

The objection may be put in a form, like so:

(1) Every inductive inference assumes the principle of uniformity.
(2) Inductive inferences that rest on unknown assumptions are not valid
    inferences.
(3) Therefore, inductive inferences are valid only if there is some valid ar-
    gument (deductive or inductive) for the principle of uniformity.
(4) There can be no valid deductive argument for the principle of uni-
    formity, since we can conceive that nature is otherwise than the con-
    clusion of the inductive inference states.
(5) Any inductive argument will assume the principle of uniformity, and
    so be circular.
(6) Circular arguments are not valid.
(7) Therefore, inductive inferences are not valid.

'Valid' here is to be taken in the Peircean sense of (in the case of induction) being truth-approximating in the long run. Inductive inferences resting on unknown assumptions can lead one further from the truth in the long run, which is why (2) is true. Circular arguments function to reinforce what one already believes, which may or may not be true. This is why (5) is true. If we have no rational justification for the principle of uniformity, then we do not know it. If we do not know it, then inductive inferences are not valid.

Failed Solutions

Hume's demand is that we supply some justification for the principle of uniformity, if we are to maintain that inductive inference rests on rational grounds. Peirce was familiar with several attempted solutions to the circularity objection. He rejects them all. I now turn to examining various solutions to the problem which were known to Peirce and why he rejects them.

The Humean Solution: The circularity objection shows only that induction cannot have a rational basis. But it would be wrong to conclude that the absence of a rational justification for the principle of uniformity implies making inductions is unjustified tout court. On one account, to be justified is but to have done something rightly or blamelessly. Provided that inductive inferences are made rightly or blamelessly, even if not on some rational basis, they may be valid. That is, one might reject premise (2) and, with it, (3).

Hume maintained that the principle of uniformity does not need to be proven for inductive inferences to be made validly:

> These two propositions are far from being the same, *I have found that such an object has always been attended with such an effect,* and *I foresee, that other objects, which are, in appearance, similar, will be attended with similar effects.* I shall allow, if you please, that the one proposition may justly be inferred from the other: I know, in fact, that it always is inferred. But if you insist that the inference is made by a chain of reasoning, I desire you to produce that reasoning. The connexion between these propositions is not intuitive. There is required a medium, which may enable the mind to draw such an inference, if indeed it be drawn by reasoning and argument. What that medium is, I must confess, passes my comprehension; and it is incumbent on those to produce it, who assert that it really exists, and is the origin of all our conclusions concerning matters of fact. (2007, 35–36)

Notice that Hume allows the conclusion is justly inferred from the premise. Inference to the conclusion is logically acceptable, only Hume denies it is justified by a chain of reasoning which proves the principle of uniformity.

The task is to find some nonrational basis on which induction may be logically acceptable. One possible nonrational basis is that which Hume proposes: "Custom, then, is the great guide of human life. It is that principle alone which renders our experience useful to us, and makes us expect, for the future, a similar train of events with those which have appeared in the past" (2007, 45). Having seen two events successively conjoined, we begin to anticipate that the second event will occur whenever we see the first. Hume holds that all of our ideas are copies of impressions. Our idea of cause and effect is a copy of the inner impression of anticipation:

> [W]hen many uniform instances appear, and the same object is always followed by the same event; we then begin to entertain the notion of cause and connexion. We then *feel* a new sentiment or impression, to wit, a customary connexion in the thought or imagination between one object and its usual attendant; and this sentiment is the original of that idea which we seek for. . . . [W]e could not, at first, *infer* one event from the other; which we are enabled to do at present, after so long a course of uniform experience. (2007, 71–72)

Hume's solution to the circularity objection faces a serious problem: Appealing to acquired habits requires experience to validly infer causes from effects. Yet inferences of cause and effect are made even at a very young age, before one has developed the requisite habits. Moreover, Hume admits that "animals as well as men learn many things from experience, and infer, that the same events will always follow from the same causes" (2007, 93). Hume also attributes this ability to make such inferences to custom or habit: "It is custom alone, which engages animals" (2007, 94). The problem is that animals make such inferences before they could possibly acquire the relevant experiences. Such inferences as are made by infants and animals are surely logically acceptable. However, they are made on the basis of instinct. They are not made on the basis of experience and consequent to the development of habits.

The Instinctualist or Commonsensist Solution: Hume holds that the principle of uniformity is nonrationally yet customarily acceptable because we have acquired the habit of regarding nature as uniform. Considerations regarding animals and infants suggest, however, that even children and animals make inferences on the basis of this habit without having had the relevant experiences to acquire the habit. Perhaps, then, we have an instinctual disposition to generalize and to make inductive inferences. Dugald Stewart makes the objection to Hume and states the instinctualist (or, as it is more commonly called, commonsensist) position quite clearly in his *Elements of the Philosophy of the Human Mind*:

> Without this principle of expectation [viz. of the continuance of the laws of nature], it would be impossible for us to accommodate our conduct to the established course of nature; and, accordingly, we find that it is a principle coeval with our very existence; and, in some measure, common to man with the lower animals.
>
> . . . [L]ong before the use of artificial signs, and even before the dawn of reason, a child learns to act upon both of these suppositions [that fire scorches and unsupported heavy bodies fall]. In doing so, it is influenced merely by the instinctive principle which has now been mentioned. (1792, 204–205)

Thomas Reid had made the same objection to Hume's position earlier. Noting that the parallel motion of the eyes (i.e., their tracking of an object) cannot be explained as a consequence of custom because then we "should see children, when they are born, turn their eyes different ways" (1785, 237), Reid claims that "previous to custom, there is something in the constitution, some natural instinct, which directs us to move both eyes always the same way" (1785, 238). With respect to "belief of the continuance of the present course of nature," Reid also attributes this to "the effect of instinct, not of reason" since "children and idiots have this belief as soon as they know that fire will burn them" (1785, 457).

One significant problem with both the Humean and instinctualist views is that they appeal to the contingencies of human experience or of human nature in their accounts of why inductive inferences are logically acceptable. However, once the core insights of the theory of evolution are brought to bear on logical questions, one is compelled to conclude that human experience and human nature are insufficient grounds for accepting the principle of uniformity. At best they can only exculpate the reasoner for inferring what nature has led her to infer. Peirce concedes that humans are "in the main logical animals" (ILS 50, 1877) and that "[l]ogicality in regard to practical matters is the most useful quality an animal can possess" (ILS 51, 1877). However, this evolutionary argument extends only to practical matters. Peirce claims that beyond practical matters "it is probably of more advantage to the animal to have his mind filled with pleasing and encouraging visions" and so "natural selection might occasion a fallacious tendency of thought" (ILS 51, 1877).

Peirce worries that our commonsense convictions or instincts, though they be universal, may be mistaken or pathological. He had made this objection as early as 1864, writing, "there is no criterion by which it may be determined whether a given conviction is normal or not. The test of universality and necessity only determines whether a proposition may be derived from real observation, but clearly not whether it is true" (W 1:154, 1864). We may be psychologically or instinctually compelled to make some inferences or accept some principle. It does not follow that such inferences are rightly made nor that the principle is rightly assumed. This is part and parcel of Peirce's unpsychological view of logic, explained in Chapter 1. To be sure, these habits of thought might be inculcated by natural selection. Further, we might discover what these habits are by reflection on the actual processes of reasoning. Moreover, they may be good habits of thought. But a defense of their rightness—and of our blamelessness in using them—demands an answer different from appeals to human experience or to human nature.

The Kantian Solution: Kant forcibly makes a non-Darwinian but no less devastating line of objection to those who would appeal to experience or human nature in accepting the principle of uniformity. In *Prolegomenon to Any Future Metaphysics,* Kant accuses Hume of making what should be objective causal relations into subjective ones. Whatever causality may be, we take it to be a feature of reality or of the way the world is constituted and not of our own minds. Kant writes that Hume

> indisputably proved that it is wholly impossible for reason to think such a connection [of cause and effect] *a priori* and from concepts, because this connection contains necessity. . . . From this he concluded that reason completely and fully deceives herself with this concept, falsely taking it for her own child, when it is really nothing but a bastard of the imagination, which, impregnated by experience, and having brought certain representations under the law of association, passes off the resulting subjective necessity (i.e., habit) for an objective necessity (from insight). (1997, 7–8 [4:257–258])

Notice that Kant concedes Hume has indisputably proved that positing necessary causal relations which hold not just here but everywhere and at all times is not rationally justified. It is, as Kant states, wholly impossible for reason to think such a connection. A defense of the logical acceptability of inductive inference, which relies on acceptance of just such causal relations, must appeal not to reason but to some other principle.

Kant finds that other principle in the concepts of the understanding. The concepts of the understanding are not derived from reason. They are deduced from the table of the functions of judgment. Properly speaking, these functions of judgment are not judgments but the logical forms of propositions. These forms are used both in inference and to synthesize sense impressions: "The same function that gives unity to the different representations **in a judgment** also gives unity to the mere synthesis of different representations **in an intuition**" (1998, 211 [A 79/B104–105]). Kant holds that the concepts of the understanding are used to structure all experience. Their authority derives not from human nature or reason but from the very conditions for the possibility of experience. Induction is justified by the logically constitutive preconditions for experiencing. As with the Humean and Instinctualist solutions, Kant rejects premise (2) and with it (3).

Peirce does not abide Kant's transcendentalism.[7] He rejects Kant's appeal to the constitutive preconditions of experience. Peirce criticizes transcendentalism quite early in his philosophical development. He claims it is but a "peculiar disease of psychology," characterizing Kant as holding that since "[t]here is no possible way of answering such a demand [viz., the demand of saying what validity for truth the mental element has] from *external* studies . . . we must study the action of consciousness in order to solve the problem *quid juris*" (W 1:72, 1861). Peirce objects to such a strategy on three grounds. First, the study of consciousness must itself proceed on the assumptions employed in any study, and so circularity threatens any such inquiry. Second, transcendentalism requires a priori support for claims that cannot possibly be given a priori support, such as why we should trust "the testimony of our impressions" (W 1:73, 1861). Requiring such a priori support "overthrow[s] all science" (W 1:74, 1861). That is, whether the testimony of our impressions on some occasion is to be trusted cannot be determined transcendentally but must be found out in the normal course of inquiry. Indeed, that is partly what science is for: To sort out those sense impressions that are misleading from those that are not. Third, and related to the second point, our concepts are themselves abstractions from sense impressions. Just as it is the function of science to sort through our impressions, it is the function of science to sort through our conceptions.

Justification by Abduction?: A fourth way of addressing the argument is to change the parenthetical claim in (3) that the justification for the principle of uniformity must be a deductive or inductive argument. We have already seen in Chapter 2 that Peirce holds there are three genera of inference. Those three genera of inference are deduction, induction, and abduction. Perhaps the argument for the principle of uniformity is abductive. If there is an abductive argument for the principle of uniformity, and if abduction does not assume the principle of uniformity, then perhaps we can know that nature is uniform.

I shall have more to say about abduction in the next chapter, but this attempted defense of the principle of uniformity faces two problems. First, on Peirce's theory of abduction, abduction does not justify claims at all. It only suggests hypotheses that might be put to the test. Peirce states the conclusion of the form of an abductive inference as "there is reason to suspect that [hypothesis] A is true" (EP 2:231, 1903). He waffles on whether it consists also in adopting the explanatory hypothesis for further probation. Second, if we do adopt the hypothesis for probation, we again face Hume's problem of induction. Suppose we adopt the hypothesis that nature is uniform for further probation. If we do, then further probation of that hypothesis will employ inductive inferences to confirm the hypothesis. Those inductive inferences, however, will employ the very hypothesis we are endeavoring to support. If we put the hypothesis that nature is uniform to the test, we will be assuming it in our inductions.

The Millian Solution: John Stuart Mill proposes a different response to the circularity objection, viz., to reject premise (6), that circular arguments are not valid. He concedes that every inductive inference uses the principle of uniformity. After defining induction as generalization from experience, Mill proceeds to note, "there is a principle implied in the very statement of what Induction is . . . namely, that there are such things in nature as parallel cases; that what happens once, will, under a sufficient degree of similarity of circumstances, happen again, and not only again, but always" (1869, 183–184). He states the principle as "the proposition that the course of nature is uniform" and describes it as "the fundamental principle, or general axiom of Induction" (1869, 184). He follows the logician Richard Whately in treating it as the suppressed "ultimate major premiss of all inductions" (1869, 185).

Though Mill variously describes the principle of uniformity as a suppressed premise, a maxim, and an axiom, he also claims that the principle of uniformity is proven by induction. He maintains that while it "would yet be a great error to offer this large generalization as any explanation of the inductive process," it is nevertheless "itself an instance of induction, and induction by no means of the most obvious kind. Far from being the first induction we make, it is one of the last, or at all events one of those which are latest in attaining strict philosophical accuracy" (1869, 184). Mill contends that the principle of uniformity is both (a) used in every induction and (b) in the process of being used is proven by induction as well. The principle of uniformity is ultimate in two senses. Logically, it is initially assumed in every induction. Temporally, it is finally supported inductively by our observations and scientific successes.

Even though the principle of uniformity is ultimate in these two senses, patently the argument for it will be circular. The inductive argument based on our observations and scientific successes will assume that that nature is uniform. But can circular arguments be valid? If not, then this reply to the circularity objection cannot succeed.

Peirce denies that circular arguments are valid. He claims that treating the principle of uniformity as a suppressed major premise is an "unpardonable blunder" and criticizes Mill by noting that "if this were not inconsistency enough, he goes on to say that this Law is proved by Induction; that is, by a syllogism of which it is itself the major premiss" (ILS 221, 1909). I will say more about that unpardonable blunder later. What I wish to stress here is Peirce's insistence that such a circular argument is unacceptable.

Peirce classifies such an argument as committing the fallacy of petitio principii, or "assum[ing] as a premiss what no intelligent man who doubted the conclusion could know to be true" (CP 2.614, 1901). In 1865, Peirce skewers Mill, expressing his surprise that "a sturdy English matter of fact logician should seriously attempt to show that a *petitio principii* is not a fallacy" (W 1:219, 1865; see also W 1:414, 1866). Inferences proceed from the (presumed to be) known to the unknown. Now imagine a person doubts the conclusion that nature is uniform. If so, that person would doubt the ultimate major premise of the inductive inference, for that ultimate major premise is none other than the conclusion itself. In such a case, the inference in question would be from the unknown to the unknown. This is not much of an inference. In fact, Mill himself glosses inference as "progress from the known to the unknown" (1869, 122). By his own lights such a circular argument as proving the principle of uniformity by using the very principle could not satisfy his conception of inference. If the premise is known, so, too, will the conclusion be known. If the premise is unknown, then so, too, will the conclusion be unknown. Here is Peirce putting the matter to rest:

> We ask Mr. Mill what is the ground of Induction. The law of Causality. And what is the ground of causality? Induction. But how could the law be acted upon before induction established it; in other words what ground had men for making inductions before induction was proved valid by induction? He answers they had the ground of induction.
>
> It is all the reason he can yield. It seems to me to amount to *none.* (W 1:223, 1865)

The Sigwartian Solution: Thus far, we have considered two strategies for replying to the circularity objection. The first strategy was to deny premise (2) and with it (3) by holding that we can accept the principle of uniformity on some grounds, only those grounds do not supply a reasoned justification for the principle. The second strategy was to deny that circular arguments are invalid, that is, to reject (6). A third strategy is to accept that we have no justification for the principle of uniformity. Therefore, the conclusion of the circularity objection is correct in the sense that we do not know inductive inferences to be valid. But that we do not have a justification for the principle of uniformity does not imply the principle is false. We might be warranted in using the principle of uniformity nonetheless, to act as if we do have a justification for the principle of uniformity. We may not know the principle of uniformity to be true, if knowledge requires some argument concluding that nature is uniform. But even if we do not know the principle of uniformity to be true, we might still be warranted in using it and in making inductive inferences.

Peirce was not familiar with the many different and nuanced ways in which contemporary philosophers have endeavored to defend induction by regarding use of the principle of uniformity to be warranted. He was, though, familiar with the work of Christoph von Sigwart, who holds a similar position. In a review of Sigwart's *Logic*, Peirce notes that Sigwart holds the principle of uniformity is a regulative principle of inquiry. Peirce likens such regulative principles to assumptions made toward the end of the card game whist. Toward the end of the game, presumptions are made "not because we possess any evidence of their truth, but because if such a proposition be true, there is a way of winning a trick that, under the supposition of its falsity, there is no way of winning" (CN 2:95, 1895). Similarly, we may hold that nature is uniform just because assuming it gives us a way of winning the truth, whereas not assuming it gives us no way of winning.

Peirce criticizes this account on the grounds that "such a regulative principle cannot, *as such*, be universal, because it applies in that capacity only to the single case in hand" (CN 2:95, 1895). The principle of uniformity is supposed to underwrite all inductive inference. Yet if the principle is but a regulative principle, then it cannot underwrite all inductions. It can only underwrite those inductions for which assuming the principle is requisite for us to secure the truth, whereas if we do not assume it, we will lose out on those truths. If treated as a regulative principle, the use of the principle of uniformity will be restricted to instances in which it is pressing in the short run that we make the induction. The result is that if the principle of uniformity is a regulative principle, then "all ground for asserting [it] to be absolutely without exception is removed. Now that would profoundly modify not only our theory of the universe, but also our philosophy of life and death" (CN 2:95, 1895). Moreover, Peirce denies that in pure scientific inquiries there is ever such a pressing need: "But pure science has nothing at all to do with *action*. The propositions it accepts, it merely writes in the list of premises it proposes to use. Nothing is *vital* for science; nothing can be" (RLT 112, 1898).

Whately's Suppressed Premise: That the principle of uniformity might be used with warrant only in a specific case of inquiry raises the question of whether such a broad principle is even called for in a specific case. This brings us to Richard Whately's *Elements of Logic*. Whately makes two claims. The first is that every inductive inference can be stated in the deductive syllogistic form Barbara "with the major Premiss suppressed; that being always substantially the same," as it asserts, that "what belongs to the individual or individuals we have examined, belongs to the whole class under which they come" (1850, 208–209). The second is that the exact statement of the relevant principle will vary depending on the field of inquiry and so "[w]hether the Premiss may fairly be assumed, or not, is a point which cannot be decided without a competent knowledge of the *nature of the subject*" (1850, 210).

Peirce objects to the first claim because it wrongly makes all inductions into deductions. He insists that "*Barbara* particularly typifies deductive reasoning; and so long as the *is* is taken literally, no inductive reasoning can be put into this form" (ILS 168, 1878). In Whately's example 'is' is not taken literally as ascribing a predicate to a class or individuals. Rather, the 'is' has the sense of 'is like.' That is, the individuals sampled are like the individuals of the class more broadly. Moreover, deduction differs from induction in that the former "rest[s] solely upon the facts observed," whereas the latter takes account of "the manner in which those facts have been collected" (CP 2.766, 1905, and see ILS 146, 1878). It matters to quantitative induction how my samples are collected. But if inductions could be treated as deductive syllogisms in the form of Barbara, how I collect my sample would be a matter of indifference. The conclusion would follow from my premises regardless.

Although Peirce objects to Whately's first claim, it is the second claim which is more pressing in the present context. Whately holds that the exact statement of the principle of uniformity varies by inquiries. Whately writes that an inquirer's "*skill as a Naturalist* is to be shown in judging whether these animals [sheep, cows, etc.] are likely to resemble in the form of their feet all other horned animals; and it is this exercise of this judgment, together with the examination of individuals, that constitutes what is usually meant by the *Inductive process*" (1850, 211). Whately is of two minds. On the one hand, he regards the principle of uniformity as "substantially the same" in every inductive inquiry. This suggests the principle is generic. On the other hand, he indicates that the exact formulation of the principle will vary depending on the area of inquiry. This suggests the principle is a specific guiding principle.

However, the specific guiding principle used in an inquiry cannot be generalized willy-nilly, as if the principles used in specific inquiries were merely specifications of the principle of uniformity. Consider Peirce's example that *the magnetic properties of this sample of copper are the properties of all samples of copper*. This cannot be generalized willy-nilly, for the magnetic properties of copper are not the same for all of its alloys, such as brass. Neither can the principle of uniformity be willy-nilly specified. From the principle that nature is uniform, we would be wrong to specify that what is true of the properties of one metal is true of every other metal. Although Whately does not acknowledge this directly, an example he gives suggests as much. He notes that if a person meets with ill luck on Fridays, the person may conclude that Fridays are unlucky days. Whately claims that though one would object to the inference "it would not be, as an *argument illogical*; since the Conclusion *follows* fairly, *if* you *grant* his implied Premiss, that the events which happened on those particular Fridays are such as must happen on all Fridays; but we should object to his *laying down this* Premiss" (1850, 211). That is, the specific guiding principle about Fridays is false. But the specific guiding principle would not be false if we could willy-nilly specify the claim that nature is uniform. These considerations bring us to Peirce's solution to the circularity objection.

Peirce's Reply to the Circularity Objection

Peirce rejects premise (1) of the circularity objection and with it (3). Peirce makes a point similar to Whately's. Although we may maintain that what is true of the magnetic properties of one sample of copper is true of all samples of copper, we cannot generalize this to the claim that what is true of the magnetic properties of some metal or alloy is true of all metals or alloys. A rotating copper disk will stop spinning if placed between the poles of a magnet, but a brass disk will not (see ILS 51, 1877). This is Peirce's conception of a guiding principle. The guiding principle in induction will be specific to a topic of inquiry. Peirce credits William Whewell for making this point explicit. It is not the principle that nature is uniform upon which induction proceeds but "upon the student's coming to his subject provided in advance with appropriate ideas—a view to which the history of science since 1837 (particularly Darwinian ideas and those of physiological psychology) has brought much additional support" (CN 2:176, 1898). It is the specific guiding principles, the appropriate ideas to the question or topic of inquiry, that are employed in induction. A broad commitment to the uniformity of nature plays no important role in scientific inquiry.

Note that Whately concedes the one who infers from his unlucky Fridays to the conclusion that Fridays are unlucky days makes that inference logically. That is, his argument is valid. His argument is not strong, however. Given the evidence he has considered, the man has concluded Fridays are unlucky. But he has not taken care to survey the Fridays of others or to reflect on all the Fridays of his life. The inference he makes from the sample considered in isolation is valid. The problem is that the sample is biased, and that bias vitiates his inference. The inference is valid relative to the information he uses, but it is weak.

The generic principle that nature is uniform does not do the work in inductive inference, as the example of unlucky Fridays suggests. Rather, the specific guiding principles do the work. Specific guiding principles link the sample in question to the population as a whole. However, they are not generic since they do not concern nature as a whole. Specific guiding principles, principles specific to the matter under consideration, do the work in scientific inquiries. Moreover, they are subject to refutation in the course of those investigations. The community of inquirers, in the course of their investigations, will find what are the magnetic properties of copper and its alloys, and similarly for other contexts of inquiry. In so doing, they will sort through the guiding principles used in various inductive inferences and either eliminate, corroborate, or confirm them in the course of inquiries.

Yet one principle that they will not use in such inquiries is the generic principle that nature is uniform. In fact, Peirce argues that the principle of uniformity is false. If a person accepts premise (1) of the circularity objection, so much the worse for induction. Recall that the claim *nature is uniform* is importantly different from the claim *there are uniformities in nature*. Specific guiding principles asserting uniformities in nature—such as that what is true of the magnetic properties of one sample of copper is true of any other sample of copper—will be used in inductive inferences. The principle that nature is uniform will not be used. Peirce's argument against the principle of uniformity has three steps. First, he argues that nature is not uniform. Second, he argues that even if it were uniform, we could never discover it is. Third, he argues that even if nature were uniform and we could discover that nature is uniform, it would be useless for inquiry.

Nature Is Not Uniform: First, Peirce argues that nature is not uniform. In "Grounds of Validity of the Laws of Logic," Peirce argues, "[n]ature is not regular. . . . It is true that the special laws and regularities are innumerable; but nobody thinks of the irregularities, which are infinitely more frequent" (W 2:264, 1869). Peirce is affirming that there are specific uniformities, or special laws. He is denying that nature is uniform overall. His comment about the innumerable uniformities and infinitely more frequent irregularities may incline us to think his argument depends on the universe being infinite. If it did, his argument would rest on a doubtful assumption. As I shall show, his argument does not turn on the assumption nature is infinite.

There are two ways to think about Peirce's argument. The first way is more modest and proceeds from straightforward reflection on the world. The second, I think, is what Peirce intended to be his main argument. It proceeds on a more mathematical basis, the foundation of which I have already laid in the earlier discussion of the chance-world objection. One way of understanding the claim that nature is uniform is that "[e]very fact true of any one thing in the universe is related to every fact true of every other. But the immense majority of these relations are fortuitous and irregular" (W 2:264, 1869). Take any two facts. There is not always some uniformity or regularity in their relation. It rains in Portugal, and a woman stubs her toe in New Zealand. A cat meows in Sweden, and a flower blooms in the Falkland Islands. As Peirce writes, "[a] man in China bought a cow three days and five minutes after a Greenlander had sneezed" (W 2:264, 1868). These events may occur and may occur successively, but there is no uniform relation between them. Therefore, nature is not uniform in the sense that every fact is related to every other fact.

A second way to understand the claim that nature is uniform is that the uniformities of nature outstrip the nonuniformities. Having rejected the first conception, Peirce remarks, "[t]he orderliness of the universe, therefore, if it exists, must consist in the large *proportion* of relations which present a regularity to those which are quite irregular" (W 2:265, 1869). The problem is that the number of nonuniformities is so large that that proportion will be exceedingly small. This is evident from straightforward reflection on our experience. The relation between any two arbitrarily selected facts—such as a cat's meow in Sweden and a blooming flower in the Falkland Islands—will likely not be uniform.

We can present Peirce's argument more mathematically. If the uniformity of nature consists in the proportion of uniformities to nonuniformities, nature is not uniform. The greater the number of objects, the more quickly will nonuniformities outpace uniformities. Our universe consists of a great number of objects. The more objects we introduce into the world, while uniformities will be found, their proportion to nonuniformities will increase at a faster rate. The formulae given earlier show why this is so. The proportion of unique arrangements of properties among three objects (for three properties, 27) to total possible groups of objects (for three objects, 256) shrinks when the number of properties increases. For three properties, it is 27/256, or roughly .105. For five properties, it is 243/~4.3 billion, or roughly .000000057. The proportion only shrinks with more properties. So, if the uniformity of nature consists in the proportion of uniformities to nonuniformities, this proportion shrinks as we increase the number of properties, objects, and events in the world. Since our universe consists of very many properties, objects, and events, the proportion is "as small as it can be; and therefore, the orderliness of the universe is as little as that of any arrangement whatever" (W 2:265, 1869). Consequently, nature is not uniform, if the uniformity of nature consists in the proportion of uniformities to nonuniformities. Nature is not uniform, but that does not entail there are no uniformities in nature.

Furthermore, if nature is collectively uniform, then there will be some grand plan to its design. For example, if God has providentially ordered the universe to a particular end and is bringing that end about according to an exact blueprint in the divine mind, then nature is collectively uniform. If the universe is infinite, Peirce holds that we could never discover that nature is uniform because we cannot know how much of it we have observed. If the universe is finite, Peirce claims, "[t]he universe ought to be presumed too vast to have any character" (ILS 162, 1878). He notes that assertions as to the benevolence of the universe are ill-founded because "beneficences, justice, etc., are of a most limited kind—limited in degree and limited in range" (ILS 163, 1878). Moreover, if proposals that the universe is divinely ordered be made, the presumption ought to be against them. Instead, we should first examine "whether such relations are susceptible of explanation on mechanical principles, and if not they should be looked upon with disfavor as having already a strong presumption against them; and examination has generally exploded all such theories" (ILS 163, 1878). We ought to proceed on the presumption that the universe can be explained by known laws rather than invoke a supernatural divinity to explain it.

Although Peirce is here inclining against religious belief, he does make a hedge: "It would be extravagant to say that science can at present disprove religion; but it does seem to me that the spirit of science is hostile to any religion except such a one as that of M. Vacherot" (ILS 163, 1878). Vacherot holds that religious symbols are syntheses of the imagination and intellect. Those symbols should be interpreted as standing for a moral vision that underlies the major monotheistic religions. As a consequence, he contends, the objections of science to religion and of religion to science rarely hit home. The objections of science strike at the high metaphysics of religion, which are but symbols of the more vital underlying moral message. The objections of religion fail to appreciate that science is aimed at describing the world as it is and not as it should be. Consequently, religion should have no fear that science will undermine its basic moral vision. Peirce was a theist, and much of his emphasis on religion is on its moral message of love. He claims that the moral message of Christianity "is contracted to a rule of ethics, it is: Love God, and love your neighbour" (CP 6.441, 1893). Furthermore, loving God is "accomplished by each man's loving his neighbour" (CP 6.443, 1893).

In sum, if nature is uniform, then we should find (a) that every fact of nature is uniformly related to every other, or (b) that the uniformities in nature quite outstrip the nonuniformities, or (c) there is some design to the whole. But we find none of (a)–(c). A Swedish cat's meow is not uniformly related to a blooming flower in the Falklands. The numerosity of such nonuniformities quite outstrips the uniformities we do observe. Moreover, there is no evident plan to the universe as a whole that has gained scientific support. Therefore, nature is not uniform, or at least there ought to be a scientific presumption against it.

Even If Nature Were Uniform, We Could Not Discover It: Let us suppose some person were interested in coming to know whether nature is uniform. Would they be able to discover whether nature is uniform? In "Grounds," Peirce argues that no, they would not. He remarks, "even if there were such an orderliness in things, it never could be discovered" (W 2:265, 1869). Although he touches on the matter in his later "The Order of Nature," in which he discusses related themes, he seems disinclined to pursue it fully.

In "Grounds," Peirce argues that such uniformity would either belong to the system of nature as a whole (collectively) or to each part of it (distributively). The distinction between collective and distributive predicates is easily grasped by considering how predication works differently between these claims:

- These sheep are a flock.
- These sheep are cloven-hoofed.

The first is an example of collective predication, for being a flock does not belong to each sheep individually but to them as a whole or as a collection. The second is an example of distributive predication, for being cloven-hoofed belongs to each sheep individually.

Suppose that uniformity belongs to nature collectively. Peirce argues that in order to know that nature is uniform, we would have to see "some considerable proportion of the whole" (W 2:265, 1869). The problem is that we cannot know "how great a part of the whole of nature we have discovered" (W 2:265, 1869). Consequently, we cannot know whether the whole is uniform or just the part we have observed.

Suppose that uniformity belongs to nature distributively. In that case, we might know that the diverse parts of nature we have observed are uniform even without having observed a considerable proportion of the whole. Peirce argues that were nature uniform, we could not know it by observing parts of it because properties are known to belong to things only by contrast: "a character can only be known by comparing something which has it with something which has it not" (W 2:265, 1869). If each part of nature were uniform, we would not have a nonuniform part of nature with which to compare it. Consequently, we cannot discover that nature is uniform.

I doubt that we should put much stock in these arguments. For this reason, I submit, Peirce does not pursue the matter as fully in "The Order of Nature." There are three reasons for doubting his arguments from "Grounds." First, if we are only warranted in using the principle of uniformity, then we do not need to know or discover it is true. Second, if nature is distributively uniform, the parts of nature may be governed by different laws. *Nature is uniform* will be true since each part of nature is uniform. Nonetheless, we could contrast the ways in which some part of nature is uniform with the uniformity of other parts because they will not all be uniform in the same way. Third, Peirce provides no argument for his claim we cannot know how much of nature we have discovered. He seems to be proceeding on the assumption that nature is infinite. A comment from "The Order of Nature" suggests as much. He indicates that if the universe is "quite boundless in space and in time," then "since the proportion of the world of which we can have any experience is less than the smallest assignable fraction, it follows that we could never discover any *pattern* in the universe except a repeating one; any design embracing the whole would be beyond our powers to discern" (ILS 161, 1878). Yet he also acknowledges that the universe may be "of limited extent and finite age" (ILS 161, 1878). He concedes the evidence "rather favors the idea of a beginning than otherwise" (ILS 162, 1878). Worries about the universe being infinite in scope, and consequently about our ability to discover real uniformities in nature, are addressed by the fact that the universe is finite.

The Uselessness of the Principle of Uniformity: All of nature is uniform—is such a principle utile in the sciences? Peirce answers it is not: "It would not explain how knowledge could be increased (in contradistinction to being rendered more distinct), and so it would not explain how it could itself have been acquired" (W 2:265, 1869). There are three problems concerning the utility of such a generic principle as that nature is uniform.

The first problem is that it is ill-adapted to explain how statistical induction actually works. Statistical inductions are proportionate. They do not assert that nature is everywhere uniform. In fact, the claim that nature is uniform "hardly seems adapted to the explanation of this *proportionate* induction [of sampling beans from a bag], where the conclusion, instead of being that a certain event uniformly happens under certain circumstances, is precisely that it does not uniformly occur, but only happens in a certain proportion of cases" (ILS 157, 1878). A similar consideration applies to the example of it raining when the clouds are dark and heavy and the atmospheric pressure drops. If nature were uniform, then it should always rain when those conditions are met. When one makes such an inference and says that rain is likely, the person is acknowledging precisely that nature is not uniform. Sometimes it does not rain when the clouds are dark and heavy and the atmospheric pressure drops. For this reason, Peirce typically prefers the terminology that certain events in nature are regular rather than uniform. It is a uniformity that it rains when the clouds are dark and heavy and the atmospheric pressure drops. Nature, however, does not always conform to this uniformity. Rather, nature conforms to it regularly, say 90% of the time.

The second problem is that even if nature is uniform, all such a statement permits us to infer is that nature everywhere follows regular laws or patterns. It does not tell us what those laws or patterns are or which parts of nature are governed by which laws. Scientific inquirers are not interested in showing that nature is uniform as a whole. They want to know what laws govern observed natural phenomena. The mere fact nature is uniform is of no aid in discovering those laws.

Even worse, some apparent uniformities may be mere coincidences not explicable as laws of nature. The mere fact that a scientist has found a uniformity of nature is no evidence that it is a law of nature. In an undated manuscript, Peirce writes, "at one time Mill speaks of 'uniformities' as if he meant what others call laws of nature, . . . but [at other times] that they are merely the bringing together by the mind of similar facts. But if the phenomena explained as due to laws of nature had no other bond of connection than our own classing them together, their resemblance would be merely fortuitous" (PSR 38). Cornelius Delaney has noted that Peirce "is perfectly willing to grant that there are regularities in nature, but insists that it is only in the uninteresting sense in which there are some regularities in any collection" (1973, 438).

What the scientist aims to do is to sort out those uniformities that explain observed phenomena from those which do not because they are merely coincidental. As Peirce states in a later work, "[t]he problem of how an accidental regularity can be distinguished from an essential one is precisely the problem of inductive logic" (CP 3.605, 1903). He identifies two characteristics we can use to sort laws from accidental regularities. First, a law expresses a "generalization from a collection of results of observation *gathered* upon the principle that the observing was done so well as to conform to outward conditions; but not *selected* with any regard to what the results themselves were found to be" (EP 2:67, 1901). Second, a law expresses "neither a mere chance coincidence among the observations on which it has been based, nor is it a subjective generalization, but is of such a nature that from it can be drawn an endless series of . . . predictions, respecting other observations not among those on which the law was based; and experiment shall verify those [predictions] . . . in the main" (EP 2:68, 1901). These quotations suggest a point noted in the previous chapter: The sorting of mere correlations or coincidences from laws pertains to the strength of induction rather than its validity.

The third problem with using the principle of uniformity is that scientists do not in fact employ the principle that nature is uniform. Rather, they employ more specific principles, such as that the sample of copper under consideration is representative of all coppers, and they employ the techniques of statistical induction. To continue with Whately's example of the naturalist studying horned animals, the guiding principle in question is not the uniformity of nature but that the horned animals sampled are representative of the population of horned animals as a whole. It is specific guiding principles, supplemented by sound practices of inquiry, that do the work. The generic principle that nature is uniform is inutile in inquiry. As Peirce writes, "to say that inductions are true because similar events happen in similar circumstances . . . is to overlook those conditions which really are essential to the validity of inductions" (ILS 158, 1878). We draw inductive inferences not using the claim that nature as a whole is uniform but that that into which we are inquiring has some uniformity or regularity, which may be the frequency with which one event follows another. This guiding principle is specific, not generic. The specific guiding principle is used in inductions, not an appeal to the uniformity of nature as a whole. Peirce is careful to point out that though the inference may be drawn on the basis of the specific guiding principle and be valid, our inquiries need to be supplemented by sound practices of inquiry such as random sampling and predesignation. "If we limit ourselves to such characters as have for us any importance, interest, or obviousness," he writes, "then a synthetic conclusion may be drawn, but only on condition that the specimens by which we judge have been taken and random" and "induction only has its full force when the character concerned has been designated before examining the sample" (ILS 158, 1878). Then the inference gains in strength. The validity of induction, however, does not repose on the truth of the specific guiding principle but on the fact that if the specific guiding principle is wrong the community of inquirers will discover that it is wrong.

Summary: Peirce's reply to Hume's circularity objection is to contend that inductive inference does not depend on the principle of uniformity. Rather, persons who draw inductive inferences will use specific guiding principles related to the subject under examination. Those specific principles can be eliminated, corroborated, or confirmed by the community of inquirers in the process of inquiry itself. Moreover, if inductive inference did assume the principle of uniformity, so much the worse for induction. Not only is the principle false, it is useless.

The End-of-Community Objection

The Objection

I have just explained Peirce's argument for the conclusion that even if nature were uniform, we would not be able to discover it. One reason for this is that we can never be sure how large of a proportion of the whole of nature we have observed. Yet this objection would seem to generalize to any sort of guiding principles, even the specific ones that Peirce claims are used in inductive inquiries. Perhaps even these are undiscoverable by us, given how vast the universe is and how limited the scope of our inquiries is. In that case, perhaps inductions are not truth-approximating in the long run just because our observations will be limited.

In the previous chapter, I explained Peirce's reply to this objection in his discussion of the unlimited community of inquirers. Peirce denies that the worth of an induction consists in the inquiries of a finite inquirer at a given time. Rather, the value of our inferences in the here and now consists in how they contribute to larger research projects into the future. Those larger research projects are those of an indefinitely extended community of inquirers embracing all rational life. Consequently, even inductions based on very limited testing and sampling in the short run may be valid. We may draw inductive conclusions in the short run (rather than suspend judgment indefinitely into the future) for the purpose of aiding future inquirers. Although we may be wrong with respect to any given inference at any given time, the community of inquirers will be able to correct our errors. Moreover, provided the induction is strong, we may be confident that the conclusion approximates the truth.

Yet the objection might be pressed further on the grounds that the community of inquirers will come to an end. Surely humans will one day go extinct. We have yet to find any other rational life in the universe, and this fact supports the crude induction that there is no other rational life in the universe. Moreover, on some theories, as the universe expands, any rational life is bound to go extinct. Accordingly, Peirce's appeal to a community of inquirers is at variance with the facts: There is and will be no such infinite community to address the fact that our observations are limited in scope. In the long run, we are all dead. Therefore, we can have no assurance that induction will be truth-approximating in the long run.

Peirce's Reply

Peirce admits human inquiries may come to an end. He denies that there is a good reason to think rational life and inquiry will come to an end. As he states in "The Doctrine of Chances," "there exist no reasons, and a later discussion will show that there can be no reasons, for thinking that the human race, or any intellectual race, will exist forever. On the other hand, there can be no reason against it" (ILS 117, 1878). Late in his life, amid his musings about God and the universe, Peirce reports:

> I look at the stars in the silence, thinking how each successive increase in the aperture of a telescope makes many more of them visible than all that had been visible before. The fact that the heavens do not show a sheet of light proves that there are vastly more dark bodies, say planets, than there are suns. They must be inhabited, and most likely millions of them with beings much more intelligent than we are. (CP 6.501, 1906)

Peirce was a practicing astronomer; the sole monograph he published in his life was *Photometric Researches*. His reference to a sheet of light hints at his solution to Olbers's paradox. Olbers's paradox is that if space is infinite and static and stars are uniformly distributed, then the night sky ought to be a sheet of light. Every ray we draw outwardly from a point on the earth should reach a star, and so light from that star should also reach the earth. Peirce's solution to the paradox is, evidently, that planets occlude some of the light from stars. (Today, the paradox is typically explained by holding that the universe is finite and expanding.) As this quotation indicates, Peirce suspects that there are exoplanets (scientists only confirmed there are in 1992). Given their sheer quantity, Peirce muses some of them harbor intelligent life. The community of inquirers extends to those alien rational creatures, too.

Yet more importantly, Peirce's reply to the worry that our observations are limited is not that there is or will be some such community. His reply is rather that we must hope there will be some such community, have faith in it, and love it. The objection only requires that we postulate some community of inquirers. But one will rightly wonder on what grounds we may postulate such a community. That there can be no rational justification for belief in the community of inquirers is evident. Any such argument would be inductive. But the validity of induction depends on a community of inquirers. Hence, an inductive argument for a community of inquirers will assume that there is a community of inquirers to correct the error. This is patently a circle, a Peircean version of the circularity objection. It applies just as directly to the conclusion that the community of inquirers will come to an end. An inductive inference is valid only if there is a community of inquirers that will be able to correct the error in the long run. But any argument against the continued existence of the community of inquirers purports to show that there will be no such community. Consequently, the inductive argument is valid only if induction is invalid. For these reasons, there can be an argument neither for nor against an enduring community of inquirers.

Peirce denies that induction's validity is underwritten by some argument for the claim that there will be an infinitely expansive and enduring community of inquirers. Postulation of an infinitely expansive and enduring community of inquirers is not rationally justified. Rather, logical considerations drive us to hope that there will be some such community. Logical considerations compel us to put our faith in our fellow inquirers. Logical considerations require that we endeavor, so far as we can, to imitate the heroic inquirer in setting aside our personal interests and trivial concerns. These are logical sentiments, which logical considerations lead us to adopt. They are not rational justifications for belief in the community of inquirers. In "The Doctrine of Chances," Peirce writes, "as the whole requirement is that we should have certain sentiments, there is nothing in the facts to forbid our having a *hope*, or calm and cheerful wish, that the community may last beyond any assignable date" (ILS 117, 1878). In "Grounds," he remarks, "this very assumption involves itself a transcendent and supreme interest, and therefore from its very nature is unsusceptible of any support from reasons. This infinite hope which we all have . . . is something so august and momentous, that all reasoning in reference to it is a trifling impertinence" (W 2:271–272, 1869).

Peirce expresses these logical sentiments as logical parallels of the theological virtues of faith, hope, and love. He claims that "interest in an indefinite community, recognition of the possibility of this interest being made supreme, and hope in the unlimited continuance of intellectual activity" (ILS 117, 1878) are adjacent to "that famous trio of Charity, Faith, and Hope, which, in the estimation of St. Paul, are the finest and greatest of spiritual gifts" (ILS 118, 1878).

First, Peirce holds that we must hope the community of inquirers never ends, that rational intelligence throughout the universe is never extinguished entirely. With respect to statistical inductions, for our inferences to hew more closely to the actual probabilities such as they are in nature, we will need an increasingly large sample size and repeated sampling. The enormity of the universe will require an enormously large, unlimited, indefinitely extended community of inquirers. Peirce makes the point in a book review: "Suppose, then, that there be an indefinitely great *series of series* of observations of the same quantity, each lesser series consisting of n observations, and each having the same mean residual. Then, there being an infinite number of such series, the mean of their mean results may be taken as the true value, by definition. For the ultimate result of indefinitely continued observation is all that we aim at in sciences of observation" (W 3:379, 1878). Provided our observations continue indefinitely into the long run, the result of inquiries will gradually hew more closely to the facts of the matter (though at no time may we claim that our results in fact align with the actual uniformities in nature). The initial conclusions may well be wrong, but continued sampling will modify the result, so that "on continuing the drawings the inference will be, not *vindicated* . . . but *modified* so as to become true" (W 4:417, 1883). Note that this hope regarding the community of inquirers is different from other hopes that we may have. We may also hope that our question is answerable by following sound methods of inquiry. Such a hope concerns inquiry generally, not induction specifically. Also, we may hope that the true hypothesis is among those we can conceive. This hope concerns abduction, not induction.

Second, we must have faith in our fellow inquirers. To have faith in them is to put our trust in them. This faith will operate along several axes. For one, we will have faith in their work and results. Their researches will supply information for present and future inquiries. For two, we will have faith that future inquirers will take up our research projects. If a hypothesis has many more consequences than we could possibly put to the test in our own lifetime, we must put our faith in future inquirers to put them to the test. For three, we will have faith in those inquirers beyond our own geological epoch. Evidence suggests the human race will one day be as extinct as the saber-toothed tiger is now. We will also have faith in those inquirers we may never know.

Third, we must have love for our fellow inquirers. Love will require putting away our local and trivial concerns. The desire for fame, for wealth, and for other rewards must be set aside so far as we can. And we must be open to criticism for the improvement of our ideas. If we fail to have such love, we may be led to fudge data. We may be subject to confirmation biases. We may become dogmatic and dismissive of evidence against our hypotheses. As I mentioned earlier, Peirce does not require that inquirers in fact be selfless. He only requires that they imitate the heroic inquirer. We must strive to have love for our fellow inquirers so much as we can. Nevertheless, future inquirers may also correct our errors. If we have fudged data, they will find out. If we have designed our experiments poorly, they will discover this. And if the evidence weighs against our theories, they will appreciate the objections in ways we individually do not.

Hope, faith, and love are logical sentiments. They are not beliefs, presumptions, or assumptions. Several scholars have maintained that Peirce accepts that some logical principles—such as the principle of bivalence—are regulative assumptions of inquiry. Some scholars indicate that these logical principles are hopes.[8] Acceptance of the principle of bivalence or some other principles of logic sometimes are assumptions. Peirce certainly thought that inquirers proceed on "the assumption that things are intelligible, that the process of nature and the process of reason are one" (W 6:392, 1890). As noted earlier, some logical principles may be accepted as regulative assumptions in specific contexts of inquiry. They are adopted as regulative assumptions when accepting them will help us win the answer to our question, as when we make assumptions in the card game whist. However, Peirce maintains that regulative assumptions are warranted only in specific circumstances.

Nonetheless, Peirce holds "much that is generally set down as presupposed in logic is neither needed or warranted. The true presuppositions of logic are merely *hopes*" (HP 2:1028, 1902). We take should take care to distinguish hopes, on the one hand, and assumptions or presuppositions, on the other. An assumption or presupposition is a proposition taken to be true without proof but used for further inference (see CP 3.632 and 3.365, 1901). What the logical sentiments are about, however, need not be taken to be true. The inquirer need not assume that inquiry will continue indefinitely into the future. Rather, she need only hope that it will. The inquirer need not assume that the community of inquirers embraces all rational life. Rather, the inquirer need only have faith that it will. The inquirer need not assume her inferences are selflessly made. She need only have love for the truth and the community of fellow inquirers. These are sentiments. They are not beliefs, not suppositions, not assumptions. Because these are hopes and not assumptions, Peirce claims that "we cannot condemn scepticism as to how far they may be borne out by facts" (HP 2:1028, 1902).

Logic presupposes or assumes these sentiments. But the propositional content of the sentiments need not be presupposed or assumed to be true. That there will be some such community of inquirers is not presupposed in logic. What is presupposed is that one hopes for, has faith in, and loves that community. Induction reposes on these sentiments. It does not repose on the truth of the propositional contents of those sentiments. As Peirce remarks, "it is requisite that there should be, if not *belief,* yet *hope,* that the particular investigation they [inquirers] have in hand at the moment, may concern a matter sufficiently regular to admit of some generalization exact or approximate" (HP 1:202–203, 1893). He stresses, "we hope that this [viz., there is some single truth], *or something approximating to this,* is so, or we should not trouble ourselves to make the inquiry. But we do not necessarily have much confidence that it *is* so. Still less need we think it is so about the *majority* of the questions with which we concern ourselves" (CP 3.432, 1896).

In fact, the inquirer may believe precisely the opposite of these sentiments. She may believe that humanity will bring about its own destruction by wars or environmental degradation. She may be convinced that the likelihood of humanity being destroyed as a consequence of asteroid strikes is far greater than the likelihood of discovering alien species in the vastness of space. Peirce admits that he "once thought the indications were that the human race would become extinct before any great number of future centuries" (EP 2:466, 1913). Yet to assume, to believe, to suppose, is one thing. To hope, to have faith, to love, is another. The inquirer may be an optimist about inquiry even if she is a pessimist about humanity's survival prospects. The logical sentiments are not beliefs, which on Peirce's view are dispositions of the mind. The logical sentiments are "dispositions of heart" (ILS 118, 1878). For his part, Peirce is pessimistic about the survival of humanity but optimistic about the continuance of the community of inquirers. He notes that though we "may take it as certain that the human race will ultimately be extirpated . . . it [is] as certain that other intellectual races exist on other planets . . . [so that] intellectual life in the universe will never finally cease" (W 5:227, 1885).

These sentiments ground the validity of induction. They do not ground the scientific enterprise as a whole. The validity of deduction, we have seen, rests on quite different grounds. The validity of abduction, we shall see in the next chapter, rests on quite different grounds as well. The logical sentiments alone do not guarantee that science will make progress. They do not even guarantee that inductive inquiries will make progress. There are no guarantees in these matters; all life may be extinguished by morning. The logical sentiments only tell us why induction is valid. It is valid because, though we may be wrong on this or that occasion, whatever error we have presently made will be corrected by the community of inquirers. The error will be corrected by the community of inquirers provided that community continues indefinitely, takes up our projects of inquiry, and has as its members individuals who set aside their self-interested and trivial concerns in pursuit of the truth. There is no rational argument for belief in the community of inquirers. Logic compels us to hope that the community will continue indefinitely. It requires us to have faith in our fellow inquirers. And logic requires us to love the community's members. This is Peirce's logical sentimentalism.[9]

Sentiments are dispositional evaluative feelings evoked in response to something and that incline us to act in ways that will bring to pass what we take to be a greater good (or that will prevent a greater harm).[10] In the case of scientific investigation, the logical sentiments are evoked in response to our inductive inquiries. They incline us to act in ways that serve the community of inquirers. We hope that that community will endure forever. We put our faith in that community to continue our researches after the truth. And we love that community, such that we endeavor to be rid of bias and put away our own local and trivial concerns in pursuit of the truth.

As noted earlier, Peirce likens the logical sentiments to religious sentiments. It is not hard to see why. T. L. Short states the matter nicely when he remarks, "the goal [of inquiry] is not a final fixation [of belief] for oneself but for the community of inquirers—a community that one hopes will continue after his death and forever. That is part of the religious dimension of science: it vanquishes egoism, it subordinates self to something greater than self" (2002, 274). Moreover, the final fixation of belief is something that may never be attained. This does not preclude our approximating it. We are to think of the indefinitely deferred final opinion counterfactually or subjunctively. It is what would be believed once "all possible evidence has been gathered, preserved, and exploited" (Short 2002, 272).

We are now in a position to better understand Peirce's conception of the community of inquirers. In the conclusion of *Cambridge Pragmatism*, Cheryl Misak states that one pressing issue with which pragmatists must grapple is "how to demarcate the community of inquirers. . . . Is the community in which our beliefs are forged and our epistemic standards located local or are we all part of the same community as far as knowledge or truth is concerned" (2016, 282–283)? Whereas Peirce and Ramsey regard the community expansively, Misak avers James and Wittgenstein waver on whether it "is like a 'tribe'" or "what is best for a particular inquirer to believe" (2016, 283). Misak sides with Peirce and Ramsey. She remarks, "the best kind of pragmatism is one that takes the community to be wide and open. Otherwise, we lose our grip on normative notions such as truth, rightness, disagreement, and improvement" (2016, 283).

Alexander Klein criticizes Misak's earlier book *The American Pragmatists* on this issue of the scope of the community of inquirers. He argues that Misak unfairly attacks James for failing to do justice to those normative notions just listed. It is rather Peirce who fails to do them justice. For Klein, "[t]he issue comes down to the size of the community in which philosophical inquiry should be conducted, according to each figure" (2013, 414). Because of Peirce's commitment to "linguistic precision," the community of inquirers turns out to be a "narrow community of trained specialists" (2013, 414). Peirce's commitment to developing a rigorous, unambiguous vocabulary for philosophical inquiry means that "his ethics of terminology . . . creates a tremendous restriction on who can participate in inquiry *in practice*" (2013, 417). In contrast, "James's Darwinian account of inquiry requires him to practice philosophy publicly" (2013, 414). (This suggests that Peirce does not hold a Darwinian view of inquiry, which is correct.[11]) Klein holds that "inquiry conducted in a larger community with a greater variety of temperaments is proportionately more likely to produce a consensus that deserves to be called 'objective'" (2013, 414) since "we can only hope to reach a consensus that approaches this ideal of bias-free objectivity if the community of inquiry is maximally diverse" (2013, 415).

Let us take up each account in turn to see how it compares to the role that the community of inquirers is designed to play in Peirce's theory of induction. Writing of Peirce and Ramsey, Misak maintains,

> everyone with whom we can talk, disagree, and so forth must be thought
> of as a member of the community of inquirers. Indeed, they argue that any

particular disagreement makes sense only against a background of agreement. Their concern for truth and objectivity leads them to use this premise to argue that, since we can meaningfully disagree about questions with anyone we can recognize in the most minimal sense as another rational agent, the class of those we perceive as members of our most basic form of life must be no narrower than the class of those we perceive as rational agents. (2016, 271)

Misak's characterization of the community of inquirers in this passage is too narrow in one respect and too broad in another. With respect to being too narrow, whether we perceive that other creatures are rational agents makes no difference to whether we or they are part of the community of inquirers. Neither does it matter that we engage in rational discussion or debate with them. Humans may go extinct before encountering any other rational, inquiring creatures. Nonetheless, should rational orcas or rational alien life forms discover and use our researches, we are all part of the community of inquirers. As Peirce states, the community of inquirers reaches beyond all geological epochs and being a member of it requires only mediate contact with other rational life forms.

On the other hand, Misak's conception of the community is too broad in another respect. By making membership in the community of inquirers subject only to being a rational agent, Misak does not require that the person endeavor to inquire in conformity with the logical sentiments. To be a member of the community of inquirers, one must be willing to make the sacrifice of being wrong. One must be willing to spend one's entire life pursuing an ultimately fruitless line of research. One must put away selfish concerns for fame or wealth. But an individual can be a rational agent without manifesting the logical sentiment of love. Scientists have fudged data to gain tenure or to get funding. They are rational agents. Yet so far as their inquiries have not conformed to the logical sentiments, they are not members of the community of inquirers. Similarly, there may be researchers who, while self-sacrificing, either never communicate their results or actively prevent other researchers from seeing or using them. Such a researcher has no faith in present or future inquirers taking up her lines of research. So far as this is so, she is not a member of the community of inquirers. Likewise, a researcher may abandon her work out of despair over humanity's prospects for survival. Such a researcher has no hope that her work will be continued. So far as she lacks such hope, she is not a member of the community of inquirers. Christopher

Hookway rightly remarks, "[w]hen we participate in scientific activity, we do subordinate ourselves to the wider community. Our inferences are evaluated in terms of what they contribute to science, not how far they satisfy our selfish desires" (1985, 215–216). The community of inquirers is broad, but it is restricted to those who endeavor to be members of that community. It is restricted to those who imitate the heroic inquirer, the inquirer who displays the logical sentiments.

At this juncture, a distinction is in order. In the preceding paragraph, I have moved between referring to members of the community of inquiries as those whose inquiries are animated by the logical sentiments and those whose inquires conform to the actions of one who is animated by the logical sentiments. The reason for this, as we shall see more clearly in the next objection, is that Peirce is concerned that if membership in the community of inquirers requires that our inquiries be animated by the logical sentiments and if all human actions are ultimately motivated by self-interest (psychological egoism), then no humans are members of the community of inquirers. Peirce rejects psychological egoism, but he also holds even if were true, it is sufficient for one's actions to conform to what the "heroic" inquirer would do. Accordingly, we might distinguish between those members of the community of inquirers who are full-fledged and dues-paying members and those who are honorary members. The former are those whose inquiries are animated by the logical sentiments, and the latter are those whose inquiries are in conformity with what the full-fledged, dues-paying member would do.

In either case, Klein is correct that the community of inquirers is limited. It is limited to those who exemplify the logical sentiments in their researches, whether as full-fledged, dues-paying members or as honorary members. If supplying ourselves with a technical vocabulary in which we conduct our research is requisite to make advances and clearly communicate our results (as it surely is), then the community of inquirers can and must do so. It makes no difference whether the run-of-the-mill person can participate in that particular line of research. It makes no difference whether the theory gains universal assent. What should it matter if a self-seeking person rejects the theory of evolution so that he can win the praise of a group of religious zealots?

Klein states a preference for James's Darwinian theory of inquiry. On this theory, articulated in the preface to *The Will to Believe*, theories are to be tested in the public sphere so long as a fair field is shown. The true theory will be the one that wins in the sense of gaining champions. Peirce rejects this view. He argues that a theory can gain champions through tortures and inquisitions. Whereas theories do need to be put to the test, the test is not that of gaining champions. Generations of self-seeking people may embrace a false theory if it wins them praise, feeds their egos, or eases their consciences. For Peirce, the better theory is the one that accounts for the diversity of phenomena observed. A good theory reduces them to unity, as Newton's theory of gravity unified the falling of an apple and the orbits of planets. The true theory is the one that leaves none of the experimental phenomena it is designed to explain unexplained.[12] Newton's theory left unexplained the precession of Mercury's perihelion; Einstein's theory of relativity did not. The feeling of conviction we may have in claiming some theory is true depends on other factors, such as the scope of our observations, the instruments we use to gather data, whether there are phenomena currently unexplained but possibly explainable by the theory, and so on. These are concerns regarding the strength of an induction. None of these—whether a theory is better, whether it is true, and whether we may confidently claim it is true—depends on the number of people who champion the theory. What induction does depend on is future research by members of a hopeful, faithful, and loving community of inquirers. In short, Peirce's defense of induction's validity does not commit him to the claim that there will be an infinitely expansive community of inquirers. Rather, it only commits him to the hope that there will be some such community in which we can put our faith and which we love.

The Egoism Objection

The Objection

Peirce's requirement that we love the community of inquirers may pose a problem. Some theorists have argued that all of our actions are ultimately motivated by self-interest. If this be so, then inquirers cannot possibly endeavor to be self-sacrificing or to be rid of their biases. Consequently, there can be no loving community of inquirers. If the validity of induction requires the possibility that there be some such community, it follows that induction cannot be valid.

Peirce's Reply

Peirce rejects the thesis "that man cannot act without a view to his own pleasure" for (a) people make enormous self-sacrifices, (b) they care what happens after they die, and (c) we often use 'we' even when "no personal interests at all are involved" such as when we talk about our national interests (W 2:271, 1869). Furthermore, Peirce argues that even if psychological egoism were true, what is important is not that inquirers actually be unselfish but that they imitate the unselfish person. He claims, "it is not necessary for logicality that a man should himself be capable of the heroism of self-sacrifice. It is sufficient that he should recognize the possibility of it" (ILS 117, 1878). One need not be solely motivated by love for the truth and for the community of inquirers, or even primarily motivated by it. One need only imitate the person who is so motivated by love for the truth and for the community. That is sufficient, at least, for honorary membership in the community of inquirers, even if full-fledged, dues-paying membership requires one's inquiries in fact be animated by the logical sentiments. As Susan Haack remarks, "[p]rovided that you can achieve your ulterior purpose *only by getting to the truth of the matter*, the fact that you are motivated by something other than the love of truth for truth's sake doesn't necessarily make you any the less serious and effective an inquirer" (2018, 222). I have examined Peirce's critique of psychological hedonism in depth elsewhere,[13] and so I shall not pursue the matter further here.

The Lost Evidence Objection

The Objection

A fifth objection to induction's validity is that some evidence will have been lost to time. There will be facts such that it is impossible for the community of inquirers to ascertain the truth regarding them. Or there may be events that are extremely rare such that they may occur only once in all of the universe yet be unobserved on that one occasion. In such a case, one might make the crude induction that the occurrence of such an event has never been established; therefore, it has never occurred. Yet if a case has occurred, the community of inquirers will never know this.

Peirce's Reply

To be certain, the problem of lost evidence shows that our inductive inferences may be vitiated. That is, they may not be strong. But that evidence can be lost does not undermine our inductive inferences or prohibit us from drawing inductive conclusions. Though the evidential grounds on which we might hold some opinion may seem forever lost, we are not precluded from hoping that they may be recovered. Neither are we prevented from exerting our energies to ascertain the truth. Peirce makes this reply to Royce, who complains that the truth we would approximate may never come to pass. "It may be he is right in this criticism;" Peirce remarks, "yet to our apprehension this 'would be' is readily resolved into a hope for *will be*" (CP 8.113, 1900). Some evidence may be forever lost, and the answers to some questions may forever elude our grasp. That does not preclude us from hoping to find the answer, provided the question has any definite meaning. It does not prevent us from exerting our energies to get closer to the truth. As Peirce claims, "it is unphilosophical to suppose that, with regard to any given question (which has any clear meaning), investigation would not bring forth a solution of it, if it were carried far enough" (W 3:274, 1878). The truth, pragmatically understood, is what would be the result of our inquiries in the long run. We hope that we shall approximate that truth by way of our inquiries. We need not assume that we will reach that truth. With respect to any given unanswered question to which scientific methods may be applied, we have no grounds for asserting that we have reached the utterly final opinion about it. But neither do we have grounds for asserting that we cannot reach (or at least approximate) the truth of the matter.

The Irrational Numbers Objection

The Objection

An objection similar to the lost evidence objection is that there may some frequencies in nature which are to be represented by irrational numbers. However, our data will be based on observations which give us rational numbers. It follows that the ascertained probabilities will never be the same as the actual frequencies in nature.

Peirce's Reply

The objection is a lamentable catch. First, it does not follow from the fact that our observational data are discrete that the results cannot involve irrational numbers. We may find, for instance, that the curve which best fits our data plot is $y = (1/\sqrt{2})x$. Second, induction only requires that we approximate the truth in the long run. With respect to irrational numbers, the result is not obtainable with ascertained frequencies, but that does not prevent us from exerting our energies to approximate it. Moreover, because of his conception of the community of inquirers as unlimited, Peirce can account for our ability to approximate uniformities in nature involving irrational numbers. It is true the actual frequencies ascertained will all be rational numbers. However, an increasingly large sample size will lead to a closer and closer approximation of the irrational number. Peirce is aware that the real probability of an event may be an irrational number (see ILS 277, 1910). His theory of probability addresses the problem by appeal to the community of inquirers.

The Short Run Objection

The Objection

Peirce's appeal to the long run, however, raises a problem concerning the short run. If discerning which hypotheses are true rests with the community of inquirers in the long run, if ensuring our probabilities ultimately hew closely to the real uniformities of nature rests with the community of inquirers in the long run, and if eventually alighting on the true hypothesis rests with the community of inquirers in the long run, then why should we trust our current scientific theories in the short run? Why should we trust the probabilities we have calculated in the here and now? And why should we presume that the true hypothesis has even been conceived let alone confirmed?

One species of the Short Run Objection is the stopping problem in probability theory. We will often use inductive inferences in order to reach a conclusion that will inform our actions. When will we have investigated enough to draw a conclusion on which we can confidently act? The stopping problem is sometimes illustrated by the secretary problem. Suppose a CEO wishes to hire a secretary. She screens the applicants randomly and makes a decision about the applicant immediately after each interview. What is the optimal time for the CEO to stop interviewing and decide on an applicant?

More generally, our theories will, in the long run, be improved upon or replaced. Why then should we trust them in the transient short run? Why should we stop and conclude that such-and-such is the case? As Rescher notes, the question "why trust science?" can be taken in two ways. In the first, it is "why accept what science tells us, rather than rely on other sorts of procedures"; in the second, it is "why should we accept what science tells us today" (1995, 103–104)? Peirce's theory of scientific inquiry and his logical sentimentalism answer the first question. Inductive procedures are truth-approximating in the long run; necromancy is not. The Short Run Objection concerns the second question.

Peirce's Reply

Rescher holds that we should trust science because it is "the optimal means of making thoroughgoing sense of our cognitively relevant experiences overall" (1995, 111). But this answer only gives us a reason to trust science in the long run. He claims the methods of science have "legitimating virtues." These are the virtues of being self-corrective, of being hypothetically effective (i.e., if any method can lead us to the truth, the methods of science can), and of being theoretically comprehensive. All of these points are well-taken. But science is self-corrective in the long run. Science is most effective in the long run. Science is comprehensive in the long run. Rescher has dodged the hard question of trust in the short run by appealing to the long run.

Christopher Hookway insists, "[t]here is no logical guarantee that induction will be of any practical or short run value at all" (1985, 210). He claims "there is no reply to skeptical doubts about [the scientific method's] short-run effectiveness" (2000, 142). But this overstates matters. If 'logical' is to be understood as pertaining to the validity of induction, Hookway is surely correct. But logic may also concern how to make our inductions strong in the short run. Even if we can have no long-run logical guarantee of being close to correct now, we can at least be as approximately correct as possible under the current circumstances. And being approximately right can have practical and short-run value. There is always room for skepticism that we are exactly right. There is much less room for positive (as opposed to Cartesian) grounds for doubting our strong inductions are approximately right.

In the last several years, a goodly amount of research has been done on the question of why we should trust science. If Peirce is correct, we need to distinguish between the questions of why we should trust science in the short run and why we should trust science in the long run. These questions admit of different answers. We should trust it in the long run because we put our hope and faith in the community of inquirers, which we love. Their continued inquiries will alight on the truth and root out error. But in the short run, the work of the community of inquirers is never completed. So we cannot put our trust in science in the short run for the same reason that we put our trust in science in the long run.

Rather, the reasons we should trust science in the short run will be determined by various marks that the science holds good, at least for now. On the Peircean view, we must distinguish between concluding for now and concluding for all time. For any course of inquiry that involves inductive inference, we never make conclusions for all time. Peirce regards inductive inference as ampliative inference, and it is ampliative in two senses. First, it increases, or amplifies, our knowledge. Second, final judgment is ampliated, or deferred. Indeed, even the infinite community of inquirers may only approximate the truth in the long run, as is evident from the Irrational Numbers Objection discussed earlier. Further, even if the truth has been reached on any given matter, it is not possible to be assured that the truth has been reached now. This is why Peirce claims that belief as a possession for all time "has no place in science at all" (EP 2:33, 1898). Science feels that "its position is only provisional. It must then find confirmations or else shift its footing. Even if it does find confirmations, they are only partial. It still is not standing upon the bedrock of fact. It is walking upon a bog, and can only say, this ground seems to hold for the present. Here I will stay till it begins to give way" (CP 5.589, 1898).

Nonetheless, scientists will reach tentative conclusions in the short run. Although inquiries involving induction do not ultimately conclude, inductive inference involves drawing conclusions for now based on the evidence we have. Jeff Kasser rightly insists that there is "room for *scientists* to believe, conclude, and commit in the name of a process that does not do these things" (2016a, 304). Scientists can, and often must, trust theories in the short run. To decide which theories to trust, they will rely on certain indicators that the theory is trustworthy in the short run. These are indicators which, while not perfectly reliable, provide grounds for trusting scientific results to different degrees. There are at least six.

First, perhaps the most important of these marks is whether the inquirers have done their due diligence to ensure that their inductions are strong. Strong inductions are more trustworthy than weak ones. Have the authors recorded their successes as well as their failures? Have they considered and tested alternative hypotheses? Have they used methods of randomization appropriately? Did they predesignate the characteristics for which they were testing? If the answers to these questions are negative, it is grounds for having less trust in the results.

Second, an indicator of trustworthiness is whether the researchers have any personal interest in their results, be those interests with respect to wealth, to fame, or to tenure and promotion. Researchers, like any other humans, are subject to biases. Those biases may subtly (or not so subtly) influence their inquiries. This is part and parcel of why Peirce requires induction to be grounded in the logical sentiment of love. Stephen J. Gould has made much of this fact in his *The Mismeasure of Man* (1996), as have Naomi Oreskes and Erik M. Conway in *Merchants of Doubt* (2010). When there is such a personal interest, we have grounds for being suspicious of the results. The researchers' reasonings may be motivated. They may weigh evidence in questionable ways that favor their preferred hypotheses. They may suppress pertinent data.

A third indicator of trustworthiness is whether the work has been through blind peer review. Research which must win the consent of other scholars when those other scholars have no personal stakes in approving it is indicative of good research. This is one of the reasons that tenure and promotion committees place emphasis on the process of peer review. Peer review is not a perfect process. Sometimes the reviewer knows who the author is. Other times the reviewer is careless or hurried. Nevertheless, it is generally a reliable indicator of the trustworthiness of the research.

A fourth indicator is whether there is scientific consensus on the matter in question. If the vast majority of inquirers have embraced a particular hypothesis, that is a good indicator of the trustworthiness of the results. Once again, this is not an infallible indicator. There may be sociological factors that influence who gets what and which posts where. Such factors may also influence which perspectives prevail in the academy. Nevertheless, consensus is a good indicator of trustworthiness in many contexts. Both Helen Longino (1990) and Oreskes (2019) have stressed the importance of consensus for trusting science.

A fifth indicator of trustworthiness is one on which Peirce laid much emphasis. It is convergence on a hypothesis. Convergence is different from consensus since it does not require that a large majority of theorists embrace some hypothesis. Rather, convergence occurs when two or more scholars or groups of scholars independently arrive at the same results. "How many times," Peirce exclaims, "have men now in middle life seen great discoveries made independently and almost simultaneously" (W 8:204, 1892)! He proceeds to list a series of such convergences on a hypothesis, including the prediction of Uranus by Le Verrier and Adams, the mechanical theory of heat by Rankine and Clausius, and the theory of evolution by Wallace and Darwin. Convergence may occur when theorists are working from the same or similar data or out of the same sources. More remarkable still is when convergence occurs from scholars working out of different data, studying different objects, or even in different fields altogether. Perhaps most remarkable is when theorists from hostile points of view arrive at the same conclusion. However it may occur that independent researchers converge on a conclusion, it is a good indicator of the trustworthiness of the result.

A sixth and final reason for trusting a scientific theory in the short run is simply that it works or has worked in analogous situations to our own. We often need to adopt or use a scientific theory because we have some practical end in view. Such ends may be to build a boat, erect a house, construct a boiler, shoot a satellite into orbit, design a vaccine, or what have you. Although some of these abilities were gained in the past using trial and error, we also have physical, chemical, and biological theories that aid us in attaining these ends. Those theories have worked in the past. Those past successes give us good practical grounds for trusting the theories now.

The Resources Objection

The Objection

An eighth objection to induction's validity is that there may be questions which are impossible to answer just because our resources are limited. Rescher notes that Peirce recognizes in science there may be increasing costs and diminishing returns such that "there is a limit, ultimately an economic limit, to the questions about the phenomena that we can even answer. Many questions of transcending scientific importance will remain unresolved because their resolution would demand a greater concurrent deployment of resources than will ever be marshalled at any one time in a zero-growth world"

(1978, 34). In such cases where limited resources entail that we cannot pursue inductive investigations, induction will not be truth-approximating in the long run, for we are not able to perform the requisite experiments to approximate the truth.

The Peircean Reply

From such economic considerations, one might be tempted to draw the conclusion that inquiry cannot ultimately root out error and alight on the truth. But 'cannot' may be understood practically or aspirationally. Practically, there are truths we will never alight on and errors we will never root out. Moreover, there are some questions such that continued inquiry brings diminishing marginal returns, such that a further deployment of resources investigating them is imprudent. But aspirationally, we may hope that the community of inquirers will marshal the needed resources. We may believe (dispositions of the mind) that we will never ascertain some truths and root out some errors while still hoping (dispositions of the heart, our sentiments) that the community of inquirers will ascertain those truths and root out those errors. The validity of induction rests on our sentiments, though the economy of research may justify beliefs contrary to our sentiments. Even when we have economic grounds for being pessimistic about the progress of science, we must do what we can now by sacrificing our own interests to the larger community in pursuit of the truth. We are to put our hope and faith in future inquirers to ascertain those truths and root out those errors that economic considerations now prohibit us from pursuing.

The Ethics Objection

The Objection

A ninth objection is that there may be theories we cannot test because to do so would be unethical. Note that this differs from the Resources Objection: We may have the resources to perform the tests, only doing so would be morally reprehensible. Peirce has a very expansive conception of science. He holds that ethics, politics, and religion might also be subjects for scientific investigation. Yet Peirce also believes that these sciences—like much in philosophy—are in a rather underdeveloped and imperfect condition. He holds "Philosophical Science" may "ultimately influence Religion and Morality" (EP 2:29, 1898). But he also maintains that "because of this utterly unsettled and uncertain condition of philosophy at present . . . I regard any practical applications of it to Religion and Conduct as exceedingly dangerous" (EP 2:29, 1898). Were some ethical, political, and religious theories well-supported, we might apply them to life. Yet ethical, political, and religious theories are hard to put to the test. It is unclear what sorts of tests might be appropriate. When it is clear, performing those tests may be grossly unethical. Also, some tests may be practically impossible to conduct. Accordingly, induction will not be truth-approximating in the long run for ethical and political sciences, or if it is, it will require grossly immoral experiments.

The Peircean Reply

Notice that there are in fact two objections here. The first objection is that an inquirer may wish to ascertain some truths (say, in biology) by engaging in grossly immoral experimentation. The second is that scientifically investigating ethical questions may itself require grossly immoral experimentation. The first concerns what the researcher may do. The second concerns what scientific investigation requires.

The first issue pertains to restrictions that we ought to place on the conduct of researchers. The objection does not affect the fact that were such experiments performed, they would enable us to ascertain the truth. Consequently, the objection does not threaten the validity of induction as a truth-approximating genus of inference. It only prohibits the researcher from attaining the truth in an immoral manner with respect to that matter. Yet even here, it may be that there are natural experiments which provide data that can help the researcher eliminate, corroborate, or confirm hypotheses. That is, there may be events which researchers do not cause but from which they may glean information to evaluate the truth of various hypotheses.

Similar considerations apply to the second objection. Scientific investigations may be prohibited by ethical restrictions we place on researchers. That does not imply, however, that induction would not ascertain the truth were there no such restrictions. But Peirce has more to say about inquiry into ethical and sociopolitical matters. Peirce holds that scientifically investigating ethical matters is itself a sort of immorality, stating, "ethics, which is reasoning out an explanation of morality is—I will not say immoral, [for] that would be going too far—composed of the very substance of immorality" (CP 1.666, 1898). Rather than trust to our philosophical theories in the conduct of life, Peirce maintains that we ought to trust to our ethical (not logical) sentiments. Furthermore, we ought to give those ethical sentiments that are instinctual (rather than merely culturally inculcated) greater weight in our practical deliberations. Peirce thus disagrees with Mill, who claims that "if the principles and rules of inference are the same whether we infer general propositions or individual facts; it follows that a complete logic of the sciences [of induction] would be also a complete logic of practical business and common life" (1869, 172). To be sure, we ought to engage in philosophical inquiry about matters of ethics, politics, and religion. Moreover, we will do well to deliberate about various courses of action. But we ought to be reticent to apply our ethical, political, and religious theories directly to life. Instead, Peirce maintains such inquiries should affect life by a "slow process of percolation of forms, as . . . researches into differential equations, stellar photometry, the taxonomy of echinoderms, and the like, will ultimately affect the conduct of life" (EP 2:28, 1898).

Peirce's sentimental conservatism is the twin theses that (a) we ought not to trust too much to our ethical, political, and religious theories in the conduct of life and (b) we should instead mainly trust to our sentiments and instincts. Conservatism, in this sense, is not ethical, political, or religious conservatism. It is theoretical conservatism: that our theoretical beliefs ought not precipitously to inform how we conduct our everyday lives. For instance, if I endorse utilitarianism as an ethical theory, it does not follow that I should live my daily life in light of its precepts.

It is true that in another work, Peirce appears to embrace political and religious conservatism. He writes, "if [my readers] were to come to know me better, they might learn to think me ultra-conservative. I am, for example, an old-fashioned christian [sic], a believer in the efficacy of prayer, an opponent of female suffrage and of universal male suffrage, in favor of letting business-methods develope [sic] without the interferences of law, a disbeliever in democracy, etc. etc." (R 645:21, 1909). First, this sort of political and religious conservatism is different from theoretical conservatism. Second, notice that Peirce does not say he is ultra-conservative, only that persons who know him better might think he is. Third, it is hard to know how seriously to take Peirce's comments here. What makes a Christian old-fashioned? Peirce does believe in the efficacy of prayer. But he denies it is efficacious because it moves God to act. Rather, prayer is efficacious because it brings about "great spiritual good and moral strength" (CP 6.516, 1906). He is critical of prayer that asks the day's bread "may be better baked than yesterday's" for being "childish, of course; yet innocent" (CP 6.516, 1906). That is not the conservative view of the efficacy of prayer. His comments about suffrage, while deserving no defense, are hardly distinctive or surprising for 1909. In whatever case, this sort of political and religious conservatism forms no essential part of Peirce's theoretical conservatism.

If our theoretical beliefs ought not to precipitously guide our daily conduct, what should? This is where Peirce's ethical sentimentalism comes in: Our sentiments and instincts ought to guide how we conduct our daily lives. Peirce's sentimental conservatism admits of a stronger and a weaker interpretation. On the weaker interpretation, sentimental conservatism may be construed as a quite modest and low-key proposal. All Peirce is doing is giving us a bit of advice about how to live our lives. We should not trust to our ethical, political, and religious theories, which have led people to engage in rather peculiar behavior on quite tenuous evidence. Instead, we should trust to our sentiments and instincts, which have been forged in the crucible of natural selection. On the stronger interpretation, sentimental conservatism may be construed as a theory about our epistemic access to ethical, political, and religious truths. Because our sentiments and instincts have been forged in the crucible of evolutionary pressures, they are good guides to the moral and religious facets of the world in which they have been forged. Peirce appears to have endorsed both positions.

Two additional points of clarification are in order. First, Peirce does not deny that our theories can and should influence how we conduct our lives. His claim is that, generally speaking, they should do so only with long and considered reflection, "with secular slowness and the most conservative caution" (EP 2:29, 1898). Peirce also acknowledges that there are times when we might act on our theories rather than our sentiments and instincts. He writes, "it is of the essence of conservatism to refuse to push any practical principle to its extreme limits,—including the principle of conservatism itself" (EP 2:32, 1898). Generally, we should rely on our sentiments and instincts in the conduct of life. Nonetheless, there may be times when we should rely on our theories.

The second point of clarification is that Peirce draws a distinction between sentiments and instincts. Our sentiments are affective patterns correlated with habitual ways of acting for a greater good or the avoidance of a worse harm. Feelings of love lead one to care for one's offspring. Religious sentiments lead one to treat others with compassion and dignity. Instincts, in contrast, are inherited habits. An instinct is a "way of voluntary acting prevalent almost universally among otherwise normal individuals of at least one sex or other unmistakable natural part of a race . . . which action conduces to the probable perpetuation of that race" (EP 2:464–465, 1913). Some instincts are allied with certain sentiments. The feelings of love leading one to care for one's offspring are an example. We do, though, have instincts that are not allied with sentiments. Our reflexes are instincts but not allied with sentiments. Also, we have sentiments that do not rest on instincts, such as sentiments inculcated as a consequence of the culture in which we are raised. A love for certain artistic or literary forms is an example of such a sentiment. Rather than being instincts, these may be mere affective prejudices. Peirce recognizes that some of our sentiments are mere prejudices inculcated from our cultural traditions. Such prejudices ought to hold no or little weight in our deliberations.[14]

I have given thorough treatment to Peirce's doctrine of sentimental conservatism elsewhere.[15] What I am claiming here, which I did not explicitly claim there, is that Peirce's sentimental conservatism is a consequence of his theory of induction's validity. Peirce's doctrine of sentimental conservatism is not explicitly articulated until 1898, twenty years after the ILS Series. However, as Mathias Girel (2011) has argued, the themes are already present in Peirce's 1885 review of Royce's *The Religious Aspect of Philosophy*. They are also found in 1892's "Evolutionary Love."

Some scholars have been deeply suspicious of Peirce's sharp distinction between theory and practice. To some, it has seemed inconsistent with Peirce's pragmatism and his emphasis on the conceivable practical consequences of a hypothesis. But having misgivings about his sentimental conservatism on these grounds is a mistake. Hypotheses have diverse conceivable practical consequences. Research projects employing inductive inferences to confirm those hypotheses must be entrusted to the unlimited community of inquirers. Because (a) such researches must be left to the unlimited community of inquirers and (b) putting some ethical, political, and religious theories to the test poses significant ethical and practical problems, we are well-advised to trust to our sentiments and instincts in the conduct of life. Our ethical, political, and religious theories are too tentative, too unproven, to deserve our trust. In contrast, our sentiments and our instincts have proven their mettle in helping us navigate the pressures of natural selection and social life. Moreover, those sentiments indicate that we ought not to pursue certain research projects because they involve grossly unethical behavior. Therefore, we ought not to pursue those research projects. It does not follow, however, that induction is invalid. If we were to pursue such projects, then we would discover the truth. Only we won't pursue them, or at least we should not do so.

Nevertheless, Peirce's sentimental conservatism is fully consistent with advocating for radical reforms. In spite of his aforementioned claim to be thought an ultra-conservative by some, Peirce's conservatism is not political conservatism. He advocates for a radical reform of the prison system in his essay "Dmesis." He remarks that Pasteur was also a practical man who "*must* see the thing [his strongly supported inductive conclusions, such as we find in his work on pasteurization] accepted and acted upon, so that here would begin those struggles with all the agencies of conservatism and of night that brought Pasteur to his comparatively early death" (CN 3:65, 1902). Those agencies of conservatism are political conservatives who would opt for the status quo over scientifically supported, trustworthy conclusions. When our conclusions are well-supported, Peirce has no qualms about advocating for radical reforms on their basis. Neither does he have any qualms about advocating for radical reforms on the basis of our sentiments and instincts, as "Dmesis" shows. Furthermore, were our ethical, political, and religious conclusions well-supported, Peirce would have no qualms about advocating for radical reforms on their basis. Yael Hungerford (2016) has aptly described Peirce as a conservative progressive. Oxymoronic as the moniker might initially sound, there is no threat of inconsistency once we recognize that theoretical conservatism is consistent with political progressivism.

The Underdetermination Objection

The Objection

At this juncture, I turn to consider three objections to induction's validity which Peirce does not clearly address but which have received significant attention in the recent literature. The first is the problem of underdetermination. There are at least four different kinds of underdetermination.[16] One kind of underdetermination is equivalence underdetermination, wherein two theories have the same consequences but differ in some important respect. Equivalence varieties of underdetermination are easily generated by proposing, for example, that there is an evil demon that is the cause of all we observe. These are instances of global equivalence underdetermination. Peirce holds that hypotheses are (a) distinguished from one another by their predictions and (b) demonstrable as true or false from their predictions. Let hypothesis H_1 have a set of predictions P_I and H_2 have a set of predictions P_{II}. If $P_I = P_{II}$, then H_1 and H_2 are equivalent hypotheses. Moreover, if H_1 is false, there will be some member of P_I that, if put to the test, will be found to be not true. The evil demon scenario would seem to show this presumption is false. But there may also be cases of local equivalence underdetermination, in which case two hypotheses have all the same consequences. The hypothesis H as well as the hypothesis H or Q, where Q is some arbitrary addition that does not affect the set of predictions P of H, might be equivalent. In such a case, it is undetermined which of H and H or Q is true.

In addition to equivalence underdetermination, there are holist, transient, and practical underdetermination. Worries about holist underdetermination arise if evidence alone cannot direct us to accept or to reject a theory because modifications may be made to the whole web of belief. This is the sort of underdetermination we find in Quine's "Two Dogmas of Empiricism." A third kind of underdetermination is transient underdetermination. Transient underdetermination arises from the fact that our present theories are liable to be overturned or improved upon by future inquiries. Transient underdetermination is supported by the skeptical meta-induction over the history of science that our scientific theories, even those which gain wide acceptance, have often ultimately proven to be false. A fourth kind of underdetermination is practical underdetermination. Practical underdetermination arises when the evidence available does not suffice to justify us in accepting one theory over its rivals.

The Peircean Reply

Peirce has responses to worries about induction arising from each of these kinds of underdetermination. Peirce would object to such varieties of global equivalence underdetermination on two grounds. First, evil demon scenarios are not hypotheses that suggest themselves to us. As such, they are not among the hypotheses that we will gather by the armfuls and subject to scientific scrutiny. They will be rejected at the abductive stage of inquiry.

Although global equivalence underdetermination might haunt philosophers in their armchairs, it doesn't haunt practicing scientists in their laboratories. I shall have more to say about abduction and evil demon scenarios in the next chapter.

Second, Peirce claims that even if the whole of experience were the consequence of an evil demon endeavoring to deceive us, still we would find uniformities in that experience. Those uniformities would be, so far as inquiry is concerned, the extent of reality. He remarks, "the existence even of an illusion is a reality; for an illusion affects all men, or it does not. In the former case, it is a reality according to our theory of reality; in the latter case, it is independent of the state of mind of any individuals except those whom it happens to affect" (W 2:269, 1869).[17]

The matter of local equivalence underdetermination needs to be treated differently. In such a case, either Q has no predictions, its predictions are a subset of the set of predictions P from H, or its predictions are identical to P. In the former case, Q is a conjecture that fails to meet one of the four criteria for admissibility to scientific inquiry. As this is a matter which concerns abduction, I must table it for further consideration in the next chapter. If the predictions of Q are a subset of P or equivalent to P, Peirce would claim that either Q is a specification of H, H and Q are equivalent, or we have failed yet to find a prediction that distinguishes between them. Recall that we do not always know all of the predictions of a hypothesis once it is conceived. Some predictions of a hypothesis will be discovered in a course of inquiry. Moreover, Peirce holds that we ought to break up our hypotheses into their most basic items so as to test each one individually (see HP 2:761, 1901).

Holist underdetermination has received significant scholarly attention. In my judgment, it has been satisfactorily addressed elsewhere (see Laudan 1990 and Mayo 1996, Ch. 6), and so I shall not take up the matter here. Suffice it to say for the present that defenders of holist underdetermination have failed adequately to support their ambitious claim that any hypothesis may be as well supported by the evidence as its rivals. Possibilities are not probabilities, and the mere fact it is possible to rejigger one's concepts or web of belief does not imply it is reasonable to do so.

As to transient underdetermination, Peirce embraces the objection. Indeed, his account of the community of inquirers is designed not to respond to transient underdetermination but to embrace it. We in fact hope that our present theories will be improved. If our theories are wrong, we hope that better theories will be developed by future inquirers and put to the test. Yet nothing about transient underdetermination requires us to deny that our theories hold good for now or to deny that they approximate the truth. Moreover, the longer our theories hold up to testing, the stronger the grounds we have for thinking they will continue to hold.

Lastly, practical underdetermination does not affect the validity of induction. It is of concern only if we must decide on a theory for the purposes of belief or action. In the scientific context, Peirce regards this worry as insignificant. "Nothing," he remarks, "is *vital* for science; nothing can be. Its accepted propositions, therefore, are but opinions, at most; and the whole list is provisional" (RLT 112, 1898). Scientists may well disagree about the conclusions they have reached from the evidence they have. We hope that further inquiry will settle such disagreements.

The Paradox of the Ravens Objection

The Objection

The paradox of the ravens turns on the logical equivalence of *All ravens are black* and *all non-black things are non-ravens*. If we wish to test whether all ravens are black, we could instead search out non-black things and examine them for non-ravenness rather than search out ravens and examine them for blackness. Consequently, it would seem as though I can have evidence for *all ravens are black* just by observing my white tennis shoes and finding they are not ravens.

The Peircean Reply

The Peircean has several lines of response. First, we will want our sampling procedures to be as efficient as possible. We can much more readily increase the scope of our sampling relative to the population sampled by first finding ravens and checking them for blackness than by examining all non-black things and checking them for non-ravenness. That is because the number of non-black things in existence is many times larger than the number of ravens. Accordingly, the better procedure is to go about sampling ravens than to go about sampling non-black things. Note that the issue here is not that sets of things with negated qualities are generally larger. It is much more efficient to sample non-colored things for non-materiality than it is to sample material things for whether they are colored, even though *all material objects are colored* is equivalent to *all non-colored things are non-material objects*.

Second, whichever way we go about sampling, the community of inquirers will eventually find a non-black raven (viz., one that is albino or leucitic). Recall that for Peirce we are to hope that intellectual activity never ceases. In the short run, we may find that all ravens are black. The hope is that in the long run, if we are wrong, the community of inquirers will discover our error. Similarly for the example of material objects being colored: Whichever be tested, the community of inquirers will eventually find a non-colored material object (e.g., clear glass).

Third, costs should be considered when we rank hypotheses to put to the test. As Rescher points out, it is much more cost-effective to check ravens for blackness than non-black things for non-ravenness, since the population of ravens is smaller (see 1976, 77). As the cost of testing concerns abduction, I shall defer a fuller consideration of this issue to the next chapter.

Fourth, one point to underscore in the preceding is the role that predesignation plays relative to the question we are asking. The nature of our sampling procedure will determine which sorts of questions we can answer. As Peter Godfrey-Smith rightly notes, we may ask (1) whether all ravens are black, or we may ask (2) what proportion of ravens are black (2003a, 215). We may predesignate that we will (a) sample ravens for blackness, (b) sample non-black things for ravenness, (c) sample non-ravens for non-blackness, or (d) sample black things for non-ravenness. Only (a) will be able to answer both questions (1) and (2). Procedure (b) will be able to answer question (1) but not question (2), for it can tell us only what proportion of non-black things are ravens. Even supposing we had a complete enumeration of non-black things, we would not know how many ravens there are. Neither (c) nor (d) can answer questions (1) or (2).

The New Riddle of Induction Objection

The Objection

Nelson Goodman's new riddle of induction proposes a new predicate, grue. An object is grue just in case it was observed prior to 2100 CE and is green or if it was not first observed prior to 2100 CE and is blue. The predicate grue poses a riddle for induction because of the following inference:

(1) All praying mantises observed prior to 2100 CE have been green.
(2) Therefore, all praying mantises are green.

And this inference:

(1) All praying mantises observed prior to 2100 CE have been grue.
(2) Therefore, all praying mantises are grue.

These are formally equivalent arguments. However, we are inclined to think the first argument is valid and strong, whereas the second argument is fallacious.

One possible answer to this riddle is to claim that there is something wrong with the idea of grue. The problem is specifying what is wrong with the idea. After all, we might also have the property bleen, where something is bleen just in case it was observed prior to 2100 CE and is blue or if it was not first observed prior to 2100 CE and is green. In that case, we can make green and blue derivative properties of grue and bleen, where something is green just in case it was grue prior to 2100 CE or bleen after 2100 CE.

The Peircean Reply

Insofar as Goodman's new riddle of induction is aimed at showing induction cannot be adequately represented by formal inference procedures, Peirce is in complete agreement. As already remarked, he holds that the procedures we use will need to be tailored to the specific inquiries we are conducting.

Yet another issue raised by Goodman's riddle is the status of our scientific concepts. Could not any compound predicate such as grue be applied to an object, so long as we appropriately define it? The Peircean would point out that there is no problem with the idea of grue per se. Rather, the idea of grue makes it impossible to identify grue with a physical property in things. The way in which our idea of grue is made distinct is to define it in terms of some perceiver-dependent properties. The problem is that methods of science aim to fix beliefs in ways such that "our beliefs may be caused by nothing human, but by some external permanency—by something upon which our thinking has no effect" (ILS 66–67, 1877). However, our thinking—our observations—do have an effect on whether something is grue or bleen.

Grue and bleen may well be perfectly acceptable ideas. The problem is that they will be inutile as scientific conceptions. Even if our ideas of green and blue are forged from our ideas of grue and bleen, it will be our beliefs about things being green or blue that are capable of being fixed by an external permanency. This is because whereas we may identify green and blue with microphysical properties, with wavelengths of light, or some other physical property, we cannot identify grue or bleen with such properties. We cannot identify grue or bleen with these properties because the underlying structure with which we identify grue will change in the year 2100. Grue will have to be one structure prior to 2100 and a different structure following 2100. But then it cannot be identified with either physical structure simpliciter. I take this to be in line with Godfrey-Smith's position that observation becomes a confounding variable, such that "it is not just the emerald's physical nature that is affecting how color predicates apply to them, but the process of observation as well" (2003b, 580).

Perhaps surprisingly, Peirce proposes a concept somewhat like grue. He recommends that we pragmatically develop the idea of hardness by claiming that it "will not be scratched by many other substances" (ILS 91, 1878). However, he proceeds to claim that "[t]here is absolutely no difference between a hard thing and a soft thing so long as they are not brought to the test" (ILS 91, 1878). He suggests we may even say that "all hard bodies remain perfectly soft until they are touched, when their hardness increases with the pressure until they are scratched. . . . [T]here would be no *falsity* in such modes of speech. They would involve a modification of our present usage of speech" (ILS 91, 1878). Peirce is here proposing that hardness could be defined so that something is hard just in case it resists being scratched in proportion to the effort exerted in scratching it, to a certain limit. This is a grue-like property because hardness, on this alternative conception, becomes perceiver-dependent. An untouched diamond at the bottom of the ocean will not be hard as nothing is exerting pressure to scratch it.

Peirce ultimately rejects this account. He claims that it denies the reality of would-bes, or laws that are operative in nature. The problem is not, however, in casting the *idea* of hardness in these earlier terms. We might call such a conception of hardness touch-hardness. The problem is that hardness can no longer be identified with or explained by anything intrinsic to the object studied. We cannot explain the hardness of the diamond, for example, by its crystalline structure. That is because the structure of the diamond remains the same whether anyone touches it or not. We might, for example, observe the diamond at the bottom of the ocean and not touch it. If the touch-hardness of the diamond could be explained by that structure, then once we do touch the diamond, we should observe changes in its crystalline structure.

It would be puerile to claim that we will find such changes. As Peirce notes, "[f]rom some of these properties [e.g., being carbon, being a transparent crystal, having a certain chemical structure] hardness is believed to be inseparable. For like it, they bespeak the high polymerization of the molecule" (EP 2:356, 1905). The consequences of touch-hardness are different from the consequences of hardness. The latter can be made into a scientific conception that tracks externally permanent, observer-independent features of objects. The former cannot. And the same considerations apply to grue. Grue cannot track externally permanent features of objects, such as microphysical properties. The same cannot be said for the idea of green, even if our idea of green is constructed from the ideas of grue and bleen. The result is that so far as scientific inquiries are concerned, scientists will find it utile to discuss the greenness of objects rather than their grueness. This is so even if the idea of greenness is derivative of grue and bleen.[18]

Second, the Peircean has an abductive reply along the same lines as her reply to the paradox of the ravens. That reply is two-pronged, where the first prong concerns the suggestion of hypotheses and the second the economy of research. The first prong is that those hypotheses we take up in inquiry as plausible and testable will be the ones our abductive instinct suggest. The hypotheses that suggest themselves are not those involving grue and bleen. Rather, the hypotheses that suggest themselves are that such-and-such objects are green, or that such-and-such objects are blue. It is these hypotheses we will put to the test. It is these hypotheses our experimental researches confirm or disconfirm.

Rescher draws our attention to the second prong of the reply. It is that the economy of research recommends we adopt green rather than grue as a predicate to put to the test. This abductive consideration is that the costs of adopting a language of green and blue are much less. They are less because in our language "color-talk can be carried on wholly in an ostensively taught language where all that matters is the surface appearance of things" but in the language of grue and bleen "a two-factor mechanism . . . adds to this ostensively accessible phenomenology . . . a layer of chronometric issues at the level of learning, teaching, explanation, and application" (1976, 80).

Abductively, the hypothesis we will select to put to the test will draw on our language of green and blue rather than grue and bleen. Once our hypothesis is settled on, we will perform our experimental researches. Our data gathered will confirm that hypothesis rather than some arbitrary hypothesis concocted after the fact. Cheryl Misak notes, "the choice between the 'green' and the 'grue' hypotheses is a matter for abduction, not induction. The inference to 'all emeralds are green' is not an induction from the premise 'all observed emeralds have been green'. Rather, we would, if we were interested enough in the matter, infer 'all emeralds are green' as an abduction" (2004/1991, 97–98). Accordingly, if these be the hypotheses we are taking up into inquiry, we are delimiting that our investigation will concern objects with respect to their being green or objects with respect to their being blue and then putting *those* hypotheses to the test. As David Wiggins remarks, "[i]nduction itself . . . can *support* generalizations but . . . it does not license us, in or of itself, to go from positive instances of an arbitrary putative generalization toward the assertion of that generalization. Before that can happen, the generalization has to enjoy the status of a hypothesis" (2004, 104). We must first select the hypothesis to test. Only then do we gain inductive support for it. Our evidence supports *that* generalization qua hypotheses-we-are-putting-to-the-test and not any arbitrarily selected generalization we might concoct after we have the evidence. This does not prohibit us from postulating a hypothesis such as that praying mantises are grue. But these are not the hypotheses we have selected to put to the test in our inquiries. Our evidence does not arbitrarily favor them independently of the context of inquiry.

Conclusion

In this chapter, I have shown that Peirce has resources to respond to twelve different objections to his account of induction's validity. Peirce maintains that induction is truth-approximating in the long run because the community of inquirers will be able to correct our errors. However, he does not aver there is any such community. Ultimately, Peirce's defense rests on a commitment to logical sentimentalism. We are to hope that there will be an infinitely enduring community of inquirers in which we put our faith and which we love. The Peircean is not committed to the claim that there is such a community, only she hopes there will be. Moreover, Peirce responds to various objections raised against the validity of induction, including that there are no uniformities in nature, Hume's circularity objection, and the problem of why we may draw conclusions in the short run when the validity of induction rests in the long-run work of a community of inquirers.

7

Abduction

Abduction is that genus of inference which introduces new ideas: "Abduction is the process of forming an explanatory hypothesis. It is the only logical operation which introduces any new idea" (EP 2:216, 1903). Abduction does not merely generate new ideas. Peirce writes that its function is the "invention, selection, and entertainment of the hypothesis" (HP 2:895, 1901). Peirce does not identify abduction as a distinct genus of inference until 1900. In 1903, he admits he "confused Abduction with the Second Kind of Induction, that is the induction of qualities" (PPM 277n3). In a late letter to Paul Carus, he confesses these were mixed up in "almost everything I printed before this century [i.e., before 1900]" (ILS 279, 1910). Abduction should not be conflated with what, in Peirce's early work, he calls hypothesis. Hypothesis is rather a species of induction. It is qualitative induction, the species of inference which eliminates, corroborates, or confirms hypotheses. Although both abduction and qualitative induction deal with hypotheses, they face different directions. Qualitative induction puts hypotheses to the test. Abduction suggests hypotheses to put to the test. "Abduction . . . is merely preparatory," Peirce writes; "[i]t is the first step of scientific reasoning, as induction is the concluding step" (CP 7.218, 1901).

In this chapter, I turn to examining Peirce's account of the validity and strength of abduction. I proceed, as with deduction and induction, by first considering the nature of abductive inference and why Peirce maintains it is valid. In the case of abduction, the truth-producing virtue of abduction is that it is truth-conducing, provided there is any truth to be ascertained. This contrasts with deduction, which is truth-preserving, and induction, which is truth-approximating in the long run. I argue that abduction is truth-conducing in two ways, first by generating worthwhile hypotheses to put to the test and second by devising a plan by which to go about testing hypotheses. These are inference to the best explanation given the evidence we have and inference to the best explanation to pursue, and these two sorts of inference come apart. Following a discussion of validity, I turn to Peirce's account of abductive strength. Because Peirce distinguishes between

Peirce on Inference. Richard Kenneth Atkins, Oxford University Press. © Oxford University Press 2023.
DOI: 10.1093/oso/9780197689066.003.0008

inference to the best explanation given the evidence we have and inference to the best explanation to pursue, these will be strong in different ways. The former are strong depending on whether the hypothesis is suggested by *il lume naturale*, whether it is plausible, whether it can be put to the test, and whether we have countervailing reasons to disbelieve it. The latter are strong depending on considerations of efficiency. What connects these two species of inference? I argue that both sorts of inference are inference to perceptive judgments. In the first case, we make perceptive judgments as to which hypothesis is likely to be the correct one. In the second case, we make perceptive judgments as to which hypothesis, if pursued, is likely to lead us to the truth in the most efficient way. Lastly, I consider objections and replies to the validity of abduction.

The Validity of Abduction

I begin by examining the nature and validity of abduction. Abduction is frequently conflated with inference to the best explanation. This is a mistake, as other scholars have shown.[1] The mistake arises because 'inference to the best explanation' is used in many ways. To be clear about both the validity and the strength of abduction, we need to home in on the relevant senses in which abductions may be regarded as inferences to the best explanation.

Abduction as Inference to the Best Explanation

Although abduction is often treated as inference to the best explanation, we should distinguish among at least six different senses in which we may make inferences to explanations. Those senses are (1) inference to an explanation, (2) inference to a plausible explanation, (3) inference to a scientifically admissible explanation, (4) inference to the best explanation for us to pursue, (5) inference to the best explanation given the evidence we presently have, and (6) inference to the best explanation simpliciter.

I begin with (6) inference to the best explanation simpliciter. In Chapter 2, I distinguished between hypothetic induction as a mode of inquiry and hypothetic induction as a species of inference. As a species of inference, hypothetic induction eliminates, corroborates, or confirms hypotheses. As a mode of inquiry, hypothetic induction includes the preparatory stages of framing hypotheses to put to the test and deducing their predictions. When we regard hypothetic induction as a mode of inquiry aimed at explaining some phenomenon, it is a process in which we gather together as many hypotheses as we can to explain the phenomena. We then make predictions, which state what we expect to observe under certain experimental conditions. Through the process of testing (hypothetic induction as a species of inference), we sort through those explanatory hypotheses until we are ultimately left with one hypothesis, provided the true hypothesis was among those gathered at the start.

The entire procedure of forming and arguing for explanatory hypotheses may be regarded synchronically (through time) or diachronically (at the end of the processes). When we consider the process synchronically, we are in the thick of the process of inquiry. We may have started, for example, with six hypotheses and eliminated three. We set about testing the other three and, eventually, eliminate two of those three. Yet if we view the process diachronically, at its end, then we will have settled on some single hypothesis, the hypothesis which explains the phenomena. In this last case, the explanatory hypothesis on which our investigations have settled is the best explanation. It is in this sense in which we have made an inference—or rather, inferences as a mode of inquiry—so as to arrive at the best explanation simpliciter.

As should be plain, inference to the best explanation simpliciter is not what Peirce means by abduction, for it includes abduction as well as other modes of inference. Inference to the best explanation simpliciter is rather a diachronic perspective on the processes of inquiring into the truth of various explanatory hypotheses. At the end of the process of inquiry, we should be left with the true explanatory hypothesis, provided we have conceived it. We have inferred to the best explanation.

In contrast with (6), all of (1)–(5) are facets of what Peirce considers to be abductive inference. At its most basic, abduction is (1) inference to an explanation of phenomena. Peirce states that abduction "merely suggests that something *may be*" (EP 2:216, 1903) and that it is "nothing but guessing" (EP 2:107, 1901). At its most basic, abduction is nothing more than the forming of hypotheses or explanations.

Hypotheses may be formed in one of three ways, only two of which Peirce thinks are deserving of any respect. The unrespectable way of forming a hypothesis is by mere conjuring. Perhaps, for example, an event has been caused by "the aspect of the planets" or by "what the dowager empress had been doing just five hours previously" (EP 2:108, 1901). This is no more than "mak[ing] random shots at the determining conditions" (EP 2:108, 1901), and it is contrary to good sense to put stock in them. In contrast, the ways in which a hypothesis may be proposed and be respectable are "of two kinds, the purely instinctive and the reasoned" (EP 2:108, 1901). First, a hypothesis may be suggested by instinct. Peirce regards Galileo's doctrine of *il lume naturale*, the natural light of reason, to be the doctrine that those hypotheses which instinct suggests are to be given consideration. Galileo claims that the simpler hypothesis is to be preferred. Peirce confesses he previously misunderstood Galileo to be claiming that the logically simpler hypothesis is to be preferred. He later realizes that by preferring the simpler hypothesis, Galileo means that it is "the more facile and natural, the one that instinct suggests, that must be preferred; for the reason that unless man have a natural bent in accordance with nature's, he has no chance of understanding nature, at all" (EP 2:444, 1908). Second, a hypothesis may be suggested by reasoning, that is, by careful excogitation. Such reasoning to a hypothesis will often occur when one is in the process of inquiring into other hypotheses. In such a process, a new hypothesis may suggest itself based on what one has discovered. Peirce's example is Kepler's efforts to reason to the orbit of Mars based on only three or four data points (see EP 2:83 and 108, 1901). Frederick Grinnell (2019) describes the process of forming a hypothesis in this way as being akin to finding a puzzle piece that finishes the puzzle. Several surprising observations have been made (the puzzle pieces one has), and the hypothesis is the piece that fits them all together. Peirce remarks that the "reasoned marks of truth" are two (EP 2:108, 1901). First, positive facts already known may recommend hypotheses. Second, whether the hypothesis accords with our preconceived notions may recommend the hypothesis. He cautions against the second, though, on the grounds that "experience shows that likelihoods are treacherous guides" because researchers being wedded to preconceived ideas has "caused so much waste of time and means" (EP 2:108–109, 1901). The distinction between instinctual and reasoned hypothesis formation is not sharp. Our instinctual guessing instinct can be honed through training and practice.

Although at its root abduction is no more than the forming of hypotheses, such guessing by instinct or by reason is rarely merely guessing. It is associated with feelings of conviction as to those hypotheses at which we guess. Peirce states that "we cannot help accepting the conjecture at such a valuation as that at which we do accept it; whether as a simple interrogation, or as more or less Plausible, or, occasionally, as an irresistible belief" (EP 2:443,

1908). When some hypothesis *H* is formed, we may merely ask: *Might it be the case that H?* In such a case, we have no or little conviction that the explanation proposed is the true explanation. This may occur when we lack much data on the basis of which to form explanatory hypotheses. For instance, we might try to solve a cipher by conjecture. Noticing some symbol appears frequently, we may ask: *Might it be the case that such-and-such symbol is the letter E?* We may not think it likely or plausible; the symbol could be the letter T or A, or it could be a cipher for a different alphabet altogether. Nonetheless, we might give the conjecture a try.

One important point to bear in mind is that inference to an explanation does not require that the explanation be consistent with what one already takes herself to know. Otherwise, it would be difficult to explain how we ever change our minds about a matter. In fact, recognizing that some phenomenon is inconsistent with we already take ourselves to know may be precisely what spurs us on to frame new hypotheses. Those hypotheses may be alternatives to other hypotheses we have already adopted and inconsistent with them.

Nonetheless, sometimes hypotheses strike us as being plausible. We might conjure up any old explanation for a phenomenon, for example, that the price of tea is determined by the transit of Jupiter. But we do not take any old explanation to be a plausible explanation for the phenomenon. For example, that (a) the price of tea is determined by the labor put into its production or that (b) the price of tea is determined by supply and demand are more plausible explanations for the price of tea than is the transit of Jupiter. Generally, we will want to separate those abductions which are mere conjurings from those abductions which give us plausible explanations. In such cases, the abduction is (2) inference to a plausible explanation.

A hypothesis will be a plausible explanation only if it explains the known phenomenon it is designed to explain given what an inquirer already takes herself to know. It is for this reason the hypothesis strikes us as plausible at all. Yet again, however, consistency and coherence with our beliefs are not necessary conditions for plausible explanations. While whether a hypothesis strikes us as plausible will be greatly affected by whether it is consistent and coherent with what we already take ourselves to know, one may doubt some hypothesis and yet admit it is a plausible explanation. Disagreements in science on the basis of scant evidence sometimes have this feature. One scientist will admit that some alternative hypothesis—say, that supervolcanoes caused the extinction of the dinosaurs—is plausible. Yet she may doubt that the hypothesis is true, instead affirming that an asteroid strike is the more plausible explanation. Similarly, disagreements in particle physics may have this feature. The mathematics may work out well for two competing theories, yet physicists may disagree about which theory is the more plausible.

Factors that may influence whether a hypothesis is plausible may include the so-called theoretical virtues. For instance, an elegant theory—one in which the math is simple and beautiful—may be regarded as more plausible than an inelegant theory. On the other side, a theory that requires fine-tuning (precise adjustment of a model's parameters to fit observations) may be regarded as less plausible. Metaphysically parsimonious theories may be preferred to unparsimonious theories. But there is a question of how seriously and rigorously we should take these theoretical virtues. On the one hand, they may simply record our preferences, in which case they are of little value in tracking whether a hypothesis is true or even pursuit-worthy. We need frequencies to supplement and monitor our degrees of conviction. On the other hand, they may track the preferences of the honed instinct of inquirers in a domain. If inquirers in some domain have found that theories which are elegant are the more pursuit-worthy, then it is the honed instincts of the inquirer that are sensitive to the plausibility of a hypothesis. A preference for elegance is only a symptom of that instinct. Also, in other domains, it may be the case that pursuing elegant hypotheses is not truth-conducing. Moreover, insofar as scientists aim to frame models for some phenomena, and as such models typically require some degree of abstraction and simplification for otherwise the calculations would be too complex, they will not perfectly fit the data.

In some cases, abductions render hypotheses that strike us as so plausible the hypothesis may be irresistibly believed. Peirce thinks the hypothesis that God is real is just such a hypothesis for some people, a claim he defends in his late essay "A Neglected Argument for the Reality of God." Yet there are many hypotheses which we recognize are nearly irresistibly believed once they are called into account. For instance, suppose that I am asked whether I have a spleen. Having never considered the matter before, I may well form the hypothesis that, in fact, I have a spleen, though I know that the spleen is not a vital organ. So plausible is the hypothesis I have a spleen, I hardly need to subject it to further inquiry, even though I could subject the hypothesis to further inquiry by getting a CT scan. I irresistibly believe I have a spleen.

Both (1) inference to an explanation and (2) inference to a plausible explanation allow that we could form explanatory hypotheses that we have no conceivable way to test. Notice that having no conceivable way to test a hypothesis is not the same as having no way to test a hypothesis given our current abilities. Neither is a hypothesis having no conceivable way to test the same as a hypothesis having no tests of which we can presently conceive. It may be that we have not fully understood the hypothesis or ascertained its conceivable practical consequences. When we make (3) an inference to a scientifically admissible explanation, the hypothesis we infer must have conceivable practical consequences which we can conceivably put to the test.

When there is a hypothesis we have no conceivable way to test, either the hypothesis has no experimental predictions or there are no predictions from the hypothesis that would differentiate it from other hypotheses. For example, the hypothesis that there are entities which cannot be observed either directly or indirectly is a hypothesis that has no experimental consequences. The hypothesis that all of our experiences are caused by an evil demon is one that cannot be differentiated from its contradiction, that all of our experiences are not caused by an evil demon. In such a case, neither hypothesis is admissible to scientific inquiry. In order to see why, it may be helpful to note that both will support a crude induction. On the one hand, it has never been established that we are deceived by an evil demon. Therefore, we are not deceived by an evil demon. On the other hand, it has never been established that we are not deceived by an evil demon. Therefore, we are deceived by an evil demon. The crude induction in both cases will be supported because there is no way to establish either hypothesis. Therefore, neither hypothesis has been established. Crucially, note that this does not preclude one hypothesis being more plausible than another, even if neither is admissible to scientific inquiry. Indeed, that we are not deceived by an evil demon strikes us as so plausible, it is irresistibly believed.[2] Worries about global equivalence underdetermination, such as prodded by the evil demon scenarios, are addressed at this abductive stage of inquiry. We cannot admit such questions to further inquiry. Instead, we settle on the hypothesis that is the most plausible, viz., that we are not deceived by an evil demon. The evil demon scenario is a mere conjuring, whereas instinct irresistibly prompts us to believe that there are rocks, we have two hands, and the like.

These considerations bring us to (4) inference to the best explanation to pursue and (5) inference to the best explanation given the evidence we have. It is important to realize that these come apart: We might decide to pursue a hypothesis that we think is not the best explanation the evidence supports. Such a decision can be made on the basis of economic considerations. Suppose we have just two hypotheses, H_1 and H_2, to explain some phenomena. The cost of testing some prediction P_1 from H_1 is millions of dollars, whereas the cost of testing prediction P_2 of H_2 is hundreds of dollars. Suppose, moreover, that P_1 and P_2 are independent of each other, but these are the only testable predictions we have for the hypotheses. Even if we think H_1 is the better explanation, we might test H_2 just because it is far less expensive to do so. If H_2 is eliminated, and given that H_1 and H_2 are our only hypotheses, the rejection of H_2 will lead us to conclude that H_1 is most likely to be the true hypothesis. Although we might seek confirmation of H_1 by testing for P_1, the better course of action was to test H_2 first. Peirce notes an investigator will "not even refuse to entertain a grossly improbable hypothesis, so long as it possesses the one merit of being the theory which is at the moment most conveniently and economically compared with observation" (HP 1:478, 1896). Peirce refers to this as the economical consideration of cheapness (see EP 2:113, 1901), and it bears on the question of which is the best explanation to pursue.

What is (4) the best explanation to pursue will be determined in the main by the economy of research. Peirce holds that whether we think a hypothesis is the best to be pursued will largely be determined by cheapness, intrinsic value, and the three considerations of "Caution, Breadth, and Incomplexity" (EP 2:109, 1901). Cheapness is the cost, as just mentioned. Intrinsic value is how natural or likely the hypothesis strikes us. As to caution, breadth, and incomplexity, he illustrates the first by the game of twenty questions. In the game of twenty questions, one is asked to ascertain of which object another person is thinking by asking no more than twenty questions. Peirce notes that the "secret of the business lies in the caution which breaks a hypothesis up into its smallest logical components, and only risks one of them at a time" (EP 2:109, 1901). In such a game, we will want to start at the broadest set of objects possible and then gradually narrow in, for if "each question could exactly bisect the possibilities, so that *yes* and *no* were equally probable, the right object could be identified among a collection numbering 2^{20}" (EP 2:109, 1901).

Next, economic considerations also recommend testing hypotheses which have breadth. Hypotheses have breadth when they are able to explain a wide diversity of phenomena across different areas of inquiry. Newton's theory of universal gravitation had breadth in that it could combine the motions of the planets and the falling of objects on earth under one theory. Peirce notes that the kinetic theory of gases has breadth both because it can explain Boyle's law that the pressure of a gas at a constant temperature is inversely proportional to its volume and it can be extended and generalized to explain nonconservative phenomena, that is, those phenomena which are irreversible, such as evolution (see EP 2:110, 1901; W 8:190, 1892).

With respect to incomplexity, Peirce recommends that we pursue those hypotheses which "give a good 'leave,' as the billiard players say. If it does not suit the facts, still the comparison with the facts may be instructive with reference to the next hypothesis" (EP 2:110, 1901). Peirce has in mind specifically the incomplexity of mathematical formulae. He states that it may be better first to try whether some experiments satisfy the formula $y = cx$ rather than $y = a + bx^2$, even though one suspects the latter to be the case. The former is less complex, and testing for it may leave one with data which suggest whether the latter is true. Similarly, Peirce notes that Mendeleev's periodic law is complex, but an appeal to atomic weights (rather than the periodicity of the element) might also suggest that there are undiscovered elements between those which have been discovered (see EP 2:111, 1901). It is noteworthy that Peirce's own table of elements proposed in 1869 (see W 2:282–284), when read upside down, bears some striking resemblances to contemporary versions of the periodic table.[3]

Yet as these comments indicate, though some hypothesis may be the best to pursue, it does not follow that we regard it as (5) the best explanation given the evidence we do have. Here, it is important to note that when we form a hypothesis on the basis of old evidence, we still must deduce predictions from it and put the hypothesis to the test. Nevertheless, old evidence may lead us to the position that some hypothesis is the best we have. For instance, prior to the framing of Einstein's theory of relativity, it was known that Newton's theory of universal gravitation did not explain the precession of Mercury's perihelion. Einstein's theory of relativity did explain it. That Einstein's theory explained some phenomenon that Newton's did not led some physicists (those who could understand Einstein's theory) to suspect it provided a better explanation than Newton's theory did. It was the best (or better) explanation given the evidence they had. Nonetheless, they sought to put Einstein's theory to the test.

In sum, although many scholars have claimed that abduction is inference to the best explanation, such a claim invites confusion. We must distinguish among at least six different ways in which we make inferences to explanatory hypotheses. First, we make inferences to explanations. Second, we make inferences to plausible explanations. Third, we make inferences to scientifically admissible explanations. Fourth, we make inferences to the best explanation to pursue. Fifth, we make inferences to the best explanation given the evidence we have. And, lastly, we make inferences to the best explanation simpliciter. This final case, however, is not abduction but hypothetic induction as a mode of inquiry viewed diachronically. The first five are all features of abductive inference. As I shall argue, they establish criteria for the strength of abductions. For the moment, though, I shall focus on the validity of abduction.

The Validity of Abduction

As shown in Chapter 2, Peirce did not clearly distinguish between qualitative induction and abduction until the early 1900s. He never arrives at a thoroughly satisfactory account of abduction's nature, though as we have just seen he has much to say about it. What is clear is that abduction, while related to what nowadays is called inference to the best explanation, is also importantly different from it. As Mousa Mohammadian (2021) argues, abduction and inference to the best explanation are similar in two respects. They both involve ranking hypotheses, and both may appeal to theoretical virtues in ranking hypotheses. However, unlike inference to the best explanation in the contemporary sense of that phrase, (a) abduction is a mode of inference which generates hypotheses, and (b) the ranking of generated hypotheses in abduction is done prior to testing with the aim of ascertaining pursuit-worthy hypotheses and those most likely to be true given the evidence we have.

Although the relation of abduction to inference to the best explanation is vexed, what becomes evident in Peirce's writings is that the primary function of abduction is to suggest hypotheses to us. In 1903, Peirce characterizes the conclusion of an abduction as giving us "reason to suspect that [hypothesis] A is true" (EP 2:231) and in 1910 characterizes an abduction as delivering a "conjecture" (ILS 283). Abduction gives us conjectures. We gather those conjectures by the armfuls. We then sort the conjectures, make them more explicit and definite, and put them to the test.

Accordingly, the truth-producing virtue of abduction is not that it is truth-approximating in the long run (as induction is) but that it is truth-conducing, provided there is any truth to be ascertained. The function of abduction is not to tell us which hypothesis is true. Rather, the function of abduction is to generate hypotheses and to tell us which to go about testing. Although we will want to abduce the true hypothesis (for otherwise the true hypothesis will not be among those we put to the test), the function of abduction is not to tell us which is true. The function of abduction is to introduce new ideas and to tell us how to sort through them to get closer to the truth. An abduction will be valid just in case it in fact introduces new ideas to pursue and does so in a way that is conducive toward the truth.

Peirce's account of abduction's validity differs from his account of induction's validity. Whereas the validity of induction reposes on the logical sentiments and the community of inquirers, the validity of abduction reposes on the fact that "one *must* trust one's instincts through life or be content with a passivity that cannot content him" (ILS 283, 1910). The inquirer must trust herself to those hypotheses which abduction suggests. Otherwise, she has no hope of attaining the truth. Peirce writes:

> Animals of all races rise far above the general level of their intelligence in those performances that are their proper function, such as flying and nest-building for ordinary birds; and what is man's proper function if it be not to embody general ideas in art-creations, in utilities, and above all in theoretical cognition? To give the lie to his own consciousness of divining the reasons of phenomena would be as silly in a man as it would be in a fledgling bird to refuse to trust to its wings and leave the nest, because the poor little thing had read Babinet, and judged aerostation to be impossible on hydrodynamical grounds. (EP 2:443, 1908)

Just as the fledgling bird trusts itself to its wings without understanding how they work, the inquirer will trust herself to her abductive instincts, though how those instincts operate remains opaque to her.

Peirce's claim that abduction's validity rests on the fact that we must trust ourselves to our instincts suggests that abduction is instinctive at root. In fact, this is Peirce's position. Peirce claims that "a primary hypothesis underlying all abductions [is] that the human mind is akin to the truth in the sense that in a finite number of guesses it will light upon the correct hypothesis" (EP 2:108, 1901) and "unless man have a natural bent in accordance with nature's, he has no chance of understanding nature, at all" (EP 2:444, 1908). Every facet of abduction is informed by our guessing instinct. Not only do we guess at which hypotheses might explain the facts, we also guess at which will be the best hypotheses to pursue.

Peirce frequently characterizes abduction as instinctual, so that it has perplexed some scholars as to how or even if it may also be regarded as an inference.[4] Two considerations are important here. First, that abduction is instinctual at root does not imply it is thoroughly instinctual. Our hypothesis-generating procedures are shaped by experience. The scientist whose guesses have been informed by long hours in the laboratory is more likely to generate the true hypothesis than is a person who has never considered the matter. Moreover, the scientist who is well-practiced will be better able to discern the best way to go about testing hypotheses.

Second, the abductive procedure should be distinguished from the abductive conclusion. The abductive procedure is a hypothesis-generating and hypothesis-ranking process. As I shall explain shortly, natural selection has favored the development of such an abductive power in rational animals. The procedure is instinctive in that sense. Peirce denies that we reflect on the procedure as we draw abductions, but he avers that "afterward when one subjects one's behaviour for the day or for any other marked period to self-criticism this is how one ought to vindicate good retroductions [i.e., abductions]" (ILS 283, 1910). The abductive conclusion is a conjecture generated by the abductive procedure. As already noted, abductive conclusions come with degrees of conviction. Not all abductive conclusions are instinctive, but the force with which some conjectures strike us (they are irresistible beliefs or regarded as plausible) is an indicator of their instinctiveness. Peirce holds that one such belief is belief in God. Other of our beliefs about space and time are also innate, or instinctive.

The forcefulness with which some hypotheses strike us should not be conflated with the validity of the abductive procedure. The abductive procedure is valid just in case it is in fact truth-conducing and we adhere to it. Now the question arises as to whether it is in fact truth-conducing. No doubt, for any given person, their instinctive abductive capacities may be perverted or pathological. But inquiry is not restricted to the endeavors of a single person alone. All the members of the community of inquirers as a whole will make abductions. On this basis, Peirce argues that in fact abduction as an instinctive hypothesis-generating and hypothesis-ranking procedure is truth-conducive.

Two points are in order. The first is that human minds—any rational minds—do not exist apart from nature. They are forged in the evolutionary crucible, under the influence of nature and the way that nature operates. In "The Fixation of Belief" Peirce had worried that natural selection might fill the animal mind with "pleasing and encouraging visions, independently of their truth" (ILS 51, 1877). In "The Order of Nature," however, he avers, "[i]t seems incontestable, therefore, that the mind of man is strongly adapted to the comprehension of the world; at least, so far as this goes, that certain conceptions, highly important for such a comprehension, naturally arise in his mind; and that without such a tendency, the mind could never have had any development at all" (ILS 161, 1878). Peirce's claim in "The Order of Nature" should be read as a response to his earlier worry in "Fixation." His defense of the claim consists in the fact that nature has furnished our minds with various conceptions needed for survival, including those of time, space, and force. Though these conceptions might be exceedingly vague, they admit of incremental improvement through inquiry.

This theme emerges quite clearly in Peirce's later writings on critical commonsensism. Peirce maintains that we have "veritably indubitable beliefs" which are "*vague*—often in some directions highly so" (CP 5.505, 1905). These very vague beliefs are "of the general nature of instincts" (CP 5.445, 1905). Scientific inquiry can make these vague beliefs more definite and determinate. Vague as our ideas of time, space, and force (for examples) may be, these ideas can be improved through inquiry.[5] Peirce contends that we will be sure to find in nature something to which these very vague ideas conform because they have been instilled in us by natural selection.

The second point is that while abduction's validity does not repose on the community of inquirers but on the truth-conducing nature of our abductive instinct, we may nevertheless hope that the community of inquirers will eventually alight on those conjectures that are true. We put our faith in the community of inquirers to continue generating hypotheses and putting them to the test. In fact, the very process of testing and eliminating hypotheses may generate new hypotheses.

Peirce maintains that we know by induction that "man has correct theories; for they produce predictions that are fulfilled" (CP 5.591, 1903). He contends that this supports the primary hypothesis which underwrites all abduction, that our minds have a natural bent in accordance with the order of nature. He writes, "inductive experience supports that [primary] hypothesis in a remarkable measure. For if there were no tendency of that kind . . . if [merely conjured] hypotheses had as good a chance of being true as those which seem marked by good sense, then we could never have made any progress in science at all" (EP 2:108, 1901). We could never have made any progress because we would be swamped by any number of hypotheses and could not possibly undertake the process of eliminating them all. Even if we should get lucky once or twice, we should have for the next question a limitless number of hypotheses through which we must sort. Such is Peirce's argument that our abductive powers, when considered relative to the community of inquirers as a whole, are truth-conducing provided there is any truth to be ascertained. Our abductive capacities have been forged in the crucible of natural selection, and we must trust them as having a bent in accordance with the order of nature itself. Moreover, the fact that we have made progress in science corroborates, if it does not confirm, the fact that our abductive procedures are indeed truth-conducing.[6]

The Strength of Abduction

I have just now explained Peirce's defense of abduction's validity. Abduction is truth-conducing because our instinctive power for guessing has been honed by natural selection and improved through inquiry and practice. Moreover, the success we have had in the sciences corroborates or confirms the thesis that our minds have a bent in accordance with nature's order. Now, I turn to Peirce's account of abductive strength. As Peirce remarks, it is "only in Deduction that there is no difference between a *valid* argument and a *strong* one" (EP 2:232, 1903). In what, then, does abductive strength consist?

Mousa Mohammadian (2019 and 2021) proposes that we distinguish different phases of abduction qua hypothesis generation and abduction qua hypothesis ranking. I have been arguing, though, we also need to distinguish between hypothesis ranking relative to the best explanation given the evidence we have and hypothesis ranking relative to the best explanation to put to the test. Among the hypotheses that are proposed, we will take some more seriously than others and rank them as to which we regard as the best explanations given the evidence we have in hand. Yet there is no reason to think that such a ranking of hypotheses will be identical to a ranking of hypotheses in the order we should put them to the test.

Accordingly, Peirce's discussions of abductive strength involve two components, though they are often presented together in his writings. The first component concerns the admissibility of hypotheses to serious consideration. The second component concerns the admissibility of hypotheses relative to the order of testing. When we both give a hypothesis serious consideration and decide to test it, we adopt the hypothesis for probation. Let us take up each consideration in turn.

Admissibility for Serious Consideration

Peirce identifies four constraints on the admissibility of hypotheses for serious consideration.[7] These constraints in fact track the different sorts of inference to an explanation discussed earlier. As such, my comments here will be brief.

The first constraint Peirce proposes on admissibility for serious consideration is that the hypothesis be the one that instinct recommends. As noted earlier, we should distinguish between those hypotheses which are mere conjurings and those which are recommended by *il lume naturale*, the natural light of reason. We may conjure any number of hypotheses to explain phenomena, such as the transit of Jupiter or what the dowager empress was doing five hours ago. But we do not give serious consideration to merely conjured explanations. Such conjured explanations are not strong abductive inferences.

In contrast, we do give serious consideration to those hypotheses that instinct suggests. As we have seen, Peirce argues that we have a rational instinct for guessing well, partly instilled in us by natural selection. It is important to bear in mind that this rational instinct can be cultivated. Through learning, whether by books, by teaching, or by hands-on experience, our ability to guess well is honed. When it is honed, our logica docens (our taught logic) improves our logica utens (our logic in use). In scientific investigations on abstract matters far removed from practical life, we will often depend on our honed rational instinct for guessing.

The second constraint on admissibility for serious consideration is that the hypothesis account for the phenomena it is designed to explain. Peirce acknowledges that the better explanation will be one that explains all the phenomena it is designed to explain. He remarks, "the first rule which we should set up is that our hypothesis ought to explain *all* the related facts" (EP 2:113, 1901). But he also acknowledges this may be too demanding, and so he weakens the rule by stating, "the hypothesis cannot be admitted, even as a hypothesis, unless it be supposed that it would account for the facts or some of them" (EP 2:231, 1903). We will typically form hypotheses against some background information we want the hypothesis to explain. This background information is the old evidence used in hypothesis formation. The evidence is old because we are already in possession of it and use it to form the hypothesis; we need new evidence to eliminate, corroborate, or confirm the hypothesis. Peirce's point is that a hypothesis should be given serious consideration only if it explains all or most of the old evidence used to form the hypothesis. A hypothesis will be plausible only if it explains what it is designed to explain.[8] As noted earlier, it is because the hypothesis explains the phenomena that the hypothesis strikes us as having a claim to truth at all.

A third constraint on the admissibility of a hypothesis is that the hypothesis must have predictions which we could conceivably put to the test. Note, again, that this constraint is only that there be some conceivable test for the hypothesis, whether or not we can presently perform the test. "To assert the truth of [an abduction's] conclusion ever so dubiously would be too much," Peirce claims. Rather, "[t]here is no warrant for doing more than putting it as an interrogation. To do that would seem to be innocent; yet if the interrogation means anything, it means that the hypothesis is to be tested" (HP 2:899, 1901). Abductive inferences result in hypotheses we suspect may be true. We then ask whether the hypothesis is true, and in so doing we take it that there is some test we can perform to eliminate, corroborate, or confirm the hypothesis.

Among those tests which are conceivable, there will be (1) those which we may perform now (with more or less economy), (2) those which we may perform in the future (with some investment of time and resources), and (3) those which we see no way for us to perform at all (because, for example, it would require an instrument we cannot possibly construct, such as a particle collider the size of the solar system). These matters, however, concern how we rank hypotheses for testing rather than the admissibility of the hypotheses themselves. What is requisite for admitting a hypothesis to serious consideration is only that there is some conceivable way to eliminate, corroborate, or confirm it through "subjection to the test of experiment" (EP 2:235, 1903).

The fourth constraint Peirce places on the admissibility of a hypothesis for serious consideration is whether we have strong countervailing reasons to disbelieve it. We will regard those hypotheses which we have few or no countervailing reasons to deny as deserving of more serious consideration. I have already noted that inconsistency and incoherence with what we already believe are not grounds for ruling out hypotheses that suggest themselves. Nonetheless, they may be grounds for giving a hypothesis more serious consideration. We rightly take it that the mass of our beliefs is true or approximately true. The body of our commonsense beliefs is a consequence of our ancestors over many generations endeavoring to cope with the world and to make the world more intelligible to us. It is, in Peirce's felicitous phrase, the development of concrete reasonableness, where 'concrete' is to be take in its etymological sense of grown together. Our commonsense beliefs have grown together, developed by our ancestors in ways that enable us to make sense of and cope with the world around us. It is part of our task to continue to develop that body of commonsense beliefs, such that we are led "to the avoidance of all surprise and to the establishment of a habit of positive expectation that shall not be disappointed" (EP 2:235, 1903). Although there is much to be said about commonsense beliefs, the key point here is only that the mass of our beliefs, many of which are commonsense beliefs, do establish habits which are not disappointed. This fact is a good indicator of their truth. When a hypothesis is consistent and coheres well with the mass of our beliefs, that, too, will be a good indicator that it deserves serious consideration.

Nonetheless, Peirce recognizes that one if not many of our commonsense beliefs may be false. The commonsensist "quite acknowledges that what has been indubitable one day has often been proved on the morrow to be false" (CP 5.514, 1905). Accordingly, the inquirer ought not to restrict her considerations only to those hypotheses which cohere with or are consistent with the other beliefs she holds. It may be that the belief she uses as a touchstone to regard some hypothesis as unlikely is itself false. If she were to reject hypotheses merely on the grounds they fail to cohere with her other beliefs, then she would be robbing herself of the possibility of doubting her commonsense beliefs themselves. Even when a hypothesis does not cohere with our commonsense beliefs, it may be admissible. But we will regard those hypotheses that do cohere with and are consistent with our beliefs as better explanations given the evidence we have than those hypotheses which do not cohere with our other beliefs.

As we can see, Peirce's criteria for admitting hypotheses as deserving of serious consideration tracks the distinctions among kinds of inference to an explanation given earlier. We will prefer those hypotheses instinct suggests, that is, those which are the result of inference from reason or instinct rather than mere conjuring. We will prefer those hypotheses which explain all or most of the phenomena they are designed to explain, that is, those which are the result of inference to a plausible explanation. We will prefer those hypotheses which admit of scientific investigation, that is, those from which we can deduce predications testable in hypothetic inductions. Finally, we will prefer those hypotheses that cohere with and are consistent with the evidence we have, that is, those that result from inference to the best explanation given the evidence we have.

Adopting a Hypothesis for Probation

I have just been explaining Peirce's account of which hypotheses deserve serious consideration. Yet we have also seen that inference to the best explanation given the evidence we have does not always track inference to the best explanation to put to the test. Economic considerations for testing come apart from considerations of which hypothesis is the best candidate for being true. Accordingly, we will want to sort hypotheses both relative to those which we consider to be the best explanation given the evidence we have and relative to those which we consider the best to put to the test. Because of the costs associated with testing, Peirce states, "the admission of an abductive conclusion to the rank of an active interrogation is a concession not to be too lightly accorded" (HP 2:899, 1901).

Peirce is especially concerned with how to ascertain which hypothesis to pursue so as to make our procedure of conducing toward the truth more efficient. He writes, "the aid which a correct logic can afford to science consists in enabling that to be done at small expenditure of every kind which, at any rate, is bound to get done somehow" (CP 7.220n18). The question becomes what sorts of ways of proceeding will be most efficient. Peirce acknowledges that this will in large measure depend on the question we are asking: "There would be no logic in imposing rules, and saying that they *ought* to be followed, until it is made out that the purpose of hypothesis requires them" (CP 7.202, 1901). Nonetheless, we do find Peirce discussing the matter, and in his discussion of drawing inferences about ancient history he explicitly lays down some rules for inquiry.

Plainly, the cost of performing the test will be a major factor in considering which hypothesis to adopt for probation. Costs include such things as the construction of equipment and the materials needed to perform the tests. The inquirer may need a laboratory, which involves costs associated with space, with cooling and heating, with staffing, and the like. Costs also include the costs of time and personal effort. Peirce remarks that factors pertaining to the economy of research relate to the costs in terms of "money, time, thought, and energy" (CP 5.600, 1901) or, more generally, "relations between utility and cost" (W 4:72, 1879). As Nicholas Rescher aptly remarks, "whatever intrinsic appeal various individual items within the welter of abductively eligible hypotheses may possess, we must face up to the harsh reality of the limitations imposed by the cost of experimental research in expert time and material resources" (1976, 72).[9]

In addition to such purely economic considerations as cheapness, Peirce notes other considerations pertaining to adopting a hypothesis for probation. In a discussion of forming hypotheses concerning ancient history, Peirce provides six rules to follow when adopting a hypothesis for probation. We have already seen some of these in the earlier discussion of caution, breadth, and incomplexity as well as in the immediately previous discussion of admitting hypotheses for serious consideration. The first rule is that the hypothesis we decide to put to the test should "explain *all* the related facts" (EP 2:113, 1901). Peirce's second rule is that once we have decided to test a hypothesis, we ought to keep our hands to the plow and not look back until that hypothesis has been decisively refuted. He writes, "[t]he second rule is that our first hypothesis should be that the principal testimonies are true; and this hypothesis should not be abandoned until it is conclusively refuted" (EP 2:113, 1901). Once we put our hands to the plow testing the hypothesis which explains all the facts, we should not give up until "it becomes evident that it is quite untenable," for "[n]o practice is more wasteful that that of abandoning a hypothesis once taken up" (EP 2:113, 1901). This is especially so in cases of ancient history, when witness testimony is rejected "without any definite, objective, and strong reason for the suspicion" (EP 2:113, 1901).

Peirce's third rule is that when evaluating which hypothesis should be adopted for probation in matters of ancient history, we should give great weight to high objective probabilities but little weight to slight objective probabilities and to subjective degrees of belief, if any at all. He states:

> The third rule will be that probabilities that are strictly objective and at the same time very great, although they can never be absolutely conclusive, ought nevertheless to influence our preference for one hypothesis over another; but slight probabilities, even if objective, are not worth consideration; and merely subjective likelihoods should be disregarded altogether. (EP 2:114, 1901)

As to subjective degrees of conviction or likelihoods, in cases of ancient history, Peirce is especially concerned that our preconceived notions of what is acceptable in our culture will lead us to reject testimony that is otherwise perfectly acceptable.

As to slight objective probabilities, a distinction is in order in cases of testimony. If a friend tells you that in a game of poker she was dealt a royal flush, what she has testified to has only a slight objective probability. Nonetheless, that she has testified truly might have a high objective probability, if (say) you know her to ordinarily speak the truth. When a reliable witness testifies to improbable things, we should nevertheless concede that the objective probability of her testimony being true is high. But when we have only modest confidence in a witness, this slight objective probability of her speaking truly should not count much in favor of the truth of her testimony and should be ignored.

But in cases when ancient history is not the subject of inquiry, it is much less clear that we should follow Peirce's third rule with any great assiduity. If most of our beliefs are true, as I indicated earlier, and if our rational instincts have been honed by careful study and labor, then our subjective feelings of conviction as to the truth of a hypothesis may well be good indicators as to the truth of a hypothesis and may supply reasons for adopting it on probation. Nonetheless, Peirce is surely right that we should be especially sensitive to objective probabilities when these are known. As I explained in Chapter 5, Peirce does not deny that subjective Bayesianism and Bayesian updating are supported under abstract conditions. Rather, he maintains that the procedure must be supplemented with and monitored by sound practices of inquiry in order for our inferences to have any strength.

Peirce's rules four through six accord with his distinctions among caution, breadth, and incomplexity. As to caution, Peirce notes that we "should split up a hypothesis into its items as much as possible, so as to test each one singly" (EP 2:114, 1901). For any given hypothesis, once we do put it to the test, we will want to use severe tests as well as unique tests. For severe tests, we will want to ensure that the hypothesis would not pass the test were it not true. As Peirce states, we will be interested in "examining such of the probable consequences of the hypothesis as would be capable of direct verification, especially those consequences which would be very unlikely or surprising in case the hypothesis were not true" (EP 2:114, 1901). Some predictions may be shared by sets of hypotheses, but we will also want to identify those predictions which are unique to individual hypotheses. Once we have split up the predictions as much as possible, we will find it economical to proceed as in twenty questions. Even if we adopt some one hypothesis for probation, we are well advised to first test a prediction shared by exactly half of the hypotheses we have on the table. That way, whether the hypothesis adopted for probation passes the test or not, we will have narrowed our field of hypotheses by at least half. Gradually, we will narrow in on predictions unique to individual hypotheses. If we proceed in this way, we will have proceeded cautiously.

Peirce's fifth rule concerns breadth. Suppose we have two hypotheses, each of which has a good claim to be the true hypothesis. Peirce advises us to enlarge the "field of facts" the hypothesis is supposed to explain (EP 2:114, 1901). If a hypothesis would explain a larger field of facts, it has greater breadth. In that case, we might prefer to test it rather than an alternative hypothesis.

Peirce's sixth rule relates to incomplexity. He notes that when a test for some single hypothesis will have to be done anyway to test for another hypothesis, we may as well perform that test. He remarks, "if the work of testing a particular hypothesis will have substantially or largely to be done in any case, in the process of testing another hypothesis, that circumstance should, other things being equal, give this hypothesis which thus involves little or no extra expense, a preference over another which would require special work of no value except for testing it" (EP 2:114, 1901). Peirce's examples of the equations, mentioned earlier, exemplifies this. As he notes, testing $y = cx$ will be preferable to testing $y = a + bx^2$ because "the residuals will be more readily interpretable" (EP 2:110, 1901). The residual is the difference between what is observed and what has been predicted by the statistical model. For both equations, we will need to plot the predicted values against the actual values. That work has to be done no matter which equation we test. Having tested the equation with more easily interpretable residuals, we will have already done a good bit of the work that will be needed to test the equation with less easily interpretable residuals. It might be helpful to bear in mind that Peirce was writing before the advent of modern computing, when all such work was done with pen and paper.

Abduction as Perceptive Judgment

I have now been presenting abduction as involving two components. One component is the formation and ranking of hypotheses according to which hypothesis one thinks is most likely to be true. The other component is the ranking of hypotheses according to those which are the most pursuit-worthy. This division between these components of abduction is represented by two different camps in the literature on abduction. On one side are those who hold that abduction consists in the generation of hypotheses, and while it is not to be confused with inference to the best explanation simpliciter, some hypotheses will nonetheless be given more serious consideration than others.[10] On the other side are those who regard abduction as concerned with the ranking of hypotheses by which is most pursuit-worthy for testing.[11] A third camp of scholars maintain that there is a middle ground between these two extremes, according to which abduction consists of two stages, one the generation of hypotheses and the other the ranking of them by pursuit-worthiness.[12] And finally, other scholars are skeptical that there even is such a thing as abduction; rather, abduction is just a special kind of familiar practical inference which is either deductive or is not properly regarded as inference at all.[13]

Complicating this landscape of scholarly disagreements are Peirce's own writings on the matter. On the one hand, Peirce maintains that abduction involves different stages consisting of the "invention, selection, and entertainment of the hypothesis" (HP 2:895, 1901). Such a comment suggests that abduction involves at least three inferential components. The first is the inference that invents the hypothesis based on perception or other data. The second is the inference that selects a hypothesis as the one to pursue. The third is the practical inference that concludes in the entertainment of the hypothesis, that is, the actual commencement of testing it. This differs from the selection of a hypothesis as the one to pursue, since one might hold that a hypothesis is the best to pursue and yet, because of other pressing concerns, opt not to investigate it further. On the other hand, Peirce maintains that he has not been able to discern any species of abductive inference, unlike with respect to deduction and induction: "my connection of Abduction with Firstness . . . was confirmed by my finding no essential subdivision of Abductions" (PPM 276–277n3, 1903).

All of the positions described earlier have passages in the great corpus of Peirce's writings to recommend them. It is not my aim to resolve these disagreements. It is likely that Peirce himself was wrong or confused on various matters. As I argued in Chapter 2, Peirce only distinguishes between qualitative induction and abduction in about 1900. Having made the distinction, he wrote much on it. But to maintain he ever touched bottom and gained perfect clarity into the nature of abduction is surely too much to maintain for him, if not for any person.[14]

Nonetheless, there is another position to develop among those which have been defended. This other position is particularly noteworthy since it draws on the strengths of all of these other positions. Both inference to the best explanation given the evidence we have and inference to the best explanation to pursue are ultimately the same mode of inference, merely with different aims. Thought is directed both to explanations and to actions. Inference to the best explanation given the evidence we have is inference to an explanation. Inference to the best explanation to be pursued is inference to an action.

In both cases, we make perceptive judgments. Both inference to the best explanation given our evidence and inference to the best explanation to pursue may be classed as inferences which result in perceptive judgments. Inferences which result in perceptive judgments are inferences made on the basis of our instinctive capacity for guessing well, where that capacity has been shaped by training and practice. It is the fact that the capacity has been shaped by training and practice which makes it perceptive. In the case of practicing scientists, their ability to guess well at the best explanation and their ability to guess well at the best way of proceeding to test hypotheses are shaped by training. That training results in perceptive judgments about which explanations are the best explanations given the evidence we have and which explanations are the best explanations to pursue. Peirce suggests such a position when he writes, "[i]t is one act of inference to adopt a hypothesis on probation. Such an act may be called an *abduction*. It is an act of the same kind, when a hypothesis is merely suggested as possibly worth consideration. For even then some degree of favor is extended to it" (HP 2:912, 1901). In this quotation, Peirce is treating inference to the best explanation to pursue as abduction proper. He recognizes, though, that inference to the best explanation given the evidence we have (i.e., those which are to be given serious consideration) is essentially of the same kind. My suggestion is that the kind of inference is inference as perceptive judgment about hypotheses based on instinct and training. In one case, we make perceptive judgments about which explanation is the best to pursue. In the other case, we make perceptive judgments about which explanations are more likely than others. Nonetheless, both can be reduced to—that is, understood as—making perceptive judgments about hypotheses for the purposes of further inquiry.

The perceptive-judgment account proposed here has the benefit of explaining why Peirce would frequently appeal to perceptual judgment in his discussion of abductive inference. He notes:

All that makes knowledge applicable comes to us *via* abduction. Looking out of my window this lovely spring morning I see an azalea in full bloom. No, no! I do not see that; though that is the only way I can describe what

I see. *That* is a proposition, a sentence, a fact; but what I perceive is not [a] proposition, sentence, fact, but only an image, which I make intelligible in part by means of a statement of fact. This statement is abstract; but what I see is concrete. I perform an abduction when I so much as express in a sentence anything I see. (HP 2:899–900, 1901)

Even making a perceptual judgment is a special case of making a perceptive judgment. Trained in how to use languages and in how to describe what we see, we make perceptive judgments even in such basic cases as ordinary perceptual judgment.

Yet Peirce also has a very broad conception as to what counts as a perceptual judgment. On Peirce's account, we can perceive facts of mathematics in diagrams. He provides as an example Legendre's proof that the sum of the angles formed when one line abuts another is always 180 degrees. Legendre's proof is accomplished by drawing a perpendicular to the point of abutment. As the sum of the angles formed by the perpendicular will be 180 degrees and any other line meeting at that point of abutment must be on side of the perpendicular or another, one can perceive that the sum of the angles formed by that line must also be 180 degrees. Peirce remarks, "[t]his perpendicular must lie in the one angle or the other. The pupil is supposed to *see* that. He sees it only in a special case, but he is supposed to perceive that it will be so in any case" (EP 2:207, 1903). One sees something, Peirce claims, that is "of a *general nature*" (EP 2:207, 1903). The perception of the mathematical truth is akin to the description of seeing an azalea in full bloom insofar as both are based on individual percepts (of the azalea in bloom or of the diagram) and yet the perceptual judgment is of a general nature. In the azalea case, the words used to describe it are general insofar as they could be used to describe other things. In the mathematical case, not only are the words used general, but the truth proved by perceiving the diagram is universal and necessary.[15]

Nonetheless, in both cases the judgments are formed in the context of having trained and sharpened our skills to guess well. Even a description of azaleas is a guess about how things are which can be verified or falsified by future experience. It may be the person describing the azaleas is suffering an illusion or a hallucination. In such a case, her description will be inaccurate: There is no azalea or it is not in bloom.

In addition to explaining Peirce's frequent appeals to perceptual judgment in his explanation of abduction, the present account has the advantage of bridging the chasm among the different accounts on offer. To understand how, we need to distinguish between abduction as a mode of inference and abduction as a stage of inquiry. Abduction as a mode of inference is of one kind: perceptive judgment. Abduction as a stage of inquiry can involve many different instances of perceptive judging. In particular, it will involve perceptive judging as to which explanations are the best given our present evidence, and it will involve perceptive judging as to which explanations are the best to pursue. Further, even in this later case, abduction will involve perceptive judging as to cost, caution, breadth, and incomplexity. Accordingly, as a stage of inquiry, abduction will involve the generation, selection, and entertainment of hypotheses, as Peirce claims. Yet as a mode of inference, these are all the same, viz., perceptive judging.

In sum, on the account I have been suggesting here, abduction as a mode of inference is of one kind. As a mode of inference, abduction is at root no more than our capacity to guess well, where our guessing has been shaped by training and practice. For this reason, abduction may be described as the mode of inference which issues perceptive judgments pertaining to hypotheses. But as a stage of inquiry, abduction will involve many different components. Most notably, we will make perceptive judgments as to the best explanations given the evidence we have. We will also make perceptive judgments as to the best explanations for us to pursue. When we make abductive inferences as to the best explanations given the evidence we have, the strength of our abductions will be dependent on a variety of factors, including (a) whether the perceptive judgment is a mere conjuring or suggested by *il lume naturale*, (b) whether the explanation is plausible (i.e., the degree of force with which it strikes us as true), (c) whether the explanation can be admitted to scientific inquiry, and (d) whether we have countervailing reasons to disbelieve the hypothesis. When we make abductive inferences to the best explanations to pursue, the strength of our perceptive judgments will be dependent on other factors, such as cost, caution, breadth, and incomplexity.

Objections to the Validity of Abduction and Replies

Having explained Peirce's account of abductive validity and strength, I now turn to various objections to the validity of abduction and Peirce's replies to them. I have argued that abduction is valid because it possesses the truth-producing virtue of being truth-conducing, provided there is any truth to be ascertained. That abduction has such a truth-producing virtue is a consequence of our capacity to guess being forged in the crucible of natural selection together with it being honed by training and practice. Moreover, the fact that we have made significant progress in the sciences in just a few millennia evidences that our minds have a bent for guessing nature's secrets. Nonetheless, there are several objections which might be raised to the validity of abduction, and I turn to those objections now.

The Instinct Objection

The Objection: One objection to the validity of abduction is that validity is a norm by which we evaluate inferences. Now it makes sense to subject our actions to norms only if we have some control over the actions we perform. Yet if the account developed here is correct, abduction is rather an instinctive power for guessing well. But we have no control over that instinct, for it has been forged by natural selection. Since our instinct for guessing well is not self-controlled, it follows that it cannot be subject to norms, including the norm of validity.

Peirce's Reply: As should be plain from the account of abductive inference as perceptive judgment, this objection overstates the claim that abduction is instinctual. It is true that we have an abductive instinct for guessing well. Yet that instinct can be shaped, directed, and improved through training and practice. We do have control over how that instinct is shaped, directed, and improved. Consequently, we do have a measure of control over our abductive inferences. As Peirce writes:

> We know that the Instincts of dogs, canaries, goldfish, etc. are capable of wonderful modification by wisely conducted exercise and training; and consequently the fact that human reasoning is susceptible of improvement (and it is sufficient to read Aristotle's Organon, to say nothing of his Metaphysics and Physics, to be convinced that it is so) is no objection to the opinion that [abduction] ultimately rests upon an Instinct analogous to those of the so-called "brutes" and insects. (R 334:A3–A4, c.1909)

Furthermore, it is false that just because we lack the relevant control over our actions in our current state that those actions are not subject to norms. Richard Feldman (2001) gives the example of a beginning cyclist who in her present condition is incapable of making turns well. Her turns are nonetheless bad. Through training, she will improve them. In like manner, a person whose instinct for guessing well is only just being shaped (for instance, a scientist in training) will make bad guesses or guesses that are not as good as they could be. Given her state of knowledge, she could not have done better. Her guesses are bad, or not good, guesses nonetheless.

Moreover, I have noted in Chapter 1 that Peirce sometimes treats inference as reasoning, which requires that the course of thought in which we draw a conclusion be self-controlled and deliberate. On such a stringent conception of inference, one might object that abduction is not inference at all and, a fortiori, not valid or invalid. To this worry, Peirce remarks:

> Any novice in logic may well be surprised at my calling a guess an inference. It is equally easy to define inference so as to exclude or include abduction. But all the objects of logical study have to be classified; and it is found that there is no other good class in which to put abduction but that of inferences. Many logicians, however, leave it unclassed, a sort of logical supernumerary, as if its importance were too small to entitle it to any regular place. They evidently forget that neither deduction nor induction can ever add the smallest item to the data of perception; and, as we have already noticed, mere percepts do not constitute any knowledge applicable to any practical or theoretical use. (HP 2:899, 1901)

In this passage, Peirce acknowledges that we could define inference in such a way that abduction is excluded. However, there is no other category of mental action into which abduction falls. It is plainly not mere perception, since it involves judgment.[16] When we consider the function of abduction, which is to introduce new elements into thought, Peirce maintains that it is best classed as an inference. It is a mode of inference which, while not entirely subject to self-controlled, can be shaped, directed, and improved through training and practice.

The Conceivability Objection

The Objection: A second objection is that abduction as a stage of inquiry involves guessing at the best explanation for the phenomena we observe. We seek those explanations that we can take up into a process of inquiry. Yet one may worry that there are some explanations of which we are incapable of conceiving. Why should it not be the case that all of the hypotheses that suggest themselves to us are false? May not the secrets of nature forever elude our creativity? Peirce acknowledges this objection when he remarks:

> It is idle to say that the doctrine of chances would account for man's ultimately guessing right. For if there were only a limited number n of hypotheses that man could form, so that $1/n$ would be the chance of the first hypothesis being right, still it would be a remarkable fact that man only could form n hypotheses, including in the number the hypothesis that future experimentation would confirm. Why should man's n hypotheses include the right one? (CP 7.680, 1903)

Peirce's Reply: One flatfooted reply to this objection is to simply deny its central claim. We will eventually conceive the true hypothesis. Peirce appears to endorse just such a reply when he states, "[i]t is true that, however carelessly the abduction is performed, the true hypothesis will get suggested at last" (CP 7.220n18, 1901). But this is not representative of Peirce's more nuanced view on the matter.

To appreciate Peirce's more nuanced view, we need to distinguish among three different theses pertaining to the conceivability of hypotheses. First, one might hold that a hypothesis is conceivable, only we have not yet conceived it. On Peirce's account, such an objection is no problem. All explanatory inquiries proceed on the assumption that we will be able to hit upon the true explanation. Peirce claims, "[t]hat there is any explanation of them [viz., a given object presenting with an extraordinary combination of characters] is a pure assumption" (HP 2:899, 1901). We may hope that we will nonetheless conceive the hypothesis in the future.

Second, one might hold that some hypothesis is conceivable, only it is not conceivable by us. We are humans, limited in our cognitive capacities. There may be some explanations of which we cannot conceive just because of our cognitive makeup. But even this objection has little bite. Peirce regards the community of inquirers to extend well beyond human inquirers. As he states, the community must be conceived as "extend[ing] to all races of beings with whom we can come into immediate or mediate intellectual relation. It must reach, however vaguely, beyond this geological epoch, beyond all bounds" (ILS 116, 1878). Consequently, while we might not be able to conceive of the true hypothesis, the community of inquirers which extends beyond us may be able to conceive of the true hypothesis. We hope that they will be able to do so.

Yet a third way to think of the inconceivability objection is to hold that there are some explanations which are inconceivable simpliciter, whether by us or by the broader community of inquirers. To this worry, there are two directions of reply. The first direction of reply is to adopt Peirce's earlier thesis that we have no conception of the absolutely incognizable, a thesis Peirce endorses in the 1860s and 1870s. If our only means of cognitive access are inferential and scientific and nothing is incognizable, it follows that there are no explanations which cannot be discovered by scientific inquiry. In his *Journal of Speculative Philosophy* series from the late 1860s, Peirce appears to endorse this position. In "Some Consequences of Four Incapacities," he claims that "[w]e have no conception of the absolutely incognizable" (W 2:213, 1868). The careful reader will have already noticed that there is some slippage in the ideas just developed: The claim that we have no conception of the incognizable is quite different from the claim that there is nothing incognizable. But even aside from that slippage, Peirce's argument for the claim we have no conception of the absolutely incognizable is lamentably bad.

Peirce thinks he has shown that we have no conception of the incognizable in "Questions Concerning Certain Faculties Claimed for Man," but that passage is deeply confused. The issue at hand in "Questions" is whether a sign can have meaning if the object of the sign is something absolutely incognizable. Now the answer to that question is obviously that it cannot. We would be presuming the sign to signify something that in no way is signifiable (for if it were signifiable, the object would be cognizable). But Peirce instead argues for the claim that we have no conception of the incognizable.

He contends that "all our conceptions are obtained by abstractions and combinations of cognitions first occurring in judgments of experience" (W 2:208, 1868). However, we have no cognition of the incognizable to be abstracted from those first judgments, for if something is incognizable we will not have experienced it. The obvious objection—and Peirce, to his credit, countenances it—is that the conception of the incognizable is formed by negating the cognizable. That is, not + cognizable = incognizable, and we do have a conception of the cognizable. Peirce responds that "*not* is a mere syncategorematic term and not a concept by itself" (W 2:208, 1868). To be sure, 'not' is syncategorematic. But we have many ideas formed by adding 'not' to our original conceptions, including the ideas of the infinite, immortal, indeterminate, indefinite, inconsequential, and so on. We have conceptions of these; why do we not have a conception of the incognizable?

In fact, Peirce had argued in 1865 that Kant's distinction between infinite judgments and affirmative judgments fails because infinite judgments are affirmative judgments (see W 1:253). To claim that Zeus is immortal is not to assert that Zeus is nonmortal (rocks are nonmortal but not immortal). Peirce might respond to the objection just made against his argument for the conclusion that we have no conception of the absolutely incognizable by holding that we do have conceptions of immortality but not of incognizability because the former is an affirmative judgment whereas the latter is not. But this will not help his case, since neither do we have any cognition of immortality first found in judgments of experience. Where, then, did we get our conception of immortality as opposed to the nonmortal? So not only is Peirce wrong that we have no conception of the incognizable, but also we cannot move from this thesis to the conclusion that nothing is incognizable.[17]

There is a second direction of reply to the worry that there are some hypotheses which are inconceivable simpliciter. The reply appeals to Peirce's logical sentimentalism. I have already examined Peirce's logical sentimentalism in my discussion of Peirce's reply to the End-of-Community Objection to induction's validity, in Chapter 6. Briefly, Peirce holds that while the community of inquirers may come to an end, we have no rational grounds for thinking that it will. This, again, is because any such argument would be an inductive argument. But inductive arguments are valid only if there is a community of inquirers that will approximate the truth in the long run. The argument, however, purports to show that there is no such community. If we accept its conclusion, then we are compelled to conclude that the argument is not valid. Similar considerations apply to any argument for the conclusion that there will be such a community of inquirers; any such argument would be circular. Consequently, we cannot offer rational grounds for the conclusion that there will or will not be a Peircean community of inquirers. Instead, we are to hope that it will continue indefinitely, to have faith in fellow inquirers to carry forward our research projects, and to love the community such that we are willing to set aside our own local and trivial concerns as well as our biases.

A similar response can be developed with respect to the concern that some hypothesis is inconceivable simpliciter. It is surely possible that some hypotheses are inconceivable simpliciter. However, we have no grounds for affirming that in fact some hypotheses are inconceivable simpliciter. For if we could identify some hypothesis that is inconceivable simpliciter, we would have conceived that hypothesis. As Peirce concedes, "some finite number of questions, we can never know which ones, will escape getting answered forever" (W 5:227, 1885). The catch is that we can never know which ones will escape.

Except when our hypotheses exhaust logical space, we have no rational grounds for thinking that the true hypothesis is among those that has been or will be conceived. But neither do we have rational grounds for thinking that it will not be among them. Instead, we should have hope, faith, and love. We should hope that the community of inquirers will eventually alight on the true hypothesis. We should have faith that they will continue to seek it out. And we should love our fellow inquirers such that we endeavor to put them into a better position to discover the truth. Peirce suggests precisely this when he remarks:

> we must hope [the facts admit of rationalization], for the same reason that a general who has to capture a position, or see his country ruined, must go on the hypothesis that there is some way in which he can and shall capture it. We must be animated by that hope concerning the problem we have in hand. . . . [A]bduction is, after all, nothing but guessing. We are therefore bound to hope that, although the possible explanations of our facts may be strictly innumerable, yet our mind will be able in some finite number of guesses, to guess the sole true explanation of them. (EP 2:107, 1901)

In his 1900 review of Royce's *The World and the Individual,* Peirce remarks that inquirers have two hopes, the first that "the course of 'our' experience may ultimately compel the attachment of a settled idea to the mental subject of the inquiry; and the second is, that the inquiry itself may compel him to think that he anticipates what that destined ultimate idea is to be" (CP 8.102). The first hope gives expression to the hope that underwrites induction. It is the hope that the community of inquirers will continue our inquiries indefinitely into the future. The second hope gives expression to the hope that animates abduction. It is the hope that the true hypothesis is conceived or at least vaguely anticipated in the hypotheses we have on hand.

Peirce makes the validity of induction to rest upon the community of inquirers, for it is the community of inquirers which will (we hope!) approximate the truth in the long run. The validity of abduction, in contrast, does not rest upon the community of inquirers but on whether our perceptive judgments are truth-conducing at all, provided there is any truth to be ascertained. As Paul Forster (1989) and Ahti-Veikko Pietarinen and Francesco Bellucci (2014) have rightly stressed, the use of abduction is inescapable. We have no choice but to trust to our capacity for guessing well. Nonetheless, for both induction and abduction, we will make an appeal to the community of inquirers and to the logical sentiments in order to address objections to their validity. For induction, we pin our hopes on the community of inquirers indefinitely continuing in pursuit of the truth. For abduction, we pin our hopes on the community of inquirers eventually alighting on the true hypothesis.

The Reliability Objection

The Objection: There is another objection similar to the conceivability objection but importantly different from it. The conceivability objection is that there are some hypotheses which are inconceivable and so abduction will never be able to alight upon them. In contrast, the reliability objection is that while true hypotheses are conceivable, in certain domains of inquiry the objects of investigation are so far removed from our practical life that abduction or perceptive guessing cannot reliably get at them in any finite number of guesses. If hypotheses are generated from our experiences, and granted that our experiences are limited to this little corner of the universe and this brief span of time, what grounds are there for thinking that in abstruse matters the true hypothesis is going to be among those of which we will think?

Mohammadian gives expression to this objection. He writes, "in some areas, insight [our abductive instinct] generates less and less truthful hypotheses. . . . The problem here is rooted in the original purpose of insight. Since insight is developed for survival, it stops functioning well about matters of no survival value" (2019, 149). As Mohammadian notes, well-functioning abduction is limited with respect to our ability to grasp rightly both the breadth of nature and the depth of nature. Abduction is limited with respect to breadth because it developed to supply us with crude ideas about space, time, and the likely behaviors of other animals. It is limited with respect to depth because it developed to help us cope with our surroundings at the level of our daily interaction rather than with subatomic particles and galaxies. Accordingly, in areas of inquiry outside of these basic needs, abduction will be an unreliable, and so not a truth-conducing, guide.

Peirce's Reply: There are two replies to be made to the reliability objection. The first is that even if our abductive instincts are forged in the crucible of natural selection, we have seen that they can also be honed by learning, training, and practice. Even if our abductive instinct when unformed is unreliable with respect to abstruse matters, it does not follow that our abductive instinct when shaped, directed, and improved is unreliable in abstruse matters. We can still leverage our abductive instincts to make inquiries into facets of the universe which do not concern survival. On this score, it is important to note that while many of the hypotheses regarding abstruse matters which are abductively generated will turn out to be false, we can still sort through them in the process of inquiry so as to find which hypotheses are true. In abstruse matters, abduction does not need to be reliable in the sense of tending to generate only true hypotheses, for nothing is vital in science (see RLT 112, 1898). Nonetheless, we will want our abductive instinct to generate true hypotheses in practical matters.

The second reply makes the same appeal to the community of inquirers that was made in the discussion of the conceivability objection. We again recur to the community of inquirers and our logical sentiments. While it may be true that in abstruse matters our instincts have yet to be honed and are unreliable, there is no reason to think that the guessing instinct of members of the community of inquirers must remain forever unhoned and unreliable. To the contrary, the history of science and of logic more generally is precisely the process of educating our logical instincts: "each chief step in science has been a lesson in logic" (ILS 48, 1877). Even if in its present condition our guessing instinct is woefully inadequate to unlocking nature's secrets in certain abstruse domains, it does not follow it must remain so. We may instead hope that it will be improved, putting our faith in and loving the community of inquirers.

The Economic Objection

The Objection: The final objection to consider is one which straddles the distinction between abduction and induction. It concerns hypothetic induction as a mode of inquiry. We will both endeavor to conceive of the true hypothesis and to put it to the test. However, economic considerations raise a concern. Rescher notes that Peirce recognizes in science there may be increasing costs and diminishing returns such that "there is a limit, ultimately an economic limit, to the questions about the phenomena that we can even answer. Many questions of transcending scientific importance will remain unresolved because their resolution would demand a greater concurrent deployment of resources than will ever be marshalled at any one time in a zerogrowth world" (1978, 34). Even if we can conceive of the true hypothesis, we may not be able to test it for economic reasons. It will be conceived but not pursued. Yet lines of inquiry that cannot be pursued are lines of inquiry that can neither conduce to the truth nor approximate it in the long run. It follows that hypothetic induction as a mode of inquiry is invalid on both abductive and inductive grounds.

Peirce's Reply: 'Cannot' in the claim that some lines of inquiry cannot be pursued may be understood practically or aspirationally. Practically, there are truths upon which we will never alight and errors we will never root out. But aspirationally, we may hope that the community of inquirers will marshal the needed resources. We may believe (dispositions of the mind) that we will never ascertain some truths and root out some errors while still hoping (dispositions of the heart, our sentiments) that the community of inquirers will ascertain those truths and root out those errors. The validity of induction rests on our sentiments, though the economy of research (an abductive concern) may justify beliefs contrary to our sentiments. Even when we have economic grounds for being pessimistic about the progress of science, we must do what we can now by sacrificing our own interests to the larger community in pursuit of the truth. We are to put our hope and faith in future inquirers to ascertain those truths and root out those errors that economic considerations now prohibit us from pursuing.

Conclusion

Although abduction is sometimes regarded as inference to the best explanation, we must distinguish among different inferences to explanations. Inference to the best explanation simpliciter is rather hypothetic induction as a mode of inquiry viewed diachronically. In contrast, abduction generates, selects, and entertains hypothesis, but it does not eliminate, corroborate, or confirm them by putting those hypotheses to the test. With respect to abduction proper, we make inferences to explanations, to plausible explanations, to scientifically admissible explanations, to the best explanation given the evidence we have, and to the best explanation for us to pursue. The best explanation given the evidence we have is not always the best explanation for us to pursue, for economic considerations may lead to us pursue explanations we do not think are likely.

Abduction is an instinctive power we have for guessing well. It has been forged in us through the process of natural selection. Nonetheless, that instinct can be shaped, directed, and improved through training and practice. Although inference to the best explanation to pursue and inference to the best explanation given the evidence we have are paramount in ordinary cases of abduction, they may both be regarded as cases of inference to perceptive judgments. Abduction is valid because it has the truth-producing virtue of being truth-conducing, and its truth-conducing virtue is evident to us in the success the sciences have had in hitting upon true or approximately true hypotheses.

This book has been an essay on Peirce's theory of inference. My task has been to develop a charitable retrospective perspective on Peirce's theory of inference. Whether and to what extent I have succeeded it is for the reader and the scholarly community to judge. But it behooves me, in conclusion, to make some comments not on inference but on inquiry. Peirce sometimes uses 'inquiry' in a broad sense as the struggle to attain a state of belief (see ILS 55, 1877). Inquiry in this broad sense is compatible with adhering to the method of tenacity, keeping all counterevidence at bay so that we can cling tenaciously to what we are already inclined to believe. But more typically, we think of inquiry as a process of scientific or quasi-scientific investigation. We cannot demarcate what is science from pseudoscience except in the practice of making inferences. In steps both gradual and great, we hone our instincts to guess, and we improve our techniques of inquiry. It is the success of some guesses and techniques and the failures of others that indicate which courses of thought are truly scientific. We know science from pseudoscience by its fruits. Peirce notes that there is "no reason why, provided reasoning at all tends to be true, in a more developed stage of growth of the reasoning-power, a former reasoning should not be put on trial and be convicted of weakness or utter irrationality" (EP 2:466, 1913).

Any process of inquiry will surely involve abductions, deductions, and inductions. But as I pointed out at the start of this book, abduction, deduction, and induction as genera of inference are different from abduction, deduction, and induction as stages of inquiry. In many cases, a course of inquiry will pass through all three stages. But it would be plainly untrue to the facts to maintain that inquiry is a neat and tidy affair in which we pass from one stage to another. Throughout the entire time of a course of investigation, the investigator will make abductive, deductive, and inductive inferences. Such inferences will be mutually enforcing and mutually informing. One might deduce a prediction from a hypothesis only to realize that there is some other, even more pursuit-worthy, hypothesis to be tried. In such a case, she will not pass from deduction as a stage of inference to induction but will return to the abductive stage of inference. A test might reveal another prediction of a hypothesis, and in that case one will recur from the inductive stage of inquiry back to the deductive stage. It is not the stages of inquiry that do the work in an investigation but the inferences made in those stages. What the inquirer will aim to do is to ensure those inferences are not only valid but strong. By subjecting our inferences to logical scrutiny, we discern which procedures of inference are both valid and strong. Logically scrutinizing our inferences and the procedures we use is an ongoing affair demanding serious and sustained focus. "There is no royal road to logic," Peirce writes, "and really valuable ideas can only be had at the price of close attention" (ILS 101, 1878).

Notes

Chapter 1

1. See Murphey 1961, 325–326; Hookway 1985, 52–56, and 2012, Ch. 5; Kasser 1999; Short 2000; Wiggins 2004; Bellucci 2015b; and Tiercelin 2017.

2. Sometimes, these beliefs or commitments will be tentatively or only hypothetically adopted, but I set this complication aside.

3. But conditional probabilities are undefined when the antecedent has a probability of 0, or they are defined stipulatively. This is unlike material conditionals which are true when their antecedents are false.

4. For a discussion, see Bellucci 2018, 28–31, who is more sympathetic to Peirce's doctrine of the leading principle than I am.

5. In Peirce's earliest works from the 1860s, he does seem to identify the leading principle with just such a conditional. Doubts appear in his discussion of guiding principles in "The Fixation of Belief." There, he begins to distinguish between guiding principles that are used by different inquirers and those that are used in any inference.

6. As noted earlier, Peirce's use of 'guiding principle' is usually thought to be synonymous with 'leading principle.' But once we clarify the different meanings of 'leading principle' in Peirce's work, there is good reason for using 'guiding principle' either as an implicit commitment or as a description of the procedure used in the inference.

7. Under some conditions (if, say, it is ionized), a sample may not be representative. Note also that while this way of stating the formal guiding principle will work for copper disks, it will not work for inductive generalizations based on sampling. A sample from a bag of beans may have 50% white beans. One should not, however, infer that the individuals comprising the bag are 50% white beans, else any individual bean would be 50% of a white bean. Rather, one infers to the population as a whole, not the individuals of the population.

8. See Pietarinen 2005 and De Waal 2005. Although some scholars appear to suggest otherwise, a logica utens is not opposed to a logica docens. The logica docens, the logic learned, sharpens and improves the logica utens, the logic used.

9. Accordingly, Peirce remarks that the English associationalists were "careful never to extend it [association] to the operation or event whereby one idea calls up another into the mind, but to restrict it primarily to a *habit* or *disposition* of mind.... As for that mental event which corresponds, as we suppose, to the nervous discharge of one part of the cortex upon another ... for that they employed the term *suggestion*" (RLT 232, 1898).

10. Because of these variations, informational necessity may not always be representable by Hintikka's modal system of belief and knowledge. Whether and to what extent

Peircean views are consistent with varieties of dynamic epistemic logics and probability logics needs exploration.

11. Accordingly, Peirce is committed to some version of epistemic closure.

12. Though see Schouls 1981, Haack 1982, and MacDonald 2020, for an examination of just how anti-Cartesian his project is.

13. These heirs include not just the rationalists such as Spinoza and Leibniz but also the empiricists such as Locke.

14. I develop the preceding line of argument in more detail in Atkins 2018b.

15. Of which Peirce was a pioneer—see Fisch and Turquette 1966 and Lane 1999.

16. See also Kasser 1999, 503–505.

17. I state "at least in part" because we use 'meaning' more broadly than Peirce typically does. He usually restricts 'meaning' to the "third grade of clearness," that is, the conceivable practical consequences of an idea, whereas we usually regard the definition of an idea to be part of its meaning.

18. For a fuller examination of Peirce's various arguments for the pragmatic maxim, see Hookway 2012.

19. For a fuller discussion, see Bellucci 2021.

20. For a fuller discussion, see Atkin 2005.

21. Both definitions are also found in earlier editions.

22. What are deductive procedures? As I will explain in the next chapter, deduction is inference from premises using the axioms and rules of the logical calculus.

Chapter 2

1. As noted in the Introduction, hypothetic induction is akin to induction by confirmation. In his later work, Peirce would call hypothetic induction qualitative induction. But 'induction by confirmation' is a misnomer since hypothetic inductions may confirm no hypothesis at all. 'Qualitative induction' is better but does not adequately capture the fact that one is eliminating, corroborating, or confirming hypotheses when one makes hypothetic inductions. 'Hypothetico-deductive method' refers to an entire process of inquiry rather than a species of inference.

2. It would not be appropriate to represent this is as $P(C|A)$, since that is a number. Not all inductions, however, are quantitative inductions.

3. What is a strength-producing inductive procedure? It is a procedure which employs techniques to identify and correct for errors. I shall have more to say about strength in Chapter 5.

4. For more on Peirce's account of such inferences, see also Bellucci 2015a.

5. The reader may wonder why, if Peirce was familiar with Boole's logic, he remains beholden to the traditional logic. In his earlier works, Peirce regards Aristotelian syllogistic logic as most suited to represent inferences. But why should he not have preferred Boole's logic to Aristotle's, especially given that he had made substantial improvements to it? Peirce tells us why in 1867's "On a Natural Classification of

Arguments," writing, "[s]ince it can never be requisite that a fact stated should also be implied in order to justify a conclusion, every *logical principle* considered as a proposition will be found to be quite empty. Considered as regulating the procedure of inference, it is determinate; but considered as expressing truth, it is nothing. It is on this account that the method of investigating logic which works on syllogistic forms is preferable to that other, which is too often confounded with it, which undertakes to enunciate logical principles" (W 2:25). The logical principle is the purely formal conditional mentioned earlier, such as the argument form modus ponens (not its substitution instances!). As this passage shows, Peirce adopts the Aristotelian approach over the Boolean approach because a purely formal investigation of logical principles does not "express truth" as logical principles are "quite empty." Peirce's claim here seems rather strange. Isn't it a fact that for valid inference forms, if such-and-such premises are true, then the conclusion is true? While that is true for deduction, it is not so for induction. Induction must be regulated not merely by formally correct rules but by appropriate procedures for approximating the truth in the long run.

6. The vestiges of this Aristotelian commitment can be found as late as 1898's lectures titled *Reasoning and the Logic of Things*. However, in that series of lectures Peirce refers to the syllogistic approach merely as "scaffolding" to help in "building up this theory" and states that we can grasp the distinctions among the various sorts of inference without it (RLT 141). Furthermore, thinking of the Aristotelian presentation in light of the arrangements of consequence, antecedent, and consequent helps to shed light on Peirce's otherwise baffling presentation in the 1878 *Illustrations of the Logic of Science* series. There, he presents the distinctions as concerning arrangements of a rule, a case, and a result. But it is not at all clear from the lecture what distinguishes these categories of rule, case, and result since they all have the form of a proposition stated in standard Aristotelian logic. The examples make much better sense when the rule is regarded as the conditional or consequence, the case as the antecedent, and the result as the consequent.

7. Burks 1964 and Rescher 1978, Ch.1, explain that in Peirce's work, he uses different conceptions of probability.

8. And there is an argument to be made that even in his late work he has not distinguished among them.

9. For more on Peirce on propensities, see Fetzer 1993.

10. Both his "Note" and his letter to Carus cut off when he turns to them—see ILS 129 and 277, 1910.

11. See Burch 2010 for a fuller discussion of Peirce's rules.

12. For a fuller examination, see Van Evra 1997.

13. For a list of such axioms, see Ronan 2012.

14. See Fisch and Turquette 1966 and Lane 1999.

15. It is helpful to distinguish different ways in which we may be said to reach a conclusion. In one sense, a conclusion is an answer to a question that no competent person can reasonably doubt. In a second sense, a conclusion is the conclusion of a course or plan of inquiry. Note, though, that a course of inquiry may conclude without the question being answered in the first way just mentioned. In a third sense, a conclusion

is the conclusion of an inference, where inferences are often made in the course of inquiries. In a fourth sense, a conclusion is one of a member of a set of propositions which constitute an argument.

16. To name just two of many, Misak 2018, 28, and de Waal 2013, 63. I pick these two because otherwise they are two of the finest expositors of Peirce's ideas.

17. De Waal adds a question mark here that is not in the original manuscript.

18. What is this doubly indirect manner? Peirce does not tell us. I speculate that it is because (1) those animals which guess best are the most likely to survive and (2) the success of the sciences evidences that we have a guessing instinct attuned to the order of nature. I explore these themes in Chapter 7.

19. Compare Short 2007, 68n6.

20. In his early works, Peirce claims that he uses 'hypothesis' in the same way as Kant (see W 2:218–219fn1). He takes this usage of 'hypothesis' from the *Jäsche Logic*: "*If all the consequences of a cognition are true, then the cognition is true too.* For if there were something false in the cognition, then there would have to be a false consequence too. From the consequence, then, we may infer to a ground, but without being able to determine this ground. Only from the complex of all consequences can one infer *to a determinate ground*, infer that it is the true ground*" (1992, 559 [JL 52]).

21. Questions about how hypotheses are suggested and so end up on this list concern abduction and not hypothetic induction. The tracing out of predictions, as we shall see, concerns deduction and not hypothetic induction.

22. I here follow Mayo 1996, esp. Ch. 6

23. See Hintikka 1983 and Short 2007, 265–266, for a discussion of introducing new objects in this way.

24. In Chapter 5, I shall call this typical presentation into question.

25. On this distinction, see also Levi 1980, 136–137, and 1995, 65–66.

26. It is important not to confuse the total information available to a reasoner with the information (i.e., premises) on which an inference is based. Relative to the individual items of information used in each inference, the inferences are valid. But whether the person has reasoned well about the matter by using all of the information at her disposal is a separate question from the validity of each inference made in any given process of reasoning.

Chapter 3

1. I adopt here the widely endorsed contemporary account of validity, but there is in fact significant debate about the "nature" or correct definition of logical consequence.

2. As Francesco Bellucci (2016) rightly argues.

3. Note that while the argument is valid, the premise is not subjectively valid because it is not consistent.

4. For more on which, see Bellucci 2014 and 2018.

5. I appeal to the authority of Ian Hacking, who holds that inductive arguments "can go wrong in two very different ways. [1] The model may not represent reality well. That

is a mistake about the real world. [2] We can draw the wrong conclusions from the model. That is a logical error" (2001, 33). A mistake in the first respect is not invalidity but, one might say, unreliability since we will not do a good job of modeling how things in fact are.

6. The astute reader will note that this is a variation on a problem raised in Chapter 1: Charitably reconstructed inferences will always be valid.

7. In one earlier work, Peirce characterizes a "scientific inference" as valid when it "lends an additional probability to the proposition inferred, altho' the fact indicated by this proposition may still remain entirely unknown or even grossly improbable" (W 1:441, 1866). But it is not clear what Peirce means by 'lends an additional probability.' He cannot mean it increases the probability of the conclusion inferred. For if that were what Peirce meant, then if we have thought that 50% of the beans in some bag are white based on one sample, but then in the next sample only 25% of the beans are white, we will decrease the statistic of 50%. But surely such an inference would be valid. Alternatively, what Peirce might mean is that when a scientific inference is valid, it increases the probability that the conclusion we have inferred is correct. But even this is inaccurate since a perfectly well-drawn sample might be unrepresentative of the entire bag of beans. In that case, modifying the probability in light of the sample gives us quantity further from the truth.

8. Oftentimes, not always, since the meanings of words will also be considered when evaluating the validity of inferences.

9. The clause about the conclusion is needed because a premise might be a statistic but the statistic might be used in a statistical deduction, as in the inference *50% of the beans in this sample, which has been drawn from this bag, are white; therefore, there is a 50% probability the next bean I draw from this bag will be white.*

Chapter 4

1. Peirce will sometimes refer to just one premise, it being the conjunction of all the premises.

2. For an exposition of Peirce's graphical logic, see especially Roberts 1973, Shin 2002, and Dau 2008.

3. In his *Regulae*, Descartes uses 'induction' in this sense, too (1985, 38 [ATX 408]).

4. Peirce does not speak of mental models, per se. I borrow the idea from the psychologist Philip Johnson-Laird (see, esp., 1983 and 2006), who is influenced by Peirce.

5. This essay anticipates our modern predicate logic in interesting ways (see Burch 1997, Merrill 1997, and Van Evra 1997).

6. See Atkins 2018a, 214.

7. For more on the perception of such necessary truths, see Wilson 2012 and Legg 2014b.

8. See also Turquette 1964.

9. And see Anellis 2012.

10. See Fisch and Turquette 1966 and Lane 1999.

11. For a chronological survey of major developments in Peirce's work on deductive logic, see Dipert 2004.

12. Levi 2004, 262; see also Forster 2011, 143, and 1989, 426.

13. The simplest because, as noted in Chapter 2, Peirce draws a more fine-grained distinction between probable deductions in the narrow sense and statistical deductions. The former are the simpler cases represented in the example that follows.

14. Here is the proof. Suppose we have a general statistic about population: $P(B|A) = N$. Let B be a proposition of the form X is G and A be a proposition of the form X is F, where X refers to no definite individual of the population sampled to ascertain the statistic. Let C name an individual of the population sampled to ascertain the statistic. Substitute C for X. Let B^* and A^* name the propositions resulting from the substitution. Then $P(B^*|A^*) = N$. Since C names an individual of the population sampled, it follows that $P(A^*) = 1$. Show that $P(B^*|A^*) = P(B^*)$ when $P(A^*) = 1$. Proof: $P(B^*|A^*) = P(B^*\&A^*)/P(A^*) = P(B^*\&A^*)/1 = P(B^*\&A^*) = P(B^*) \times P(A^*|B^*) = P(B^*) \times [P(B^*\&A^*)/P(B^*)] = P(B^*) \times [P(B^*\&A^*)/P(B^*\&A^*) + P(B^*\&\neg A^*)]$, now from the denominator $P(B^*\&\neg A^*) = P(\neg A^*) \times P(B^*|\neg A^*)$, the conditional probability of which is undefined or stipulated since $P(\neg A^*) = 0$, but then $P(\neg A^*) \times P(B^*|\neg A^*) = 0$ regardless; hence, $P(B^*) \times [P(B^*\&A^*)/P(B^*\&A^*)+0] = P(B^*) \times [P(B^*\&A^*)/P(B^*\&A^*)] = P(B^*) \times 1 = P(B^*)$. Q.E.D.

15. These considerations suggest that Peirce embraces an identity theory of truth, where a true proposition when viewed from the side of reality is a fact and when viewed from the side of mind is a thought. On this matter, see also Atkins 2016b and Legg 2020.

16. See W 2:457–459, 1871, for Peirce's announcement of Babbage's death.

17. I shall have more to say about this thesis, and why it is doubtful, in Chapter 7.

18. On this point in relation to truth, see Legg 2014a, 207–208, and compare Rescher 1978, Ch. 2.

19. I examine Peirce's various replies to the liar paradox in Atkins 2011.

20. If Peirce is correct, it would follow that this solution also suffices as a reply to the truthteller paradox. A reader of my essay of 2011 will find that there I take Peirce to be further objecting to his claim that every proposition implicitly asserts its own truth. Additional reflection leads me to think that Peirce's statements in the passage under discussion are ambiguous between objecting to the claim that every proposition implicitly asserts its own truth (his early view) and objecting to the claim that there can be any such proposition as one that predicates truth or falsity of itself (the Ockhamist solution).

21. I take it that Peirce's qualm about propriety is that Aristotle's definition is only a nominal definition and not a real definition of what a proposition is. But nominal definitions can be perfectly acceptable and true.

22. Which is not to deny that it could be a proposition and be true, if say the person inferred it from the fact that it is morning in India, India is a large country, and so it is likely at least one person is eating breakfast somewhere. I develop these ideas in more detail in Atkins 2021.

23. The reader may note that this solution is somewhat different from the solution in my 2011 essay. Though both appeal to assertibility, my essay of 2011 argued for the

unassertibility of the liar proposition by drawing on Peirce's pragmatic elucidation of truth. Here, I have argued for the unassertibility of the liar proposition by drawing on the nature of propositions. By giving a different argument here, I am not thereby abjuring my argument there.

Chapter 5

1. *It is not established that p* may be glossed as *it is not known that p* ($\neg Kp$). Kp is logically equivalent to $\neg P\neg p$. So $\neg Kp$ is logically equivalent to $\neg\neg P\neg p$, or $P\neg p$.
2. The proof of this may be found in Hintikka 1962/1964, 80.
3. One may also doubt that the prediction is in fact a consequence of the hypothesis, but as that concerns deduction, I set it aside in the present context.
4. What constitutes the degree of belief (which, as we shall see momentarily, is better described as a degree of conviction) being sufficiently high? That is a difficult question to answer, and Peirce does not address it directly. One plausible answer, consistent with Peirce's psychophysics of belief, is that the person is no longer motivated to inquire into the matter for evidential reasons (rather than, say, laziness). As Peirce states in "The Fixation of Belief," the "irritation of doubt" leads us to inquire and "is the only immediate motive for the struggle to attain belief" (ILS 55, 1877). In other words, whether the degree of belief is sufficiently high will be determined by the diminishing marginal utility of further inquiry.
5. I am here using 'evidence' in a nontechnical sense. It may be evidence that one has gathered through careful investigation, or it may be whatever has been impressed on us in the stream of experience.
6. Since Peirce is here providing two psychophysical laws of belief, evidently his "should" is to be interpreted both descriptively and normatively. Descriptively, whether one believes p or not will ordinarily be determined by the ratio of favorable (f) to unfavorable (u) cases. Normatively, beliefs should track that ratio. If f/u represents the chance of p, then (a) if f>u, then we ordinarily do and should believe p, (b) if f<u, we ordinarily do and should believe not p, and (c) if f = u, we ordinarily do and should suspend belief as to whether p or not p. The normative force of the "should" is supported by standard Dutch-book arguments that our beliefs should conform to the probability calculus.
7. How do we measure the degree of conviction? First, note that Peirce does not measure it on a scale of zero to one but of zero to infinity. For this reason, his scale is logarithmic, as logarithmic scales are apt to be used when we need to compress ranges of large numbers. Consequently, degrees of conviction are not equivalent to probabilities, which are measured on scale of zero to one. Nevertheless, we could map degrees of conviction onto the real numbers on a scale from zero to one and fictitiously treat degrees of conviction as akin to probabilities, as I shall do in what follows. Second, in recent articles, Jeff Kasser has emphasized that Peirce distinguishes between a conception of net weight of evidence and gross weight of evidence. The former is the balance of feelings pro and con for a belief. The latter is "the resistance of the probability

judgment to be changed by further evidence" (2016b, 643). Consider two people. The first has drawn but a few beans from a bag and finds they are all black. She has a high net weight of evidence for the belief that the next bean drawn will be black. She is convinced of it given her sample. However, because her sample is so small, she has a low gross weight of evidence. She is not convinced of it, knowing her sample is small. The second person has drawn a large number of beans from a bag and finds that half are black and half are white. She will have a low net weight of evidence for the belief that the next bean drawn will be black but a high gross weight of evidence (see 2016b, 640, and 2018, 849). She will not be convinced she will draw a black one, but she will be convinced that there is roughly a 50% probability it will be black. Both numbers are needed to measure the degree of conviction. The degree of conviction is best thought of as a complex feeling produced by the nature and quality of the evidence. But the measure of conviction should not, as it seems to me, be conflated with the belief. A belief is settled just in case the chance of it being true is greater than one to one and the degree of conviction is high (cf. Kasser 2011).

8. As mentioned in note 6, this in fact requires mapping the logarithmic scale of feelings of conviction which range from zero to infinity onto a scale from zero to one. Moreover, it might be tempting to claim that if one's degree of conviction is n, one's degree of doubt is $1 - n$. Then, if one's degree of conviction is .6, then one believes it and yet one's degree of doubt is .4. In such a case, one may then both believe and doubt some hypothesis. But it is important to distinguish between doubt as a measure of the degree to which one is not convinced and doubt as the contrary of belief. Peirce typically uses 'doubt' in the latter sense, in which case one cannot both believe and doubt a hypothesis. But in the other sense, one can both believe and doubt a hypothesis, that is, have some degree of conviction that the hypothesis is true while still having doubts in the sense of not being fully convinced of it. For a more detailed discussion, see Kasser 2018.

9. Peirce's reference to algebraically adding the feelings of approval suggests he has in mind something akin to Jeffrey's Rule, which is a generalization of Bayes's Rule. In what follows, I shall focus on the simpler case of Bayes's Rule as nothing turns on the difference.

10. Peirce's treatment of historical testimony is much more detailed than these brief comments suggest. See HP 2:705–801 and Burbidge 1981.

11. Much has already been written on Peirce's account of science in contrast with other methods of belief fixation, and so I shall pass over the details here (for some recent and worthwhile studies, see, e.g., Lane 2018, Chs. 1 and 2; Short 2018 and 2000; Atkin 2016, Chs. 3 and 4; de Waal 2013, Chs. 6 and 8; Hookway 2000, Chs. 1–4).

12. If probabilities are treated both as frequencies ascertained by testing and as degrees of conviction, how do we avoid equivocation? The principal principle: Our degrees of conviction should reflect the frequencies ascertained.

13. In fact, statistical testing has settled on an intermingling of both frequentist and Bayesian approaches to probability, as David Spiegelhalter notes (2019, 336–337) and as the Peircean would expect.

14. The application of this procedure may be a messy affair, but the application of the procedure should not be conflated with the species of inference. That the species of

inference is truth-approximating does not imply that we will be good at applying the procedure so as to approximate the truth.

15. See also Mayo 1996, esp. Ch. 12.
16. See also CN 2:214, 1899, and Wible 2018.
17. See also EP 2:25, 1895, and CN 3:48, 1901.
18. For a discussion of this issue with respect to particle physics, see Hossenfelder 2018.
19. See also CN 3:191, 1904, and Short 2008.
20. For a fuller discussion of Peirce's astronomical work, see Lenzen 1964.
21. See also Mayo 1996, 341–359, and Levi 2004, 276–277.
22. And Whately, for reasons to be explained in the next chapter.

Chapter 6

1. I shall set aside here questions as to what individuality consists in, but for one Peircean account, see DiLeo 1991.
2. A fact, for Peirce, is something quite different. It is an abstract state of things which can be represented in a simple proposition; see EP 2:378, 1906, and Atkins 2016b.
3. The null object is simply something that is none of A, B, and C. It is technically nonexistence. The null object may be excluded from consideration without affecting Peirce's argument.
4. See McNeill 1980 and Cheng 1967 and 1970.
5. Peirce uses 'characteristic,' but he uses it in several senses that invites confusion.
6. Worries about circularity should be distinguished from worries about scope. What motivates theorists to think we must appeal to the principle of uniformity is that the scope of our observations is limited. Peirce addresses this worry by appealing to the community of inquirers. But even the observations of any given community at any given stage will be limited, and so one might think that we will need to appeal to the principle of uniformity regardless. Suppose we admit the principle of uniformity as a necessary postulate for inductive inference. Won't all inductive inference be circular? As we shall see, Peirce addresses worries about circularity by rejecting the thesis that inductions assume the principle of uniformity. Yet inductions might assume that principle even granting an infinitely enduring community of inquirers. Commentators on Peirce have sometimes conflated worries about scope with worries about circularity. I want to underscore that concerns regarding the limited scope of our observations are separate from concerns about the circularity of inductive inference. Sometimes the Peircean reply to Hume's argument is presented as primarily concerning the mere possibility that we might be proven wrong. But that is not right. That gives expression to worries about scope. Hume's objection is a worry about circularity. Moreover, it is not the mere possibility we may be wrong which gives rise to worries about scope. Rather, worries about scope arise from the fact that as compared with the mind-boggling vastness of space and time, our sampling and testing have been miniscule. The universe is about 14 billion years old. To be exceedingly generous, we've been observing it scientifically for about 5,000 years. That amounts to about .00000036 of its duration. The

observable universe is 93 billion light years across. Its objects range in size from sub-atomic particles to galaxies and cosmic webs. But those parts of the universe we can observe well are comparatively little. It is the fact that our testing and sampling are so dramatically limited in comparison with the enormity of the universe in which we live *and we know our testing and sampling to be so limited* that gives rise to worries about scope. The force of these worries does not rest on the mere possibility that we may be wrong but on the fact that we are making inferences from relatively little information as compared with the vastness of the universe. Peirce endeavors to allay these worries with his appeal to the community of inquirers. Moreover, these issues are separate from worries about circularity. Compare Madden 1981, 66; Levi 1995, 86–87n8; Wiggins 2004, 119–123; and Misak 2004/1991, 111–116, and 2016, 42.

7. I have examined Peirce's critique of Kant's table of judgments elsewhere (see Atkins 2018a, Ch. 1).

8. See, e.g., Hookway 1999; 2000, 39; and 2004, 134–136; Cooke 2005, 654; Misak 2011 and 2016, 48–51; Howat 2013; Atkin 2015.

9. To my knowledge, Rachel Herdy (2011) is the first to describe Peirce as a logical sentimentalist. Herdy avers that Peirce's logical sentimentalism applies to all sorts of inferences. I endorse the more modest thesis that Peirce's logical sentimentalism is his unique and powerful response to the end-of-community objection and its version of the circularity objection. Nonetheless, the community of inquirers does play a role in abductive inference, as will be explained in Chapter 7.

10. Peirce's logical sentimentalism is not to be confused with his sentimental conservatism, on which see Atkins 2016, 49–55, and also Savan 1981 for an account of Peirce's theory of emotion. Sentimental conservatism is an ethical doctrine; logical sentimentalism is a logical doctrine.

11. On which, see Atkins 2016, Ch. 1.

12. Peirce writes of experimental phenomena that they are not "any particular event that did happen to somebody in the dead past, but what *surely will* happen to everybody in the living future who shall fulfill certain conditions" (EP 2:340, 1905).

13. See Atkins 2016, Ch. 5.

14. See Atkins 2016, 47–75.

15. See Atkins 2016.

16. See Turnbull 2018, which I follow here.

17. See also Magnus 2005 and de Regt 1999 for a discussion from a Peircean point of view. See Lane 2018 for a fuller account of Peirce's theory of reality.

18. On these matters, see also Ullian 1995, Abrams 2002, and Mayorga 2005.

Chapter 7

1. On this matter, see Niño 2009, Campos 2011, and Mcauliffe 2015.

2. Even Descartes concedes as much when he distinguishes between moral and metaphysical certainty. We can be morally certain that we are not deceived by an evil demon, even if we never follow Descartes in his *Meditations*.

3. In particular, the first two columns of Peirce's table mix together elements found in columns 8–13. The remaining correspond well with rows 14, 16, 2, 11, 1, 17, 15, and 13. But Peirce's table does not exhibit well the periodic law.

4. See Paavola 2005; Mohammadian 2019.

5. Peirce's critical commonsensism and its connections to the commonsensist tradition have been explored by others and so I shall not pursue the matter in detail here. See Wilson 2016, esp. Ch 5.

6. It turns out that abduction does not support the inductive commitment that nature is uniform (on which, see Chapter 6), but induction supports the abductive conjecture that our abductive instincts are attuned to nature's order. Is there a worry of circularity here? No, there is not, for abduction does not justify hypotheses at all. It merely proposes hypotheses which may be put to the test.

7. I have examined these four conditions elsewhere (see Atkins 2016, 105–110). By way of a mea culpa, I should note that in the work just referenced I had not clearly distinguished between the validity and strength of abduction as I do here. In my earlier work, I treat these as conditions on validity, whereas they are better regarded as conditions on strength relative to the best explanation given the evidence we have.

8. What evidence the hypothesis must explain is a difficult question and may depend partly on the judgment of the theorist, for which reason I emphasize that the theory must explain the evidence it is designed to explain. Suppose one has hypothesis H_1 which explains X, Y, and Z but neither A nor B and hypothesis H_2 which explains X, Y, A, and B but not Z. One might prefer H_2 over H_1, even though it does not explain Z because it explains A and B and one suspects that Z is susceptible to explanation in some other way.

9. A detailed examination of Peirce's work on the economy of research is beyond the purview of this book, though see Wible 2018, de Waal 2018, Tiercelin 2018, Haack 2018, and Chiffi and Pietarinen 2019.

10. Writers who defend a position broadly in line with such a view are Niño 2009, Campos 2011, Mcauliffe 2015, and Yu and Zenker 2017, albeit with some important nuances.

11. Writers who fall into this camp are Laudan 1977 and McKaughan 2008.

12. Anderson 1986 and Mohammadian 2019 and 2021 represent such a position.

13. Kapitan 2000, for example, contends that abduction is reducible to other forms of inference.

14. For a discussion of some contemporary attempts to classify kinds of abduction, see Park 2015.

15. One might worry that my use of 'proof' is inapt; a diagram is not a proof. Two points are in order. First, Peirce regards any argument that suffices to remove all real doubt to be a proof (see CP 2.782, 1901). If the diagram does so, it is a proof. Second, Peirce notes that while the "more careful logician may demonstrate that [the line of abutment] must fall in one angle or the other," this demonstration is only a substitute diagram (EP 2:207, 1903). Admittedly, the original diagram is not a proof in the sense of 'proof' demanded by contemporary mathematicians, but while Peirce values such high standards of proof, he does not regard them as the sine qua non of proof.

16. By mere perception, I mean perception that does not involve perceptual judgments. Peirce maintains that perceptual judgment and abduction shade into one another with no clear line of demarcation between them (see EP 2:227, 1903, and the essays collected in Hull and Atkins, 2017).

17. See Wilson 2020 for an examination of related themes.

Bibliography

Abbreviated References

BD: Baldwin, James Mark, ed. 1901. *Dictionary of Philosophy and Psychology*. 3 vols. New York: The Macmillan Company.

CN: Peirce, Charles. 1975–1987. *Contributions to the Nation*. 4 vols. Ed. Kenneth Laine Ketner and James Edward Cook. Lubbock: Texas Tech Press.

CP: Peirce, Charles. 1931–1958. *Collected Papers*. 8 vols. Ed. Charles Hartshorne, Paul Weiss, and Arthur Burks. Cambridge, MA: Belknap Press of Harvard University Press.

CWJ: James, William. 2000. *The Correspondence of William James*. Vol. 8. 1895–June 1899. Ed. Ignas K. Skrupskelis and Elizabeth M. Berkeley. Charlottesville: University Press of Virginia.

EP: Peirce, Charles. 1992 and 1998. *The Essential Peirce*. 2 vols. Ed. Nathan Houser and Christian Kloesel and The Peirce Edition Project. Bloomington: Indiana University Press.

HP: Peirce, Charles. 1985. *Historical Perspectives on Peirce's Logic of Science*. Ed. Carolyn Eisele. Berlin: De Gruyter Mouton.

ILS: Peirce, Charles. 2014. *Illustrations of the Logic of Science*. Ed. Cornelis de Waal. Chicago: Open Court.

LI: Peirce, Charles. 2009. *The Logic of Interdisciplinarity: The* Monist*-Series*. Ed. Elize Bisanz. Berlin: Academie Verlag GmbH.

LoF: Peirce, Charles. 2020. *Logic of the Future: Writings on Existential Graphs*. Vol. 1. Ed. Ahti-Veikko Pietarinen. Berlin: De Gruyter.

NEM: Peirce, Charles. 1979. *New Elements of Mathematics*. 4 vols. Ed. Carolyn Eisele. Atlantic Highlands, NJ: Humanities Press.

PPM: Peirce, Charles. 1997. *Pragmatism as a Principle and Method of Right Thinking*. Ed. Patricia Ann Turrisi. Albany: State University of New York Press.

PSR: Peirce, Charles. 2016. *Prolegomena to a Science of Reasoning: Phaneroscopy, Semeiotic, Logic*. Ed. Elize Bisanz. Frankfurt am Main: Peter Lang.

R: Peirce, Charles. 1839–1914. *The Charles S. Peirce Manuscripts*. Cambridge, MA: Houghton Library at Harvard University. Citations are by manuscript number (as assigned in Robin 1967 and 1971) and, where available, page number.

RLT: Peirce, Charles. 1992. *Reasoning and the Logic of Things: The Cambridge Conferences Lectures of 1898*. Ed. Kenneth Laine Ketner. Cambridge, MA: Harvard University Press.

SS: Peirce, Charles, and Victoria Welby. 1977. *Semiotic and Significs: The Correspondence between Charles S. Peirce and Victoria Lady Welby*. Ed. Charles S. Hardwick with the Assistance of James Cook. Bloomington: Indianapolis University Press.

SWS: Peirce, Charles. 2020. *Selected Writings on Semiotics 1894–1912*. Ed. Francesco Bellucci. Berlin: DeGruyter.

W: Peirce, Charles Sanders. 1982-Present. *The Writings of Charles S. Peirce: A Chronological Edition*. Ed. The Peirce Edition Project. 8 vols. Bloomington: Indiana University Press.

Other References

Abrams, Jerold J. 2002. "Solution to the Problem of Induction: Peirce, Apel, and Goodman on the Grue Paradox." *Transactions of the Charles S. Peirce Society*, 38:4, 543–558.

Anderson, Douglas. 1986. "The Evolution of Peirce's Concept of Abduction." *Transactions of the Charles S. Peirce Society*, 22:2, 145–164.

Anellis, Irving H. 2012. "Peirce's Truth-Functional Analysis and the Origin of the Truth Table." *History and Philosophy of Logic*, 33:1, 87–97.

Aristotle. 1984. *The Complete Works of Aristotle*. 2 vols. Ed. Jonathan Barnes. Princeton, NJ: Princeton University Press.

Atkin, Albert. 2005. "Peirce on the Index and Indexical Reference." *Transactions of the Charles S. Peirce Society*, 41:1, 161–188.

Atkin, Albert. 2015. "Intellectual Hope as Convenient Friction." *Transactions of the Charles S. Peirce Society*, 51:4, 444–462.

Atkin, Albert. 2016. *Peirce*. London: Routledge.

Atkins, Richard Kenneth. 2011. "This Proposition Is Not True: C.S. Peirce and the Liar Paradox." *Transactions of the Charles S. Peirce Society*, 47:4, 421–444.

Atkins, Richard Kenneth. 2016a. *Peirce and the Conduct of Life*. Cambridge: Cambridge University Press.

Atkins, Richard Kenneth. 2016b. "Peirce on Facts and True Propositions." *British Journal for the History of Philosophy*, 24:6, 1176–1192.

Atkins, Richard Kenneth. 2018a. *Charles S. Peirce's Phenomenology: Analysis and Consciousness*. New York: Oxford University Press.

Atkins, Richard Kenneth. 2018b. "Peirce on Truth as the Predestinate Opinion." *European Journal of Philosophy*, 26:1, 411–429.

Atkins, Richard Kenneth. 2021. "A Peircean Examination of Gettier's Two Cases." *Synthese*, 199:5–6, 12945–12961.

Bellucci, Francesco. 2014. "'Logic Considered as Semeiotic': On Peirce's Philosophy of Logic." *Transactions of the Charles S. Peirce Society*, 50:4, 523–547.

Bellucci, Francesco. 2015a. "Neat, Swine, Sheep, and Deer: Mill and Peirce on Natural Kinds." *British Journal for the History of Philosophy*, 23:5, 911–932.

Bellucci, Francesco. 2015b. "Logic, Psychology, and Apperception: Charles S. Peirce and Johann F. Herbart." *Journal of the History of Ideas*, 76:1, 69–91.

Bellucci, Francesco. 2016. "Charles S. Peirce and the Medieval Doctrine of *consequentiae*." *History and Philosophy of Logic*, 37:3, 244–268.

Bellucci, Francesco. 2018. *Peirce's Speculative Grammar: Logic as Semiotics*. New York: Routledge.

Bellucci, Francesco. 2021. "Peirce on Symbols." *Archiv für Geschichte Philosophie*, 103:1, 169–188.

Bergman, Merrie, James Moor, and Jack Nelson. 2004. *The Logic Book*. 4th ed. Boston: McGraw Hill.

Bunnin, Nicholas, and Jiyuan Yu. 2008. "Argument." In *The Blackwell Dictionary of Western Philosophy*. Malden, MA: Blackwell.

Burbidge, J. W. 1981. "Peirce on Historical Explanation." In *Pragmatism and Purpose.* Eds. L. W. Sumner, John G. Slater, and Fred Wilson. Toronto: University of Toronto Press, 15–27.

Burch, Robert. 1997. "Peirce on the Application of Relations to Relations." In *Studies in the Logic of Charles Sanders Peirce.* Eds. Nathan Houser, Don D. Roberts, and James Van Evra. Indianapolis: Indianapolis University Press, 206–233.

Burch, Robert. 2010. "If Universes Were as Plenty as Blackberries: Peirce on Induction and Verisimilitude." *Transactions of the Charles S. Peirce Society,* 46:3, 423–452.

Burks, Arthur W. 1964. "Peirce's Two Theories of Probability." In *Studies in the Philosophy of Charles Sanders Peirce.* Second Series. Eds. Edward C. Moore and Richard S. Robin. Amherst: The University of Massachusetts Press, 141–150.

Campos, Daniel. 2011. "On the Distinction between Peirce's Abduction and Lipton's Inference to the Best Explanation." *Synthese,* 180:3, 419–442.

Cheng, Chung-Ying. 1967. "Charles Peirce's Arguments for the Non-Probabilistic Validity of Induction." *Transactions of the Charles S. Peirce Society,* 3:1, 24–39.

Cheng, Chung-Ying. 1970. *Peirce's and Lewis's Theories of Induction.* The Hague: Martinus Nijhoff.

Chiffi, Daniele and Ahti-Veikko Pietarinen. 2019. "Risk and Values in Sciences: A Peircean View." *Axiomathes,* 29:4, 329–346.

Cooke, Elizabeth F. 2004. "Fallibilism, Progress, and the Long Run in Peirce's Philosophy of Science." *Southwest Philosophy Review,* 20:1, 155–162.

Cooke, Elizabeth F. 2005. "Transcendental Hope: Peirce, Hookway, and Pihlström on the Conditions of Inquiry." *Transactions of the Charles S. Peirce Society,* 41:3, 651–674.

Dau, Fritzhoff. 2008. *Mathematical Logic with Diagrams.* Unpublished manuscript. http://www.dr-dau.net/Papers/habil.pdf.

De Regt, Herman C. D. G. 1999. "Peirce's Pragmatism, Scientific Realism, and the Problem of Underdetermination." *Transactions of the Charles S. Peirce Society,* 35:2, 374–397.

De Waal, Cornelis. 2005. "Why Metaphysics Needs Logic and Mathematics Doesn't." *Transactions of the Charles S. Peirce Society,* 41:2, 283–297.

De Waal, Cornelis. 2013. *Peirce: A Guide for the Perplexed.* London: Bloomsbury.

De Waal, Cornelis. 2018. "The Economics of Truth." *Transactions of the Charles S. Peirce Society,* 54:2, 162–182.

Delaney, Cornelius F. 1973. "Peirce on Induction and the Uniformity of Nature." *The Philosophical Forum,* 4:3, 438–448.

Delaney, Cornelius F. 1993. "Peirce on the Conditions of the Possibility of Science." In *Charles S. Peirce and the Philosophy of Science.* Ed. Edward C. Moore. Tuscaloosa: The University of Alabama Press, 17–29.

Descartes, Rene. 1985. "Regulae." In *The Philosophical Writings of Descartes.* Vol. I. Trans. John Cottingham, Robert Stoothoff, and Dugald Murdoch. Cambridge: Cambridge University Press, 7–78.

DiLeo, Jeffrey. 1991. "Peirce's Haecceitism." *Transactions of the Charles S. Peirce Society.* 27:1, 79–109.

Dipert, Randall. 2004. "Peirce's Deductive Logic." In *The Cambridge Companion to Peirce.* Ed. Cheryl Misak. Cambridge: Cambridge University Press, 287–324.

Feldman, Richard. 2001. "Voluntary Belief and Epistemic Evaluation." In *Knowledge, Truth, and Duty: Essays on Epistemic Justification, Responsibility, and Virtue.* Ed. Matthias Steup. New York: Oxford University Press, 77–92.

Fetzer, James H. 1993. "Peirce and Propensities." In *Charles S. Peirce and the Philosophy of Science*. Ed. Edward C. Moore. Tuscaloosa: The University of Alabama Press, 60–71.

Fisch, Max, and Atwell Turquette. 1966. "Peirce's Triadic Logic." *Transactions of the Charles S. Peirce Society*, 2:2, 71–85.

Forster, Paul. 1989. "Peirce on the Progress and Authority of Science." *Transactions of the Charles S. Peirce Society*, 25:4, 421–452.

Forster, Paul. 2011. *Peirce and the Threat of Nominalism*. Cambridge: Cambridge University Press.

Girel, Mathias. 2011. "Peirce's Early Re-readings of His Illustrations: The Case of 1885 Royce Review." *Cognitio*, 12:1, 75–88.

Godfrey-Smith, Peter. 2003a. *Theory and Reality: An Introduction to the Philosophy of Science*. Chicago: Chicago University Press.

Godfrey-Smith, Peter. 2003b. "Goodman's Problem and Scientific Methodology." *The Journal of Philosophy*, 100:11, 573–590.

Gould, Stephen Jay. 1996. *The Mismeasure of Man—Revised and Expanded*. New York: W. W. Norton.

Greaves, Mark. 2002. *The Philosophical Status of Diagrams*. Stanford, CA: CSLI.

Grinnell, Frederick. 2019. "Abduction in the Everyday Practice of Science: The Logic of Unintended Experiments." *Transactions of the Charles S. Peirce Society*, 55:3, 215–227.

Haack, Susan. 1982. "Descartes, Peirce, and the Cognitive Community." *The Monist*, 65:2, 156–181.

Haack, Susan. 2018. "Expediting Inquiry: Peirce's Social Economy of Research." *Transactions of the Charles S. Peirce Society*, 54:2, 208–230.

Hacking, Ian. 1980. "The Theory of Probable Inference: Neyman, Peirce, and Braithwaite." In *Science, Belief, and Behaviour*. Ed. D. H. Mellor. Cambridge: Cambridge University Press, 141–160.

Hacking, Ian. 1988. "Telepathy: Origins of Randomization in Experimental Design." *Isis*, 79:3, 427–451.

Hacking, Ian. 1990. *The Taming of Chance*. Cambridge: Cambridge University Press.

Hacking, Ian. 2001. *An Introduction to Probability and Inductive Logic*. Cambridge: Cambridge University Press.

Halbach, Volker. 2010. *The Logic Manual*. Oxford: Oxford University Press.

Hausman, Alan, Frank Boardman, and Howard Kahane. 2021. *Logic and Philosophy: A Modern Introduction*. Indianapolis: Hackett.

Herdy, Rachel. 2011. "Peirce's Logical Sentimentalism." https://www.academia.edu/24185249/Peirces_Logical_Sentimentalism.

Hintikka, Jaako. 1964. *Knowledge and Belief: An Introduction to the Logic of the Two Notions*. Ithaca, NY: Cornell University Press.

Hintikka, Jaako. 1983. "C.S. Peirce's 'First Real Discovery' and Its Contemporary Relevance." *The Monist*, 63:3, 304–315.

Hookway, Christopher. 1985. *Peirce*. London: Routledge.

Hookway, Christopher. 1999. "Modest Transcendental Arguments and Sceptical Doubts: A Reply to Stroud." In *Transcendental Arguments: Problems and Prospects*. Ed. Robert Stern. Oxford: Oxford University Press, 173–187.

Hookway, Christopher. 2000. *Truth, Rationality, and Pragmatism*. Oxford: Clarendon Press.

Hookway, Christopher. 2004. "Truth, Reality, and Convergence." In *The Cambridge Companion to Peirce*. Ed. Cheryl Misak. Cambridge: Cambridge University Press, 127–149.

Hookway, Christopher. 2012. *The Pragmatic Maxim*. Oxford: Oxford University Press.

Hossenfelder, Sabine. 2018. *Lost in Math*. New York: Basic Books.

Howat, Andrew. 2013. "Regulative Assumptions, Hinge Propositions, and the Peircean Conception of Truth." *Erkenntnis*, 78:2, 451–468.

Howat, Andrew. 2014. "Peirce on Grounding the Laws of Logic." *Transactions of the Charles S. Peirce Society*, 50:4, 480–500.

Hull, Kathleen A., and Richard Kenneth Atkins, eds. 2017. *Peirce on Perception and Reasoning: From Icons to Logic*. New York: Routledge.

Hume, David. 2007. *An Enquiry Concerning Human Understanding*. Ed. Stephen Buckle. Cambridge: Cambridge University Press.

Hungerford, Yael. 2016. *Charles S. Peirce's Conservative Progressivism*. Ph.D. Dissertation, Boston College.

James, William. 1896/1979. *The Will to Believe*. Ed. Frederick Burkhardt. Cambridge, MA: Harvard University Press.

Johnson, Gregory. 2016. *Argument and Inference: An Introduction to Inductive Logic*. Cambridge, MA: MIT Press.

Johnson-Laird, Philip N. 1983. *Mental Models. Towards a Cognitive Science of Language, Inference and Consciousness*. Cambridge: Cambridge University Press.

Johnson-Laird, Philip N. 2006. *How We Reason*. Oxford: Oxford University Press.

Kant, Immanuel. 1992. *Lectures on Logic*. Trans. and Ed. J. Michael Young. Cambridge: Cambridge University Press.

Kant, Immanuel. 1997. *Prolegomena to Any Future Metaphysics: That Will Be Able to Come Forward as Science, with Selections from the* Critique of Pure Reason. Trans. and Ed. Gary Hatfield. Cambridge: Cambridge University Press.

Kant, Immanuel. 1998. *Critique of Pure Reason*. Trans. and Ed. Paul Guyer and Allen W. Wood. Cambridge: Cambridge University Press.

Kapitan, Tomis. 2000. "Abduction as Practical Inference." In *The Commens Encyclopedia: The Digital Encyclopedia of Peirce Studies*. Eds. M. Bergman and J. Queiroz. New Edition. Pub. 121219-1528a. http://www.commens.org/encyclopedia/article/kapitan-tomis-abduction-practical-inference.

Kasser, Jeff. 1999. "Peirce's Supposed Psychologism." *Transactions of the Charles S. Peirce Society*, 35:3, 501–526.

Kasser, Jeff. 2011. "How Settled are Settled Beliefs in 'The Fixation of Belief'?" *Transactions of the Charles S. Peirce Society*, 47:2, 226–247.

Kasser, Jeff. 2016a. "Confidence, Evidential Weight, and the Theory-Practice Divide in Peirce." *Transactions of the Charles S. Peirce Society*, 52:2, 285–308.

Kasser, Jeff. 2016b. "Two Conceptions of Weight of Evidence in Peirce's *Illustrations of the Logic of Science*." *Erkenntnis*, 81:3, 629–648.

Kasser, Jeff. 2018. "Genuine Belief and Genuine Doubt in Peirce." *European Journal of Philosophy*, 26:2, 840–853.

Kasser, Jeff. 2019. "Normativity and Naturalism in 'The Fixation of Belief.'" *Transactions of the Charles S. Peirce Society*, 55:1, 1–19.

Klein, Alexander. 2013. "Who Is the Community of Inquirers?" *Transactions of the Charles S. Peirce Society*, 49:3, 413–423.

Klima, Gyula. 2016. "Consequence." In *The Cambridge Companion to Medieval Logic*. Ed. Catarina Dutilh Novaes. Cambridge: Cambridge University Press, 316–341.

Kyburg, Henry E. 1993. "Peirce and Statistics." In *Charles S. Peirce and the Philosophy of Science*. Ed. Edward C. Moore. Tuscaloosa: The University of Alabama Press, 130–138.

Lane, Robert. 1999. "Peirce's Triadic Logic Revisited." *Transactions of the Charles S. Peirce Society*, 35:2, 284–311.

Lane, Robert. 2018. *Peirce on Realism and Idealism*. New York: Cambridge University Press.

Laudan, Larry. 1973. "Peirce and the Trivialization of the Self-Correcting Thesis." In *Foundations of Scientific Method in the 19th Century*. Eds. Ronald N. Giere and Richard S. Westfall. Bloomington: Indiana University Press, 275–306.

Laudan, Larry. 1977. *Progress and its Problems: Towards a Theory of Scientific Growth*. London: Routledge & Kegan Paul.

Laudan, Larry. 1990. "Demystifying Underdetermination." In *Scientific Theories*. Ed. C. Wade Savage. Minneapolis: University of Minnesota Press, 267–297.

Legg, Catherine. 2008. "The Problem of the Essential Icon." *American Philosophical Quarterly*, 45:3, 207–232.

Legg, Catherine. 2014a. "Charles Peirce's Limit Concept of Truth." *Philosophy Compass*, 9:3, 204–213.

Legg, Catherine. 2014b. "'Things Unreasonably Compulsory': A Peircean Challenge to a Humean Theory of Perception, Particularly with Respect to Perceiving Necessary Truths." *Cognitio*, 15:1, 89–112.

Legg, Catherine. 2020. "Is Truth Made, and if So, What Do We Mean by That? Redefining Truthmaker Realism." *Philosophia*, 48:2, 587–606.

Lenz, John W. 1964. "Induction as Self-Corrective." In *Studies in the Philosophy of Charles Sanders Peirce*. Second Series. Eds. Edward C. Moore and Richard S. Robin. Amherst: The University of Massachusetts Press, 151–162.

Lenzen, Victor. 1964. "Charles S. Peirce as Astronomer." In *Studies in the Philosophy of Charles Sanders Peirce*. Second Series. Eds. Edward C. Moore and Richard S. Robin. Amherst: The University of Massachusetts Press, 33–50.

Levi, Isaac. 1980. "Induction as Self-Correcting according to Peirce." In *Science, Belief, and Behaviour*. Ed. D. H. Mellor. Cambridge: Cambridge University Press, 127–140.

Levi, Isaac. 1995. "Induction According to Peirce." In *Peirce and Contemporary Thought*. Ed. Kenneth Laine Ketner. New York: Fordham University Press, 59–93.

Levi, Isaac. 2004. "Beware of Syllogism: Statistical Reasoning and Conjecturing According to Peirce." In *The Cambridge Companion to Peirce*. Ed. Cheryl Misak. Cambridge: Cambridge University Press, 257–286.

Liszka, James Jakób. 1996. *A General Introduction to the Semeiotic of Charles Sanders Peirce*. Bloomington: Indiana University Press.

Longino, Helen. 1990. *Science as Social Knowledge*. Princeton, NJ: Princeton University Press.

MacDonald, Ian. 2020. "Did Peirce Misrepresent Descartes? Reinvestigating and Defending Peirce's Case." *Transactions of the Charles S. Peirce Society*, 56:1, 1–18.

Madden, Edward. 1964. "Peirce on Probability." In *Studies in the Philosophy of Charles Sanders Peirce*. Second Series. Eds. Edward C. Moore and Richard S. Robin. Amherst: The University of Massachusetts Press, 122–140.

Madden, Edward. 1981. "Scientific Inference: Peirce and the Humean Tradition." In *Pragmatism and Purpose*. Eds. L.W. Sumner, John G. Slater, and Fred Wilson. Toronto: University of Toronto Press, 59–74.

Magnus, P. D. 2005. "Peirce: Underdetermination, Agnosticism, and Related Mistakes." *Inquiry*, 48:1, 26–37.

Mayo, Deborah F. 1993. "The Test of Experiment: C.S. Peirce and E.S. Pearson." In *Charles S. Peirce and the Philosophy of Science*. Ed. Edward C. Moore. Tuscaloosa: The University of Alabama Press, 161–174.

Mayo, Deborah F. 1996. *Error and the Growth of Experimental Knowledge*. Chicago: Chicago University Press.

Mayo, Deborah F. 2005. "Peircean Induction and the Error-Correcting Thesis." *Transactions of the Charles S. Peirce Society*, 41:2, 299–319.

Mayo, Deborah F. 2018. *Statistical Inference as Severe Testing*. Cambridge: Cambridge University Press.

Mayorga, Rosa. 2005. "Diamonds Are a Pragmaticist's Best Friend." *Transactions of the Charles S. Peirce Society*, 41:2, 255–270.

Mcauliffe, William H. B. 2015. "How Did Abduction Get Confused with Inference to the Best Explanation?" *Transactions of the Charles S. Peirce Society*, 51:3, 300–319.

McKaughan, Daniel. 2008. "From Ugly Duckling to Swan: C.S. Peirce, Abduction, and the Pursuit of Scientific Theories." *Transactions of the Charles S. Peirce Society*, 44:3, 446–468.

McNeill, John W. 1980. "Peirce on the Possibility of a Chance World." *Transactions of the Charles S. Peirce Society*, 16:1, 49–58.

Merrill, Daniel D. 1997. "Relations and Quantification in Peirce's Logic." *Studies in the Logic of Charles Sanders Peirce*. Eds. Nathan Houser, Don D. Roberts, and James Van Evra. Indianapolis: Indianapolis University Press, 158–172.

Mill, John Stuart. 1869. *A System of Logic: Ratiocinative and Inductive*. New York: Harper and Brothers.

Misak, Cheryl. 1995. *Verificationism*. London: Routledge.

Misak, Cheryl. 2000. *Truth, Politics, Morality*. London: Routledge.

Misak, Cheryl. 2004/1991. *Truth and the End of Inquiry*. Oxford: Clarendon Press.

Misak, Cheryl. 2011. "American Pragmatism and Indispensibility Arguments." *Transactions of the Charles S. Peirce Society*, 47:3, 261–273.

Misak, Cheryl. 2016. *Cambridge Pragmatism*. Oxford: Oxford University Press.

Misak, Cheryl. 2018. "Peirce and Ramsey: Truth, Pragmatism, and Inference to the Best Explanation." In *Best Explanations: New Essays on Inference to the Best Explanation*. Eds. Keven McCain and Ted Posten. Oxford: Oxford University Press, 25–38.

Mohammadian, Mousa. 2019. "Beyond the Instinct-Inference Dichotomy: A Unified Interpretation of Peirce's Theory of Abduction." *Transactions of the Charles S. Peirce Society*, 55:2, 138–160.

Mohammadian, Mousa. 2021. "Abduction – the Context of Discovery + Underdetermination = Inference to the Best Explanation." *Synthese*, 198:5, 4205–4228.

Murphey, Murray. 1961. *The Development of Peirce's Philosophy*. Cambridge, MA: Harvard University Press.

Niño, Douglas. 2009. "Peircean Pragmatism and Inference to the Best Explanation." https://www.researchgate.net/publication/232607842_PEIRCEAN_PRAGMATISM_AND_INFERENCE_TO_THE_BEST_EXPLANATION

Oreskes, Naomi. 2019. *Why Trust Science?* Princeton, NJ: Princeton University Press.

Oreskes, Naomi, and Erik M. Conway. 2010. *Merchants of Doubt*. New York: Bloomsbury.

Paavola, Sami. 2005. "Peircean Abduction: Instinct or Inference?" *Semiotica*, 153:1/4, 131–154.

Park, Woosuk. 2015. "On Classifying Abduction." *Journal of Applied Logic*, 13:3, 215–238.

Pietarinen, Ahti-Veikko. 2005. "Cultivating Habits of Reasoning: Peirce on the 'Logica Utens' vs. 'Logica Docens' Distinction." *History of Philosophy Quarterly*, 22:4, 357–372.

Pietarinen, Ahti-Veikko. 2020. "Publish or Peirceish." *Transactions of the Charles S. Peirce Society*, 56:2, 261–290.

Pietarinen, Ahti-Veikko, and Francesco Bellucci. 2014. "New Light on Peirce's Conceptions of Retroduction, Deduction, and Scientific Reasoning." *International Studies in the Philosophy of Science*, 28:4, 353–373.

Prendergast, T. L. 1977. "The Structure of the Argument in Peirce's 'Questions Concerning Certain Faculties Claimed for Man.'" *Transactions of the Charles S. Peirce Society*, 13:4, 288–305.

Reid, Thomas. 1785. *An Inquiry into the Human Mind on the Principles of Common Sense.* 4th ed. London: T. Cadell.

Rescher, Nicholas. 1976. "Peirce and the Economy of Research." *Philosophy of Science*, 43:1, 71–98.

Rescher, Nicholas. 1978. *Peirce's Philosophy of Science.* Notre Dame, IN: University of Notre Dame Press.

Rescher, Nicholas. 1995. "Peirce on the Validation of Science." In *Peirce and Contemporary Thought.* Ed. Kenneth Laine Ketner. New York: Fordham University Press, 103–112.

Roberts, Don D. 1973. *The Existential Graphs of Charles S. Peirce.* The Hague: Mouton.

Robin, Richard S. 1967. *Annotated Catalogue of the Papers of Charles S. Peirce.* Amherst: The University of Massachusetts Press.

Robin, Richard S. 1971. "The Peirce Papers: A Supplementary Catalogue." *Transactions of the Charles S. Peirce Society*, 7:1, 37–57.

Ronan, Mark Andrew. 2012. "Modern Algebra." *Encyclopædia Britannica*, https://www.britannica.com/science/modern-algebra/additional-info#history

Savan, David. 1981. "Peirce's Semiotic Theory of Emotion." In *Proceedings of the Charles S. Peirce Bicentennial International Congress.* Ed. Kenneth Laine Ketner et. al. Lubbock: Texas Tech Press, 319–333.

Schouls, Peter A. 1981. "Peirce and Descartes: Doubt and the Logic of Discovery." In *Pragmatism and Purpose.* Eds. L.W. Sumner, John G. Slater, and Fred Wilson. Toronto: University of Toronto Press, 88–104.

Shin, Sun-Joo. 2002. *The Iconic Logic of Peirce's Graphs.* Cambridge, MA: MIT Press.

Short, T. L. 1998. "The Discovery of Scientific Aims and Methods." *American Catholic Philosophical Quarterly*, 72:2, 293–312.

Short, T. L. 2000. "Peirce on the Aim of Inquiry: Another Reading of 'Fixation.'" *Transactions of the Charles S. Peirce Society*, 36:1, 1–23.

Short, T. L. 2002. "Robin on Perception and Sentiment in Peirce." *Transactions of the Charles S. Peirce Society*, 38:1/2, 267–282.

Short, T. L. 2007. *Peirce's Theory of Signs.* New York: Cambridge University Press.

Short, T. L. 2008. "Measurement and Philosophy." *Cognitio*, 9:1, 111–124.

Short, T. L. 2018. "Peirce's Irony." *Transactions of the Charles S. Peirce Society*, 54:1, 9–38.

Skyrms, Brian. 2000. *Choice and Chance: An Introduction to Inductive Logic.* 4th ed. Belmont, CA: Wadsworth.

Spiegelhalter, David. 2019. *The Art of Statistics: How to Learn from Data.* New York: Basic Books.

Stewart, Dugald. 1792. *Elements of the Philosophy of the Human Mind.* London: A. Strahan and T. Cadell.

Tiercelin, Claudine. 2017. "Was Peirce a Genuine Anti-Psychologist in Logic?" *European Journal of Pragmatism and American Philosophy*, IX:1. https://journals.openedition.org/ejpap/1003?lang=en

Tiercelin, Claudine. 2018. "The Economy of Research and the Proper Defense of Knowledge and Intellectual Virtues." *Transactions of the Charles S. Peirce Society*, 54:2, 183–207.

Turnbull, Margaret Greta. 2018. "Underdetermination in Science." *Philosophy Compass*, 13:2, 1–11. doi:10.1111/phc3.12475.

Turquette, Atwell. 1964. "Peirce's Icons for Deductive Logic." In *Studies in the Philosophy of Charles Sanders Peirce*. Second Series. Eds. Edward C. Moore and Richard S. Robin. Amherst: The University of Massachusetts Press, 95–108.

Ullian, Joseph S. 1995. "On Peircean Induction: A Response to Levi." In *Peirce and Contemporary Thought*. Ed. Kenneth Laine Ketner. New York: Fordham University Press, 94–99.

Van Evra, James. 1997. "Logic and Mathematics in Charles Sanders Peirce's 'Description of a Notation for the Logic of Relatives.'" In *Studies in the Logic of Charles Sanders Peirce*. Eds. Nathan Houser, Don D. Roberts, and James Van Evra. Indianapolis: Indianapolis University Press, 147–157.

Whatley, Richard. 1850. *Elements of Logic*. Boston: James Munroe and Company.

Wible, James. 2018. "Game Theory, Abduction, and the Economy of Research." *Transactions of the Charles S. Peirce Society*, 54:2, 134–161.

Wiggins, David. 2004. "Reflections on Truth and Inquiry Arising from Peirce's Method for the Fixation of Belief." In *The Cambridge Companion to Peirce*. Ed. Cheryl Misak. Cambridge: Cambridge University Press, 87–126.

Wilson, Aaron. 2012. "The Perception of Generals." *Transactions of the Charles S. Peirce Society*, 48:2, 169–190.

Wilson, Aaron. 2016. *Peirce's Empiricism*. Lanham, MD: Lexington Books.

Wilson, Aaron. 2020. "Interpretation, Realism, and Truth: Is Peirce's Second Grade of Clearness Independent of the Third?" *Transactions of the Charles S. Peirce Society*, 56:3, 349–373.

Yu, Shiyang, and Frank Zecker. 2017. "Peirce Knew Why Abduction Isn't IBE—A Scheme and Critical Questions for Abductive Argument." *Argumentation*, 32:4, 569–587.

Index

For the benefit of digital users, indexed terms that span two pages (e.g., 52–53) may, on occasion, appear on only one of those pages.
Tables are indicated by *t* following the page number